Organic Chemicals from Biomass

Editor

Dr. Irving S. Goldstein
Professor of Wood and Paper Science
Department of Wood and Paper Science
North Carolina State University
Raleigh, North Carolina

CRC Press, Inc.
Boca Raton, Florida

53276048

Library of Congress Cataloging in Publication Data

Main entry under title:

Organic chemicals from biomass.
Includes bibliographies and index.
1.   Chemicals—Manufacture and industry.
2.   Chemistry, Organic.   3.   Biomass energy.
I.   Goldstein, Irving S., 1921-
TP247.73   660.2'82   80-11939
ISBN 0-8493-5531-1

   Direct all inquiries to CRC Press, Inc., 2000 N.W. 24th Street, Boca Raton, Florida 33431.

©1981 by CRC Press, Inc.

International Standard Book Number 0-8493-5531-1

Library of Congress Card Number 80-11939
Printed in the United States

PREFACE

Although the current energy crisis receives much attention and ranks high in the public consciousness, its corollary, the materials crisis, has been comparatively neglected. Since most of our organic chemicals and the synthetic organic polymeric materials derived from them are now obtained from petroleum and natural gas, a shortage of these fossil fuels will be immediately translated into a materials shortage as well. Alternative carbon sources will then be called upon to at first fill the gap and eventually replace exhausted oil and gas supplies. These alternatives include shale oil and coal as well as biomass, the only renewable resource and the precursor of the fossil fuels. The purpose of this volume is to describe the potential role of biomass for the production of organic chemicals.

I have placed some arbitrary restrictions and limitations on the subject matter included, first to keep the size of the book manageable, and second because much of the omitted material which could arguably be called organic chemicals or biomass has been adequately treated elsewhere. In the present context organic chemicals refers generally to currently or potentially industrially important nonpolymeric compounds. Excluded are the thousands of organic compounds found in small quantity in nature as well as the high polymers such as rubber, cellulosic fibers and plastics, starch, lignins, etc.

The biomass emphasis is on material of terrestrial plant origin, although the principles are directly transferable to aquatic plants with similar components. Products of animal origin are not included. Since animal fats and oils are not considered, it seemed logical to exclude vegetable oils as well. The inclusion of the fatty acids associated with rosin in pine trees is an apparent inconsistency resulting from their collection with the rosin. Other inconsistencies of inclusion or exclusion may also appear because of the heterogeneous nature of the biomass.

The organization of the book is by groups of related chapters. Following introductory chapters describing the availability and composition of the biomass, several chapters are devoted to processes for the conversion of biomass to chemicals by fermentation, gasification, or pyrolysis. Then the major components of plant cell walls, cellulose, hemicelluloses, and lignin, are individually considered as sources of chemicals. Chemicals obtainable by extraction from wood, bark, or foliage make up the next group of chapters, and the final chapters cover economic and related factors.

The advantages of multiple authorship in providing expert treatment of many different subjects more than compensates for any lack of uniformity of style and exposition. I believe that the conversion of biomass into chemicals will receive increasing attention in the years ahead and that this volume is only a forerunner of things to come.

Irving S. Goldstein
Raleigh, North Carolina
December 1978

THE EDITOR

Irving S. Goldstein, Ph.D., Professor of Wood and Paper Science, is a member of the Department of Wood and Paper Science at North Carolina State University, in Raleigh.

Dr. Goldstein received his B.S. degree in Chemistry from Rensselaer Polytechnic Institute, Troy, New York in 1941; his M.S. degree in Chemistry from Illinois Institute of Technology, Chicago, in 1944; and his Ph.D. degree in Organic Chemistry from Harvard University, Cambridge, Mass. in 1948.

From 1971 to 1978, Dr. Goldstein was Head of the Department of Wood and Paper Science. Prior to that he was Professor of Forest Science at Texas A&M University, College Station.

Dr. Goldstein began his career as a Teaching Assistant at Illinois Institute of Technology and then Teaching Fellow at Harvard University. He was Research Chemist for North American Rayon Corporation, Elizabethton, Tenn.; Senior Research Chemist, and, subsequently, Manager of Wood Chemistry Research, Koppers Company, Inc., Pittsburgh, Pa.; and Senior Research Chemist at Nalco Chemical Company, Chicago, Ill. From 1966 to 1968, he was Manager, Paper Research, for Continental Can Company, Chicago, Ill.

Besides pulping, bleaching, and papermaking, Dr. Goldstein's fields of research interest include chemical utilization of wood, lignin, and bark, as well as the treatment of wood for resistance to decay, insects and marine borers, chemicals, fire, and dimensional change.

CONTRIBUTORS

George M. Barton, F.C.I.C.
Western Laboratory
Manager Wood Science Department
Forintek Canada Corporation
Vancouver, British Columbia
Canada

David L. Brink, Ph.D.
Professor of Forestry
Department of Forestry and Resource
  Management
University of California, Berkeley
Forest Products Chemist
University of California Forest
  Products Laboratory
Richmond, California

Robert Detroy, Ph.D.
Research Leader
Agricultural Microbiology Research
  Fermentation Laboratory
U.S. Department of Agriculture
Peoria, Illinois

Thomas J. Elder, Ph.D.
Assistant Professor
Department of Forestry
Auburn University
Auburn, Alabama

David W. Goheen, Ph.D.
Project Leader
Pioneering Research
Central Research Division
Crown Zellerbach Corporation
Camas, Washington

Irving S. Goldstein, Ph.D.
Professor of Wood and Paper Science
Department of Wood and Paper
  Science
North Carolina State University
Raleigh, North Carolina

Richard W. Hemingway, Ph.D.
Forest Products Utilization Research
Southern Forest Experiment Station
U.S. Department of Agriculture
Pineville, Louisiana

Ed J. Soltes, Ph.D.
Associate Professor of Wood
  Chemistry
Forest Science Laboratory
Texas A&M University
College Station, Texas

Norman Storm Thompson, Ph.D.
Professor of Chemistry and Senior
  Research Associate
The Institute of Paper Chemistry
Appleton, Wisconsin

Duane F. Zinkel, Ph.D.
Research Chemist
Forest Products Laboratory
U.S. Department of Agriculture Forest
  Service
Madison, Wisconsin

TABLE OF CONTENTS

Chapter 1

BIOMASS AVAILABILITY AND UTILITY FOR CHEMICALS

Irving S. Goldstein

TABLE OF CONTENTS

I. WHY CHEMICALS FROM BIOMASS?

In the context of this book, chemicals refers to nonpolymeric organic compounds of commercial importance, and biomass refers to material produced by plants, for the most part terrestrial. Chemicals have been produced from biomass in the past, are being produced at present, and will be produced in the future wherever technical utility and economic conditions have combined or will combine to make it feasible. This is unremarkable and hardly worthy of yet another addition to an already swollen literature. However, the potential exists for biomass to assume a major role in meeting future chemical needs, and for this reason, special attention to the conversion of biomass into chemicals is warranted.

At present, almost all our organic chemicals and the synthetic organic polymeric materials derived from them are obtained from petroleum and natural gas. This dependence is so great that the term petrochemicals has become synonymous with the chemical industry. Stimulated by abundant supplies of cheap fossil hydrocarbons, the industry has grown far beyond its traditional coal tar base, and applied the concurrently emerging science and technology of polymers to meet the needs of expanding world populations for fibers, rubbers, plastics, adhesives, coatings, etc. displacing natural materials along the way.

Since 1973, oil and gas are no longer inexpensive and are becoming more expensive each year. Furthermore, there is an increasing awareness of the finite nature of these fossil liquid and gaseous hydrocarbons and their ultimate depletion. However, there is much argument over when this will occur. Two recent conferences have addressed the question of where the chemical industry will obtain its raw materials when oil and natural gas are no longer available. Similar opinions were voiced at both the Conference on Chemical Feedstock Alternatives held in Houston,[1] and at the World Conference on Future Sources of Organic Raw Materials (CHEMRAWN I) held in Toronto.[2]

For the next 10 years, oil will continue its undisputed leadership as the feedstock for chemicals, but then coal, at first, and later, hydrocarbons from oil shale and oil sands will assume greater importance. These fossil resources are more expensive to process than oil and are also finite, so ultimately, chemicals from biomass, the precursor of fossil fuels and our only renewable carbon source will have an important role as well. Whether this will occur after the year 2000 or before, will depend on the entire fossil energy picture and national and international policy decisions.

Since only the timing of a major contribution of chemicals from biomass to the raw material needs of the chemical industry is under question, the rationale for this book seems well established.

II. CHEMICALS FROM BIOMASS IN THE PAST

Before the era of petrochemicals, various chemicals were produced from biomass by such techniques as extraction, fermentation, and pyrolysis. The resinous exudates from pine trees provided the raw material for the naval stores industry, the oldest chemical industry in North America. First used for the tarring of ropes and the caulking of seams in wooden ships, the crude exudates were later distilled to provide turpentine and rosin, which have a great variety of industrial uses. Extracts from the heartwood of certain hardwoods, as well as the bark of various species, provided tannins, which as their name indicates, were important in tanning leather.

The origins of fermentation are lost in antiquity, and the production of ethanol for beverages by this method is still a major industry. However, the production of industrial ethanol by fermentation was virtually completely displaced by hydration of ethylene. Biomass substrates used for fermentation ethanol were sucrose, molasses, starch, or cellulose hydrolyzates. Commercial wood hydrolysis plants, producing glucose for fermentation, were operated in the U.S. during World War I and in Germany during World War II. Other chemicals that have been produced by fermentation include glycerol, butanol, acetone, isopropanol, lactic acid, acetic acid, oxalic acid, citric acid, itaconic acid, etc.

The destructive distillation of wood to produce charcoal was once an important industry,[3] but only vestiges remain in the U.S. to provide briquettes for outdoor recreational cooking. The gas produced in wood carbonization can be used as a low Btu fuel, and was occasionally used during World War II to power internal combustion engines when gasoline was unavailable. A number of volatile organic chemicals can be recovered from the distillate of wood pyrolysis. Acetic acid, methyl alcohol, and acetone were formerly obtained from wood distillation leading to the common name wood alcohol, for methanol. In addition, various wood tar oil fractions, used for medicinals, smoking meats, disinfectants and weed killers were recovered.

III. CHEMICALS FROM BIOMASS IN THE PRESENT

Considerable quantities of chemicals from biomass are still in use today. About 300 million lb of organic and amino acids are derived annually from a sugar or grain base by fermentation.[4] Chemicals derived from wood, sometimes called silvichemicals,[5] are for the most part by-products of the pulp and paper industry.

The chemical fragments of the cell wall polymers that end up in solution after pulping can be often be isolated from the pulping liquors and used. Sugars in spent sulfite liquor can be fermented to produce ethanol. By mild alkaline oxidation of lignin sulfonates, vanillin can be obtained for flavoring and odorant applications. Volatile products from kraft black liquor include dimethyl sulfide, dimethyl sulfoxide, and dimethyl sulfone, useful as solvents and chemical reactants.

The gum naval stores industry has for the most part been replaced by the recovery of oleoresinous wood components from the kraft pulping process. Volatiles such as turpentines are recovered from the digestor relief gases. The alkaline pulping liquor converts the fatty acids and resin acids to soaps that can be skimmed off the concentrated black liquor and acidified to yield crude tall oil. Almost a million tons of tall oil are produced annually in the U.S. Some turpentine and rosin are still obtained by steam distillation or extraction of pine stumps.

Wood hydrolysis to glucose is still in use in the U.S.S.R. Furfural is produced by strong acid treatment of pentose sugars from such agricultural residues as corn cobs, oat hulls, or sugar cane bagasse.

Phenolic acids extracted from the bark of various conifers are used as extenders for synthetic resin adhesives, and waxes extracted from Douglas-fir bark can be used for general wax applications. Rubber is obtained from latexes secreted by the inner bark of certain species, and vegetable oils are extracted from various plant seeds.

IV. POTENTIAL CHEMICALS FROM BIOMASS IN THE FUTURE

Future uses of biomass can be divided into two major categories. The first represents chiefly an extension and expansion of present practices. This category does not merely represent physical expansion of existing operations, but emphasizes extension to related opportunities in by-product and extractives utilization. These are elaborated in Chapters 9 to 11.

The second category would involve conversion of the cell wall polymers into low molecular weight chemical feedstocks. The major portion of the biomass consists of the cell wall components. This source of raw materials far exceeds the extractive components or the chemical by-products in volume, and represents a potential resource for meeting all of our chemical needs in place of petrochemicals. Large scale production of industrially important chemicals from lignocellulose can be accomplished by various routes[6] that are treated in detail in Chapters 3 to 8. These tonnage quantities of chemicals such as methanol, ethanol, furfural, ethylene, butadiene, phenol, benzene, etc. could provide the basic building blocks for conversion to synthetic polymers.[7]

V. PRODUCTION OF BIOMASS

Even before the energy crisis, ecologists had calculated the primary production of biomass in studies extending over the past 100 years. In one summary, Lieth[8] estimated total continental productivity of the world at 100 billion t/year, about 65% of the total continental and oceanic value of 155 billion t. This is somewhat more conservative than another estimate of 172 billion t/year on land, 74% of the world total of 233 billion t.[9] Both estimates attributed about 70% of terrestrial production to forests, and the latter study also calculated that 82% of the existing phytomass in the world is in forests.

Lieth[8] has also estimated the range of productivity per unit of area for various vegetation types, and these figures will be useful in arriving at specific estimates for the U.S. For temperate forests, the range of productivity is given as 600 to 2500 g/m²/year, with an average of approximately 1000 g/m²/year (2.7 to 11.2 tons/acre, with an average of about 4.5 tons/acre). For cultivated land, the range of productivity is given as 100 to 4000 g/m²/year, with an average of about 650 g/m²/year (0.5 to 17.8 tons/acre, with an average of about 2.9 tons/acre).

These generalized values may be compared with the specific calculation of 4.3 tons/acre as the potential biomass producton of east Texas woodlands,[10] and the maximum annual yield of sorghum (a C-4 plant) of 16 tons/acre.[11] Sugar cane, another C-4 plant with very limited site suitability in the U.S., has been reported to yield as much as 50 tons of biomass/acre/year.[11]

Considering that only 40% of the land area in the U.S. is classified as commercial timberland (500 million acres, or 22%) or cropland (427 million acres, or 19%), the application of Lieth's average productivity figures gives a total annual biomass productivity of 3.5 billion tons/year for the U.S., not counting unproductive or reserved forest land, pasture and rangeland, swampland, and industrial and urban areas. At 8000 Btu/lb, this is equivalent to 56×10^{15} Btu (Quads) or about 75% of the present total fuel use in the U.S.

VI. AVAILABILITY OF BIOMASS

Since the energy crisis of 1973, considerable interest has developed in the use of biomass to help meet the energy budget of the world. Projections of potential fuel material have included over-optimistic extremes based on maximum yields per acre for special crops on productive sites extrapolated to total acreages, as well as overly conservative minimum values based on residues from present harvesting and manufacturing practices. Realistic appraisals of the true availability of biomass fall somewhere in between. Recent surveys by Benemann[12] and by Tillman[13] consider these in detail.

For the purposes of this book, a simplified calculation using approximations and rounded numbers will suffice to provide a rough estimate of the biomass available in the U.S. for conversion to fuel or chemicals.

The minimum productivity criterion for classification of the 500 million acres of commercial forest land in the U.S. is not very stringent, only 20 ft³ of wood per acre per year, considerably less than 1 ton of biomass per acre. If, however, an average biomass productivity of 4.5 tons/acre is assumed, the increment would amount to about 2.25 billion tons of biomass being produced by our forests every year. However, since 140 million acres of this land is publicly owned as state and national forests, where total harvesting for energy or chemical uses would be opposed on environmental grounds, the wood produced on the remaining acreage amounts to only 1.6 billion tons. Furthermore, about 200 million tons of wood are now harvested from private and industrial lands for current production of solid wood and fiber products, reducing the maximum biomass available on a renewable basis for energy or chemicals from our forests to about 1.4 billion tons.

Our 400 million acres of cropland can often be more productive than forest land, with some sites yielding as much as 20 tons of biomass/acre. However, this land (certainly the best of it) is for the most part committed to our existing agriculture. Therefore, all we can reasonably expect in the way of biomass for energy or chemicals is the crop residues or animal wastes. These amount to[14] 130 million tons of cereal straws, 125 million tons of cornstalks, 100 million tons of miscellaneous residues, and 200 million tons of cow manure, for a total of approximately 550 million tons of agricultural residue biomass. The inclusion of 160 million tons of municipal solid waste brings the maximum amount of biomass potentially available for energy or chemicals in the U.S. to about 2.1 billion tons annually. This is equivalent to 34×10^{15} Btu.

To be realistic, we cannot expect to exploit all of this biomass. Of the private and industry forest lands, 80% is in private ownership. Many of these individuals would not allow harvesting of their lands for energy uses. The crop residues, animal wastes, and municipal wastes are often too widely dispersed for economical collection. Some availability factor must be applied to the potentially available figure, and if an arbitrary value of 20% is used, the amount available becomes about 400 million tons, and the energy equivalent about 6.5×10^{15} Btu, or just under 10% of annual needs.

VII. FUEL OR CHEMICALS

Biomass is not in an ideal form for fuel use. The heat content calculated above on a dry mass basis must be corrected for the natural water content that can reduce the net heat available by as much as 20% in direct combustion applications. Gasification to low Btu gas carries an additional net energy loss, and conversion to synthetic natural gas and liquid fuels results in still greater reductions of net energy to perhaps 30% of the original heat content. In the form of such synthetic liquid and gaseous fuels, the available biomass would yield only about 2 Quads of energy, or only about 3% of our present fuel needs. While not negligible, it would make only a minor contribution to the total energy needs.

There are sources other than fossil fuels for energy for power, heating, and transportation including nuclear fission and fusion, hydroelectric, wind, geothermal, tides, direct solar, and photoelectric. Vehicles can be powered by hydrogen or electric storage batteries. However, synthetic organic materials such as adhesives, electrical insulation, fibers, and rubbers that are so important in our economy, depend on a source of carbon for their synthesis. Presently, they are produced from natural gas and petroleum. Alternatively, they can be produced from coal, but ultimately biomass, the precursor of fossil fuels, will remain the only renewable carbon source.

Our impending materials crisis has received much less attention than our energy crisis, but the two are inseparable, because in a fossil energy economy, chemicals and energy are interdependent. A significant portion of our fossil hydrocarbons is used as feedstocks for the chemical industry. That portion converted to chemicals is not available as fuel, and, conversely, chemicals such as ethanol, derived from the processing of biomass, could alternatively be used for fuel or as the starting material for the polymers now obtained from petroleum. However, what would it profit us to have ethanol as a fuel for internal combustion engines if we did not have rubber tires?

While it has been shown that we do not have enough biomass to meet more than a small fraction of our energy needs, fortunately, the supply is adequate to meet our chemical material needs. Eventually, the carbon compounds derivable from biomass may need to be reserved for materials applications, with energy needs being met from other sources, although under present circumstances, petroleum and gas remain the chemical raw materials of choice, and biomass can only serve as an auxiliary energy source by direct combustion.

At present, 10% of our natural gas, 21% of our natural gas liquids, and 4% of our crude oil are used for chemicals.[1] On a combined basis, 7.5% of all liquid and gaseous hydrocarbon consumption is for chemicals, with about 4% (60 million tons or 2.5 Quads) directly assignable to feedstocks, and the remainder used as process fuel. This total energy consumption for chemicals is within the 6.5 Quads calculated as available from biomass. On a mass basis, the 400 million tons of available biomass would require only a net product yield of 15% to provide the 60 million tons of petrochemical feedstocks that actually end up as perhaps only 40 million tons of chemicals (only 10% of the weight of the available biomass). Conversion efficiencies of biomass to chemicals are better than that, with starch capable of yielding 32% ethylene or 22% butadiene. Even with present technology, cellulose can yield 16% ethylene or 11% butadiene.

Critics of chemicals from biomass point to the unfavorable ratio of carbon to oxygen in biomass for chemical synthesis and the low energy trajectory efficiency for producing liquid fuels from wood.[13] However, coal conversion through gasification has to pass through CO and H_2, which in equimolar mixture have the same carbon, hydrogen, and oxygen content as carbohydrates. The ready availability of reduced carbon in fossil liquid hydrocarbons and their consequent great advantage for either fuels or chemicals is a temporary situation resulting from geologic accident, and not the natural state in a world with an atmosphere containing 20% oxygen. Maintaining our chemical industry without fossil hydrocarbons will require an energy or material cost to provide equivalent materials. Although biomass can make only modest contributions to the energy needs of the U.S., it can provide the raw material to meet all our synthetic organic material needs on a renewable sustained yield basis.

VIII. THE SPECIAL PLACE OF WOOD

The biomass component with the greatest potential contribution to make is wood. It has already been shown that forests are responsible for producing over two thirds of all the dry matter that might be available for processing to energy or chemicals.

How is one to reconcile the popular conception that wood is becoming more scarce and expensive with the calculation that 1.4 billion tons of woody biomass suitable for conversion to energy or chemicals is produced in the U.S. each year? Much of our concern about our wood supplies stems from the traditional method of forest inventory, the merchantable bole concept, that counts only boles of selected species of at least a 5 in. diameter at breast height extending up to a 4 in. at the top. Despite the fact that the trend in wood utilization from solid products to reconstituted products is very marked, this inventory of sawtimber is the only one we have, and grossly underestimates the amount of wood available for new methods of utilization. Experience has shown that the tonnage yield of whole-tree chips is between two and three times the amount that would be expected from conventional cruise inventory data. Therefore, we actually have at least twice as much total wood as the U.S. Forest Service inventories indicate.[15]

That does not mean walnut logs or redwood trees, or even Douglas-fir or southern pine trees that we may need for lumber or pulp. In contrast to the traditional procurement restrictions for solid wood products of species and size, and for pulpwood of species suitability, fiber length, and fiber morphology, the form or nature of wood for conversion into chemicals is of no importance. Species or size does not matter. Mixed stands of uneven age that are otherwise not suitable (rotten trees, crooked trees, small trees, and noncommercial species) can be converted into chips in the forest with no residues. Furthermore, the potential utilization of this "green junk" (as it is alluded to by many foresters) could be the deciding factor in bringing about the use of the land for productive managed forests that would meet our increasing needs for high-quality wood for lumber, plywood, and pulp. The capital supplied by using low-grade wood for chemicals could hasten improved forest management.

This woody biomass exists in highly concentrated form, where densities of 75 tons/acre and greater are not unusual. Harvesting and collection using modern whole-tree harvesting equipment can supply raw material within a 50-mi radius at less than $0.01/lb dry basis.[16] Furthermore, there are no seasonal problems of collection and storage as with agricultural residues. Wood can be stored on the stump until needed and, even in the form of chip piles as at pulp mills, is much less susceptible to microbiological deterioration than nonwoody biomass.

REFERENCES

1. Van Antwerpen, F. J., Ed., *Proc. Conf. Chem. Feedstock Alternatives,* American Institute of Chemical Engineers, New York, 1977.
2. Krieger, J. H. and Worthy, W., CHEMRAWN I faces up to raw materials future, *Chem. Eng. News,* 56(30), 28, 1978.
3. Stamm, A. J. and Harris, E. E., *Chemical Processing of Wood,* Chemical Publishing, New York, 1953, 440.
4. Seeley, D. B., Cellulose saccharification for fermentation industry applications, *Biotechnol. Bioeng. Symp.,* 6, 285, 1976.
5. Goheen, D. W., Silvichemicals - what future, *Am. Inst. Chem. Eng. Symp.,* 69, 20, 1972.
6. Goldstein, I. S., Chemicals from lignocellulose, *Biotechnol. Bioeng. Symp.,* 6, 293, 1976.
7. Goldstein, I. S., Potential for converting wood into plastics, *Science,* 189, 847, 1975.
8. Lieth, H., Modeling the primary productivity of the world, *Cienc. Cult.,* 24(7), 621, 1972.
9. *Productivity of world ecosystems,* 5th Symp. Proc. Gen. Assembly Spec. Comm. Int. Biol. Prog., International Biological Program, National Research Council, National Academy of Sciences, Washington, D.C., 1975.
10. Byram, T. D., van Bavel, C. H. M., and van Buijtenen, J. P., Biomass production of East Texas woodlands, *Tappi,* 61(6), 65, 1978.
11. Bassham, J. A., Increasing crop production through more controlled photosynthesis, *Science,* 197, 630, 1977.
12. Benemann, J. R., Biofuels: a Survey, Special Report No. EPRI ER-746-SR, Electric Power Research Institute, Palo Alto, Calif., 1978.
13. Tillman, D. A., *Wood as an Energy Resource,* Academic Press, New York, 1978.
14. Renewable Resources for Industrial Materials, National Academy of Sciences, Washington, D.C, 1976.
15. The Outlook for Timber in the United States, Forest Resource Report No. 20, Forest Service, U.S. Department of Agriculture, Washington, D.C., 1973.
16. Goldstein, I. S., Holley, D. L., and Deal, E. L., Economic aspects of low-grade hardwood utilization, *For. Prod. J.,* 28(8), 53, 1978.

Chapter 2

COMPOSITION OF BIOMASS

Irving S. Goldstein

TABLE OF CONTENTS

I. COMMON FEATURES OF BIOMASS

If the composition of biomass were as varied as its morphology, it would be necessary to deal with many thousands of compounds in attempting to convert biomass to chemicals. Such variability is, in fact, present in the minor components of plants to the extent that individual species may be differentiated and classified by a system of chemical taxonomy. However, on a total biomass basis, the common features far outweigh the differences and only a small number of biomass components can yield almost the entire quantity of chemicals obtainable when mass, and not variety, is the criterion.

Plants are made up of individual cells, each of which has elaborated a cell wall that together serve to define the morphology of the plant, provide its structural support, and control the passage of water and nutrients. The formation, organization, and properties of these cell walls are complex[1] and beyond the scope of this book. However, they all consist predominantly of polysaccharides, of which cellulose is the most abundant, and in the case of woody plants they contain another important component, lignin. Since cell walls may comprise as much as 95% of the plant material, it is also apparent that most chemicals from biomass will be derived from only a small number of antecedents. This greatly simplifies the entire field of biomass conversion.

Other biomass components that are more variable in structure or quantity are extractives, the soluble compounds that are often species or genus specific, bark, the outer material of woody plants, nonstructural carbohydrates such as starch or sucrose, and proteins. More detailed descriptions of the cell wall components, extractives, bark, and nonstructural carbohydrates are presented in the following sections.

II. CELL WALL COMPONENTS

Plant cell walls contain skeletal polysaccharides, hemicelluloses, polyuronides, lignin, and proteins. Cellulose (the most important skeletal polysaccharide), the hemicelluloses, and lignin are the chief components of woody plant cell walls. Their structure and organization are described in detail below, since woody plants comprise 70% of the biomass produced.

Other skeletal polysaccharides include chitin, a glucosamine polymer found in most fungi; unbranched β-1,3′-linked xylan, a xylose polymer; callose, a polymer of β-1,3′-linked glucose residues; and a straight chain β-1,4′-linked mannose polymer. The latter three polysaccharides are found in a few families of mostly tropical algae.[1] Polyuronides are a minor component of mature cell walls in higher plants, but are abundant in cambial tissue and in algae. In land plants they exist as pectic acid, a polymer of α-1,4′-galacturonic acid residues, while in algae they exist as alginic acid, consisting of polymannuronic, polyguluronic and polyglucuronic acids. Proteins are also present in cell walls, especially in growing cells, but in relatively small amounts.

A. Cellulose

Cellulose is the most abundant organic material on earth, comprising approximately 50% of all biomass for an annual production of about 100 billion t. It is a long chain polymer of β-D-glucose in the pyranose form (Figure 1), linked together by 1,4′-glycosidic bonds to form cellobiose residues (Figure 2) that are the repeating units in the cellulose chain.

The β-linkage requires that the alternate glucose units must be rotated through 180°. An important implication of this structure is a marked tendency for the individual cellulose chains to come together to form bundles of crystalline order held together by hydrogen bonds as shown in Figure 3. The conformation of the pyranose rings is such

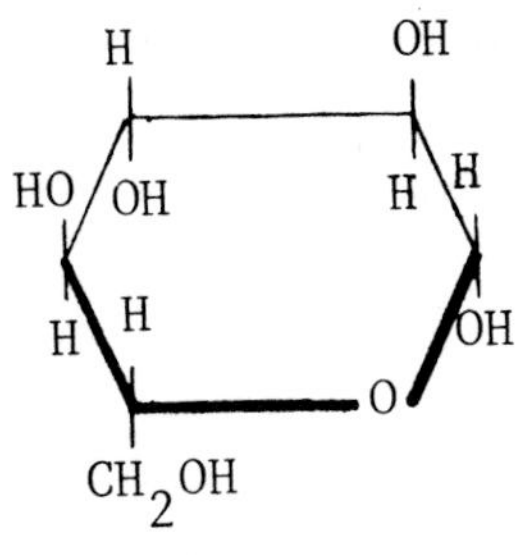

FIGURE 1. β-D-Gluco-pyranose.

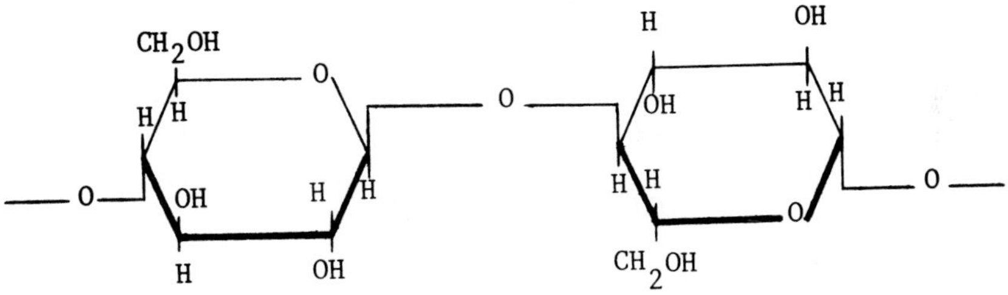

FIGURE 2. Cellobiose residue.

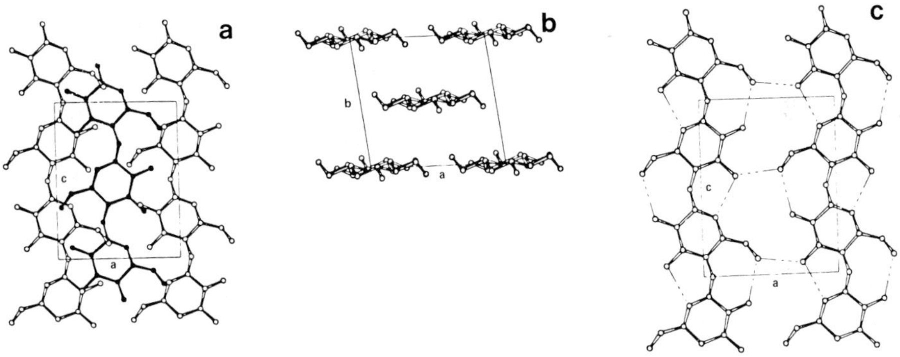

FIGURE 3. Projections of the parallel chain model for cellulose I. (a) Projection perpendicular to the ac plane. (b) Projection perpendicular to the ab plane looking along the fiber axis. (c) Hydrogen bonding network (From Blackwell, J., Kolpak, F. J., and Gardner, K. H., Structures of native and regenerated celluloses, in *Cellulose Chemistry and Technology*, Arthur, J. C., Jr., Ed., American Chemical Society, Washington, D.C., 1977, 49. With permission.)

that the total energy lies close to the minimum, a major factor in the high stability of cellulose.

Despite the presence of three hydroxyl groups on each anhydroglucose residue in the cellulose chain, cellulose is completely insoluble in water. The innumerable hydrogen bonds holding the chains together are not broken by water. Strong acids or alkalis, concentrated salt solutions, and various complexing reagents can swell and disperse, or even dissolve the cellulose, breaking up the highly ordered crystallites. On regeneration from dispersion or solution the cellulose assumes a different crystalline structure. The degree of crystallinity of celluloses varies with their origin and treatment. Cotton

(70% crystalline) is more ordered than wood pulp, which in turn is more ordered than regenerated cellulose (40% crystalline).

The chain length of cellulose as determined by measurements on cellulose solutions of different origins and treatment histories not unexpectedly shows considerable variation. Values for the degree of polymerization (DP) range from 7000 to 10,000 for wood, to as high as 15,000 for cotton.

Cellulose chemistry has been extensively treated in several publications.[3,4] For conversion into chemicals, the important features of the structure of cellulose are that it is a linear polymer of glucose with a highly ordered crystalline structure that limits the accessibility of reagents and enzymes. Cellulose utilization for chemicals will be considered in detail in Chapter 6.

B. Hemicelluloses

Closely associated with the skeletal polysaccharide cellulose in the cell wall are other structural polysaccharides collectively called hemicelluloses.[5] They differ from the cellulose in that although water insoluble, they can be dissolved in strong alkali. This property may be used to separate the hemicelluloses from the total carbohydrate fraction called holocellulose, leaving essentially pure or α-cellulose behind. Hemicelluloses are also more readily hydrolyzed by acid than is cellulose.

The hemicelluloses consist, for the most part, of sugars other than glucose, both pentoses and hexoses, and are usually branched with degrees of polymerization ranging from less than 100 to about 200 sugar units. Their greater solubility and susceptibility to hydrolysis than cellulose result from their amorphous structures and low molecular weights.

There are only a few basic hemicellulose structures found in all plants, which show small modifications in different plants, and often within the same plant in different tissues. In softwoods which contain about 25% hemicelluloses, the principal constitutent sugars in decreasing abundance are mannose, galactose, xylose, glucose, and arabinose. In hardwoods which contain about 30% hemicelluloses, the principal constitutent sugars in decreasing abundance are xylose, galactose, and mannose, with minor amounts of rhamnose and arabinose. Both types contain 4-*O*-methylglucuronic acid. In general, annual plants and hardwoods are rich in pentosans while in softwoods, the hexosans predominate.

Although linear xylans (1,4'-linked β-D-xylopyranose polymers) are found in many plants and arabinoxylans have been found primarily in certain grains and grasses, the most important hemicelluloses of hardwoods are glucuronoxylans, in which a linear or singly branched 1,4'-linked β-D-xylopyranose backbone has pyranose forms of 4-*O*-methyl-D-glucuronic acid attached by an alpha link to the xylose. Acetyl groups are also attached to this xylan system.

Arabinoglucuronoxylans, containing both arabinose and uronic acids, appear in wheat straw, softwoods, and in food crops. Arabinogalactans are water-soluble polysaccharides that are highly branched. Although more properly classified as extractives, they are also hemicelluloses by virtue of structure. They have been isolated from a number of softwoods, especially western larch where they constitute up to 18% of the weight of the wood.

The predominant hemicellulose in conifers is glucomannan of low molecular weight. The ratio of glucose to mannose varies from 1 to 1 to 1 to 4, and the structure is essentially 1,4'-β-pyranose in nature, with the length of the sequences of a single sugar dependent on the ratio of the monomers.

C. Lignin

The third major cell wall component in woody plants is lignin, comprising approxi-

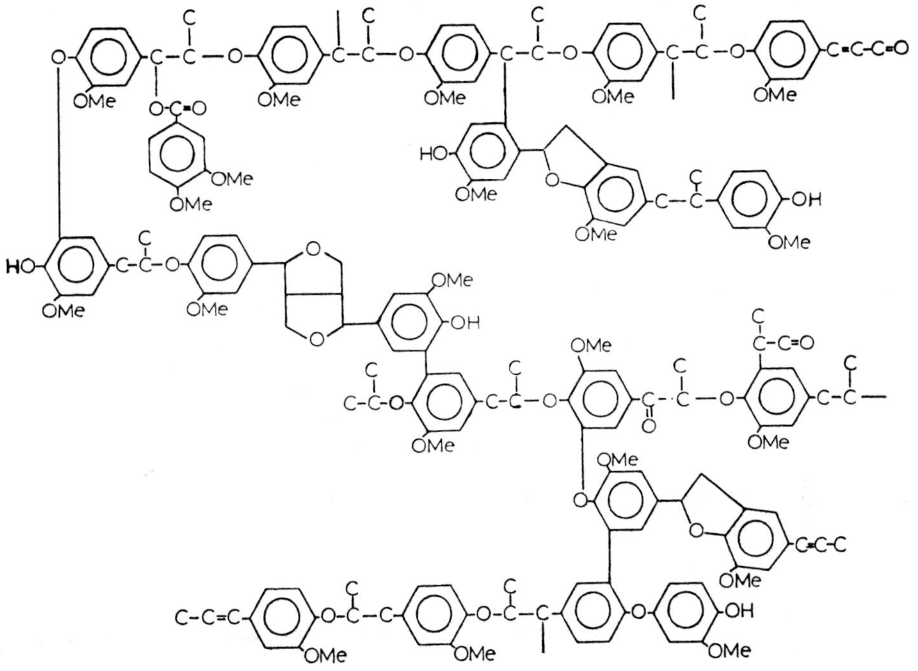

FIGURE 4. Abbreviated skeletal schematic structure of conifer lignin (From Goldstein, I. S., Chemicals from lignocellulose, *Biotechnol. Bioeng. Symp.*, 6, 298, 1976. With permission.)

mately the remaining 25% of the cell wall material. In fact, lignin is the necessary component for a plant to be classified as woody. Lignin serves as a cement between the wood fibers, as a stiffening agent within the fibers, and as a barrier to enzymatic degradation of the cell wall. In the production of chemical wood pulps, the lignin is dissolved away by various chemical processes.

Despite their importance and tremendous natural abundance, second only to cell wall carbohydrates, lignins resisted structural characterization until quite recently and still pose a major problem for their utilization. The occurrence, formation, structure, and reactions of lignins have been treated in detail.[6]

Lignins are three-dimensional network polymers formed from phenylpropane units that have randomly grown into a complicated large molecule with many different kinds of linkages between the monomers. The lignins from grasses, softwoods, and hardwoods differ somewhat in composition, chiefly in methoxyl substitution and the degree of carbon–carbon linkage between phenyl groups. However, their common structural features predominate, and the schematic structure of a conifer lignin[7] shown in Figure 4 shows the features important for conversion into chemicals. The aromatic and phenolic character is apparent, as is the covalent carbon–carbon bonding that prevents reversion to monomers by mild processing.

This random structure arises from an enzymatically initiated free radical polymerization of lignin precursors in the form of *p*-hydroxycinnamyl alcohols. In conifers the precursor is principally coniferyl alcohol (3-methoxy-4-hydroxycinnamyl alcohol) that yields the so-called guaiacyl lignin. In hardwoods and grasses, additional precursors such as sinapyl alcohol (3,5-dimethoxy-4-hydroxycinnamyl alcohol) and coumaryl alcohol (4-hydroxycinnamyl alcohol) are also present resulting in the so-called guaiacyl–syringyl lignins. Because the guaiacyl unit has an additional potential reactive site instead of the extra methoxyl group in the syringyl unit, a higher degree of crosslinking exists in guaiacyl lignins. The lower apparent molecular weight and easier dissolution of hardwood lignins are manifestations of their lower degree of cross-linking.

The molecular weight of any lignin sample under investigation will depend on its previous history. In its native state in the cell wall, the lignin probably exists as an infinite three-dimensional polymeric network by virtue of its origin from a random free-radical polymerization. However, it is not possible with present knowledge to remove lignin from the wood without some degradation. Even enzymatic removal of the carbohydrate fraction is suspect as far as modification of the lignin residue is concerned. Soluble lignin preparations exhibit wide molecular weight variations ranging from below a thousand to over a million daltons, and are polydisperse within the same preparation. The dissolved lignin fragments may be small enough to behave as simple compounds or large enough to exhibit high polymer behavior. In breaking the native lignin network, fragments of varying size are produced.

The chemical properties of lignins will also vary with the method of their isolation. Lignins which are by-products of different pulping processes are different from each other and differ in turn from the lignins in wood hydrolysis residues, whether acidic or enzymatic, and the lignins dissolved from wood by neutral solvents after ball milling or other pretreatment.

D. Cell Wall Organization

The cell wall components individually described above are intimately associated in their native state, and, as a consequence, their properties and response to processing are influenced by the close proximity of the other components as well as possible chemical combination with them.

Cellulose occurs in the cell wall in microfibrils that possess a crystalline core surrounded by an amorphous region. There is some evidence that the basic cellulose structure is an elementary fibril 35 Å wide (four unit crystallite cells) with the microfibril an assembly of four crystalline elementary fibrils. The length of the microfibrils has not been determined, but if a degree of polymerization for cellulose of 10,000 is accepted, the minimum length would be 5000 × 10 Å (the approximate length of the cellobiose residue) or 5 μm. Greater actual values would be expected because of chain overlap, and a continuous fibril is conceivable.

Microfibril widths vary from 100 to 300 Å depending on such factors as aggregation of single microfibrils to form larger assemblies and the amount of amorphous matrix material surrounding the cyrstalline areas. Hemicelluloses and lignin are the matrix polymers that surround the microfibrils as well as their constituent elementary fibrils. In addition, amorphous regions of the cellulose may also be penetrated by matrix polymers. The cell wall is then a fiber-reinforced plastic with cellulose fibers embedded in an amorphous matrix of hemicelluloses and lignin.

This composition is not uniform across the cell wall, with regions such as the middle lamella that cements the cells together containing higher proportions of lignin, especially in the cell corners. Various studies, including dissolution of either the lignin or carbohydrate portions of the cell wall leaving a carbohydrate or lignin skeleton,[8] respectively, and ultraviolet microscopy studies of the cell wall[9] have demonstrated the intimate association of lignin and carbohydrate within the cell wall at the ultramicroscopic if not the molecular level.

This molecular architecture of the plant cell wall has a great influence on the utilization of the components, whether for fibers or chemicals. The separation of the components prior to conversion to chemicals is difficult and expensive, and the application of conversion processes to the mixed components is often inhibited, as for example, the effect of lignin in preventing the access of cellulase to the cellulose.

III. OTHER BIOMASS COMPONENTS

While the cell wall structural materials constitute by far the greatest proportion of the biomass, significant quantities of other materials are also present. The most important are extractives that may comprise up to 10% of the biomass material, bark that makes up 10 to 15% of woody plant biomass, and nonstructural carbohydrates, which in some specialized cases such as cereal grains, constitute almost the entire seed. Vegetable oils are abundant in certain plant seeds, and rubber is found in the latexes secreted by the inner bark of some species. These two latter biomass components are already important commercially and have been thoroughly described in many other publications. Further treatment in this volume is unnecessary.

A. Extractives

Extractives is the term applied to the extraneous plant components that may be separated from the insoluble cell wall material by their solubility in water or organic solvents.[10,11] They include many different kinds of chemicals and a large number of individual compounds. Traditional uses of chemicals from biomass have involved such extractive components as turpentine, rosin, tall oil fatty acids, tanning materials, camphor, volatile oils, gums, and rubber.

Classification of extractives is made difficult by their great variety. Although related species often contain similar compounds, distinct differences occur even between closely related species so the total number of compounds is very large. Even within the same plant, different structures contain different extractives.

Major categories of extractives include volatile oils, terpenes (turpentines, tropolones, cymene, resin acids, and sterols), fatty acids, unsaponifiables, polyhydric alcohols, nitrogen compounds, and aromatic compounds (acids, aldehydes, alcohols, phenylpropane dimers, stilbenes, flavonoids, tannins, and quinones). Some of these are treated in detail in Chapters 9 to 11.

B. Bark

In contrast to the extensive literature devoted to the chemistry of wood and other plant cell wall material, bark, the outer part of the stem and branches surrounding the wood, has been relatively neglected. This probably results from its removal in traditional forest products processing and relegation to the status of waste. While not fundamentally different from wood in the types of components present, bark does have some special chemical properties.[12]

The structure of bark is very different from that of wood, and differences between species are much greater than in the case of wood. Extractives are much higher than in wood, amounting to 20 to 40% of the total bark. Extractive-free bark contains two components not found in wood, suberin and phenolic acids, as well as the familiar cellulose, hemicelluloses, and lignin.

Suberin is a hydroxy acid complex found in the cork cells of bark consisting of esters of higher aliphatic hydroxy acids and phenolic acids. Acids with 18 to 22 carbon atoms such as phellonic, phloionolic, and eicosanedicarboxylic acid have been identified. The suberin content varies from 2 to 8% in the barks of pine, fir, aspen, and oak, while it is much more prevalent in birch (20 to 39%) and cork oak(35 to 40%).

Phenolic acids are high molecular weight phenols that contribute to the Klason "lignin" content of bark (insoluble in 72% sulfuric acid), but differ from lignin in their lower molecular weight, alkali solubility (caused by high carboxyl content), and lower methoxyl content. They are easily extracted from bark by alkali and may comprise almost 50% of the weight of the bark of conifers.

It follows from the high contents of suberin and phenolic acids in some barks that the amounts of the typical wood components present must be considerably less. For example, the cellulose content of bark is low, usually between 20 and 30%. Of the hemicellulose components, galactose and mannose are more prevalent in coniferous tree bark, while xylose exists in much higher percentages in deciduous tree bark in agreement with their occurrence in the wood. Some of the chemicals derivable from bark are considered in Chapter 10.

C. Nonstructural Carbohydrates

Although the principal carbohydrate components of biomass are insoluble cell wall constituents, plants do produce significant quantities of more available carbohydrates that function as energy reserves for the plant or its seedlings. Most important of these are starch and sucrose.

Starch has been found in the sapwood of some species to the extent of about 5%. In seeds such as cereal grains, it is the predominant component. Like cellulose, it is a polymer of D-glucopyranose, but since it is linked in the α-1,4' configuration, the repeating unit is anhydroglucose rather than the dimer maltose, because the two glucose residues in maltose are indistinguishable and superimposable. In contrast to the stiff straight chains of cellulose that readily cyrstallize, the α-linked starch chains are more flexible and unlike cellulose, can be dissolved or dispersed in hot water because of fewer interchain hydrogen bonds.

Furthermore, starch is a mixture of two glucose polymers.[13] The linear fraction amylose that is water soluble comprises about 20% of the starch. Its degree of polymerization is commonly between about 400 and 900, although values as high as 6000 have been reported.

The other starch component, amylopectin, is swollen by water, but is insoluble. It is a branched polymer of glucose with its straight chain portions similar in structure to amylose. However, the branches are provided by α-1,6' links. The molecular weights of amylopectins are in the range of several million, resulting from thousands of relatively short chains (25 units) assembled in a tree-like structure.

By virtue of the above structural features, starch is readily accessible to hydrolysis by either acids or enzymes, and has traditionally provided the major substrate for fermentation processes (Chapter 3).

Sucrose is a dimer of glucose and fructose found in high concentrations in such plants as sugar cane (14% of juice) and sugar beets (16 to 20% of weight). It is common in the sap of trees, for example, maple sugar. In addition to its importance in fermentation, it can also be converted into useful derivatives.

IV. COMPOSITION OF VARIOUS TYPES OF BIOMASS

The biomass components described above are found in differing amounts in plant matter of various origins depending on species and other biological variations such as genetic differences within species, specialized tissues within individual plants, and growing conditions. These differences in biomass composition would be of little concern in gasification processes that nonselectively convert the biomass to carbon monoxide and hydrogen. They would be of greater concern in more selective biomass conversion processes where, for example, pentose sugars might be either a desired or unwanted product. In addition, species differences are completely determinant in using the biomass for specific components such as turpentine or rubber.

The range of variation of individual sugars in wood is shown in Table 1. Greater variations in composition among plant materials of widely different origins are apparent from Table 2. The high lignin content of most of the biomass available is evident and must be considered in the processing of biomass to chemicals.

Table 1
RELATIVE ABUNDANCE OF
INDIVIDUAL SUGARS IN
CARBOHYDRATE FRACTION OF WOOD
(%)[5, 14]

Sugar	In softwoods	In temperate hardwoods
Glucose	61—65	55—73
Xylose	9—13	20—39
Mannose	7—16	0.4—4
Galactose	6—17	1—4
Arabinose	<3.5	<1
Rhamnose	<1	<1
Uronic acids	4—7	4—7

Table 2
COMPOSITION OF VARIOUS TYPES OF BIOMASS (% dry weight)

Material	Cellulose	Hemicelluloses	Lignin	Pectic compounds	Starch	Extractives	Ref.
Algae (green)	20—40	20—50	—	30—50	—	—	1
Cereal grains	—	—	—	—	65—70	—	—
Cotton, flax, etc.	80—95	5—20	—	—	—	—	15
Grasses (including palms, bamboo, sugar-cane, etc.)	25—40	25—50	10—30	—	—	—	15
Wood							
Amazonian	48	22	26	—	—	6.4	14
Teak	34	17	30	—	—	15	14
Temperate hardwoods	45 ± 2	30 ± 5	20 ± 4	—	—	5 ± 3	16
Softwoods	42 ± 2	27 ± 2	28 ± 3	—	—	3 ± 2	16
Wastes and Residues							
Animal manure	18	—	—	—	—	—	17
Corn cobs	41	36	6	3	—	14	18
Cornstalks	29	28	3.1	—	—	—	19
Wheat straw	40	29	14	—	—	—	19
Newspapers	40—55	25—40	18—30	—	—	—	15
Chemical pulps	60—80	20—30	2—10	—	—	—	15

REFERENCES

1. Preston, R. D., *The Physical Biology of Plant Cell Walls,* Chapman and Hall, London, 1974.
2. Blackwell, J., Kolpak, F. J., and Gardner, K. H., Structures of native and regenerated celluloses, in *Cellulose Chemistry and Technology,* Arthur, J. C., Jr., Ed., American Chemical Society, Washington, D.C., 1977, 49.
3. Ott, E. and Spurlin, H. M., *Cellulose and Cellulose Derivatives,* Vols. 1—3, Interscience, New York, 1954—1955.
4. Immergut, E. H., Cellulose, in *The Chemistry of Wood,* Browning, B. L., Ed., Interscience, New York, 1963, chap. 4.
5. Schuerch, C., The hemicelluloses, in *The Chemistry of Wood,* Browning, B. L., Ed., Interscience, New York, 1963, chap. 5.
6. Sarkanen, K. V. and Ludwig, C. H., *Lignins: Occurrence, Formation, Structure and Reactions,* Wiley-Interscience, New York, 1971.
7. Freudenberg, K. and Harkin, J. M., Modified structural formula for spruce lignin, *Holzforschung,* 18, 166, 1964.
8. Côté, W. A., *Wood Ultrastructure: An Atlas of Electron Micrographs,* University of Washington Press, Seattle, 1967.
9. Procter, A. R., Yean, W. Q., and Goring, D. A. I., The topochemistry of delignification of kraft and sulfite pulping of spruce wood, *Pulp Pap. Mag. Can.,* 68, T-445, 1967.
10. Buchanan, M. A., Extraneous components of wood, in *The Chemistry of Wood,* Browning, B. L., Ed., Interscience, New York, 1963, chap. 7.
11. Hillis, E. W., *Wood Extractives and Their Significance to the Pulp and Paper Industry,* Academic Press, New York, 1962.
12. Jensen, W., Frener, K. E., Sierila, P., and Wartiovaara, V., The chemistry of bark, in *The Chemistry of Wood,* Browning, B. L., Ed., Interscience, New York, 1963, chap. 12.
13. Bolker, H. I., *Natural and Synthetic Polymers,* Marcel Dekker, New York, 1974.
14. Browning, B. L., Ed., The composition and chemical reactions of wood, in *The Chemistry of Wood,* Interscience, New York, 1963, chap. 3.
15. Cowling, E. B. and Kirk, T. K., Properties of cellulose and lignocellulosic materials as substrates for enzymatic conversion processes, *Biotechnol. Bioeng. Symp.,* 6, 95, 1976.
16. Thomas, R. J., Wood: structure and chemical composition, in *Wood Technology: Chemical Aspects,* Goldstein, I. S., Ed., American Chemical Society, Washington, D.C., 1977, chap. 1.
17. Stephens, H. R. and Heichel, G. H., Agricultural and forest products as sources of cellulose, *Biotechnol. Bioeng. Symp.,* 5, 27, 1975.
18. Foley, K. F., *Chemical Properties, Physical Properties and Uses of the Anderson's Corncob Products,* The Andersons, Maumee, Ohio, 1978.
19. Sloneker, J. H., Agricultural residues, including feedlot wastes, *Biotechnol. Bioeng. Symp.,* 6, 235, 1976.
20. Goldstein, I. S., Chemicals from lignocellulose, *Biotechnol. Bioeng. Symp.,* 6, 298, 1976.

Chapter 3

BIOCONVERSION OF AGRICULTURAL BIOMASS TO ORGANIC CHEMICALS

Robert W. Detroy

TABLE OF CONTENTS

I. INTRODUCTION

This article will deal primarily with the current methods available to generate organic chemicals via fermentation from crop biomass, starch materials, agri-residues, and agro-industrial wastes. A comprehensive analysis of the characteristics and availability of agri-residues and industrial wastes is available and will be identified by other authors contributing to this subject. Relative composition of biomass, residues, and waste materials will be identified only when necessary to define substrates for production of specific chemicals through fermentation. Extensive studies on the utilization of animal products and animal waste management by Loehr[1] cover research conducted in the past 15 years. Overviews by Sloneker et al.[2,3] on crop residues and animal wastes defines the availability of these resources in the U.S. A more recent review by Detroy and Hesseltine[4] deals mainly with both chemical and microbiological conversion of crops and agri-residues to useful by-products, i.e., animal feed supplements, biopolymers, single-cell protein, methane, and chemical feedstocks.

II. IDENTIFICATION AND POTENTIAL OF BIOMASS AND AGRI-RESIDUES

Increasing attention has been noted to the possibilities of utilizing photosynthetically active plants as natural solar energy-capturing devices, with the subsequent conversion of available plant energy into useful fuels or chemical feedstocks, such as alcohol and biogas, via fermentation. Acquisition of biological raw materials for energy capture follows three main approaches: (1) purposeful cultivation of so-called energy crops, (2) harvesting of natural vegetation, and (3) collection of agricultural wastes. Lewis[5] has recently described the energy relationships of fuel from biomass in terms of net energy production processes (Table 1). Table 1 presents data in terms of energy requirements, net energy gains and losses, and land area equivalents for a number of relevant conversion systems. Starch crops like cassava and other saccharide plants, notably sugar cane, appear to be the most favorable in terms of energy balance. More technological innovations would be required to derive a favorable energy balance for the conversion of the lignocellulosic raw materials owing to the energy intensive pretreatment requirements to render the substrate fermentable.

Biomass, or chemical energy, can serve as an energy mechanism to be harvested when needed and transported to points of usage. Land availability must be carefully evaluated in view of the potential of this energy alternative.

Since energy deficits are enormous, significant sources of biomass must be acquired. Some 95% of the field crops are planted for food grains. Since the majority of the plant residues (stalks and straw) are unused after harvest, these residues are potentially available for collection and conversion to useful energy.

The potential annual supply of U.S. cellulosic residues from domestic crops is certainly in excess of 500 million tons (dry weight). In general, cereals produce some 2 lb of straw per pound of grain harvested. Significant accumulations of major crop residues are, of course, confined to those areas of intensive cropping. The general distribution of potentially collectible cereal straws in the U.S. is depicted in Figure 1. All crops produce collectible residues; however, the distribution of straw residues increases the costs of utilization. These collectible residues from major and minor crops are depicted in Tables 2 and 3. The residues produced by the majority of these crops are left in the fields after harvest. Only with sugar cane, vegetables, fruit, and peanuts are there significant accumulations at specific processing sites.

Since the quantity of straw produced is equal to or greater than the quantity of

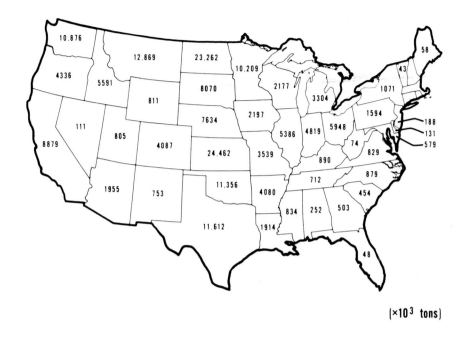

(×10³ tons)

FIGURE 1.   Geographical distribution of cereal straws (flax, wheat, rye, rice, oats, and barley).

Table 1
ENERGY REQUIREMENTS, NET ENERGY GAINS AND
LOSSES, AND LAND AREA EQUIVALENTS FOR A
NUMBER OF CONVERSION AND PRODUCTION SYSTEMS

| | | | Net energy | |
Principal substrate	Product	GER product (GJ/t)	(GJ/t)	(GJ/ha/yr)
CO_2	Energy crops	1.26	+16	+1090
Raw sewage	Algae[a]	57	−34	−850
Raw sewage	Algae[b]	18	+5	+125
Algae	Methane[b]	168	−112	−627
Livestock waste (UK)	Methane	144	−88	−0.88
Sugar cane	Ethanol	24	+3	+51
Cassava	Ethanol	61	−34	−71
Timber	Ethanol[c]	239	−212	−74[d]
Timber	Ethanol[e]	98	−71	−16[d]
Straw	Ethanol	222	−195	−138

[a]   The figures relate to current methods adopted.
[b]   The figures are estimates of what should be possible at present.
[c]   Cellulose hydrolyzed to fermentable sugars by fungal enzymes.
[d]   Figures expressed on basis of land area requirement to annually replenish the quantity of wood substrate used.
[e]   Cellulose hydrolyzed to fermentable sugars by acids. Also requires 470% manpower increase over enzyme route.

edible grain from cereal crops, its utilization is of paramount importance. Present constraints on the utilization of cereal by-products include: new technology development, residue collection, marketability, practical utility of residues, and research on

Table 2
MAJOR CROPS—CURRENT ESTIMATES

Commodity	Acres harvested (× 10⁶)	Tons/acre	Residue (dry wt) Total × 10⁶ Minimum	Maximum
Corn	65	2—3	130	195
Hay	64	3—7	192	448
Soybeans	60	1—2	60	120
Wheat	60	1—2	60	120
Sorghum	16	2—3	32	48
Oats	14	1—2	14	28
Cotton	12	1—2	12	24
Barley	11	1—2	11	22
Total	302	—	319[a]	557[a]

[a] Total yields do not include hay crop.

Table 3
MINOR CROPS—CURRENT ESTIMATES

Commodity	Acres harvested × 10⁶	Tons/acre	Residue (dry wt) Total × 10⁶ Minimum	Maximum
Vegetables	3.5	1—2	3.5	7.0
Fruit	3.3	1	3.3	3.3
Rice	2.2	1—2	2.2	4.4
Flax	1.8	1	1.8	1.8
Peanuts	1.5	1—2	1.5	3.0
Sugar beets	2.0	1—2	2.0	4.0
Sugar cane	1.5	6—10	9.0	15.0
Rye	1.0	1—2	1.0	2.0
Total	16.8	—	24.3	40.5

model bioconversions. Collection costs of important residue resources govern the economic feasibility of bioconversion processes for fermentation chemicals.

Mechanical equipment exists for harvesting corn refuse, silage, or hay, and can be readily be used for the collection and hauling of plant residues to central locations for processing. Sloneker[3] discusses types of harvesting operations that can be employed to stack, bail, windrow, chop, and transport various crop residues. Time and expensive equipment are serious deterrents to collection of crop refuse in on-the-farm operations. Any major increase in the use of cereal straws and other residues will require major efforts to collect, handle, transport, and deliver at a central location or plant so that they will be competitive with other raw materials for chemical production. Benefits from mass collection of straw residue must be balanced against the consequences of its removal from fertile crop land. Residues plowed under or left on the surface (conservation tillage) increase the tilth of the soil, aid in H_2O sorption, and reduce soil erosion; therefore, the impact that continuous residue removal will have on soil fertility must be thoroughly examined. Refractory material that remains after bioconversion of agro-residues may, if returned to the land, provide sufficient organic matter in the soil for tilth.

Table 4
GRAIN PROCESSING WASTE CHARACTERISTICS[a]

Parameter	Corn wet milling (average)	Corn dry milling (average)
Flow[b]	18.3	—
Biological Oxygen Demand (BOD)	7.4	1.14
Chemical Oxygen Demand (COD)	14.8	2.69
Suspended solids	3.8	1.62

[a] Corn wet milling, to produce corn syrup or starch. Corn dry milling —
    to produce meal and flour, water usage limited to washing, tempering,
    and cooling.
[b] Flow = 1/kkg grain processed. BOD and suspended solids = kg/kkg
    grain processed.

From Development Document for Effluent Limitations Guidelines and New
Source Performance Standards for the Grain Processing Segment of the
Grain Mills Point Source Category, EPA 440/1-74-028a, Environmental
Protection Agency, Washington, D.C., 1974.

The wet-milling process of cereal grains produces considerable quantities of grain carbohydrate waste. The waste-liquid streams that arise as a result of steeping, corn washing, grinding, and fractionation of corn yield cornstarch, corn syrup, gluten, and corn steep liquor. Increased studies are necessary on the bioconversion of these negative value carbohydrate wastes into alcohol, C_3 and C_4 chemicals, and methane, as well as on economical pretreatment of the industrial waste being produced. A summary of waste characteristics from grain processing is depicted in Table 4. No process wastewaters are produced by the milling of wheat and rice grains. However, the bran from these two cereals contains 5 to 10% oil and is rich in certain B vitamins and amino acids.

A major potential resource of the immense animal industry in the U.S. is the annual generation of over 2 billion tons of waste. Recent changes in the fertilizer and animal-feeding industries have resulted in the accumulation of animal wastes into localized areas. This localization has produced air and water pollution problems. Technological changes in large-volume cattle feeding have created a serious need for new waste technology, either through cost reductions in handling to eliminate pollution hazards or some type of bioconversion process to useful fuels or chemical feedstocks.

The utilization of animal wastes, other than land usage, as a waste management alternative has proceeded in two main areas: biological and thermochemical. Major experimentation has involved methane formation, single-cell protein production, and microbial fermentation and refeeding. Animal wastes are excellent nutrient sources for microbial development. Major constituents are organic nitrogen (14 to 30% protein), carbohydrate (30 to 50%, essentially all cellulose and hemicellulose), lignin (5 to 12%), and inorganic salts (10 to 25%).

In most biological processes, microorganisms consume nutrients present in the wastes to increase their own biomass and, through substrate utilization, release various gases and other simple carbohydrate materials. There are mainly two classes of biological processes: biogas (or an anaerobic fermentation) and biochemical hydrolysis. The biochemical processes produce primarily protein, sugar, and alcohol, whereas the anaerobic fermentation takes place under an oxygen-deficient environment to produce methane.

All of these processes have been successfully demonstrated for livestock manure.[6]

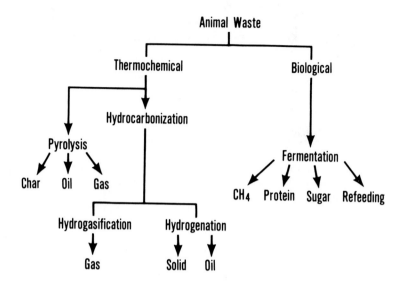

FIGURE 2.   Process alternatives for the generation of fuels from animal waste.

Table 5
MANURE PRODUCTION IN
THE UNITED STATES[a]

Animal[b]	Dry measure $\times 10^6 t$	Percent of total
Cattle	210	81.3
Swine	25	9.7
Horses	14	5.4
Poultry	6.2	2.4
Sheep	3.1	1.2
All	258.3[a]	100

[a]   Wet weight $= 1.5 \times 10^9$ t at 16.1% dry matter.

The various biological and chemical processes alternatives for the generation of renewal fuels and chemicals from animal manure is depicted in Figure 2. Total production of manure in the U.S. according to classes of animals and relative concentrations to the total, is shown in Table 5.[7-9]

The utilization of sugar cane bagasse must be considered on a country-by-country basis. Bagasse is the fibrous residue obtained after the extraction by crushing of sugar cane stalks. This roller-mill process removes 95% of the sucrose, producing a residue that contains some 50% moisture and consists of 15% lignin and 75% cellulose. Annual world production of bagasse is greater than 100 million tons. Bagasse has been used mainly as a fuel in sugar cane factories, for production of pulp and paper, and for structural materials. Extensive research has been conducted in the past few years on bagasse as a cellulosic raw material for single-cell protein production.[10,11] Cellulosic wastes, such as bagasse, have also received considerable attention as resource material for chemical processes and energy conversions (anaerobic fermentation to methane or ethanol).

The largest wastes from dairy food plants are whey from cheese production and

Table 6

RAW WASTE LOADS[a] FOR THE FRUIT AND
VEGETABLE PROCESSING INDUSTRY

Category	Flow (gal/ton)	BOD (lb/ton)	Total suspended (lb/ton solids)
Fruit			
Apple processing	690	4.1	0.6
Apple products, except juice	1,290	12.8	1.6
Citrus, all products	2,420	6.4	2.6
Olives	9,160	87	15
Pickles, fresh packed	2,050	19	4
Tomatoes	2,150	8	12
Peeled products	1,130	3	5
Vegetables			
Asparagus	16,530	4.2	6.9
Beets	1,210	39.4	7.9
Carrots	2,910	39.0	24
Corn			
Canned	1,070	28.8	13.4
Frozen	3,190	40.4	11.2
Lima beans	6,510	27.8	20.7
Peas			
Canned	4,720	44.2	10.8
Frozen	3,480	36.6	9.8
White potatoes	1,990	54.6	74.8

[a] The raw waste load is in terms of the quantity of wastewater parameter per ton of raw material processed for fruits and vegetables. Raw waste loads are those generated from canning processing.

pasteurization water. A pound of cheese produces 5 to 10 lb of fluid whey with a biological oxygen demand (BOD) of 32 to 60 g/ℓ, depending upon the process. Whey is an excellent nutrient source for microbe development, containing 5% lactose, 1% protein, 0.3% fat, and 0.6% ash.

Processing plant wastes for different fruits and vegetables vary in character and quantity. The effluents consist primarily of carbohydrates, starches and sugars, pectins, vitamins, and plant cell-wall residues. One must consider how the various processing operations affect availability and type of residues. Table 6 depicts some typical fruit and vegetable residues and characteristics based upon the quantity of material processed or quantity of material produced. Supply problems, due to various geographical locations and seasons, hinder large-scale utilization of these residues for fermentation purposes. Waste-waters and peels from potato processing also serve as an excellent starch source, but seasonal production hinders utilization of residues. The most promising end uses for potatoes involve recovery of starch for cattle feeding and for production of sugar, single-cell protein, and biogas.

The enormous amounts of spoiled, damaged, and culled fruits and vegetables are excellent sources of carbohydrate material. These materials typically are good substrates for the growth of many fungi, especially on acid fruits. However, a real problem exists in that these materials are seasonal, so that a microbial process cannot be run the year around because large amounts are available only at certain times.

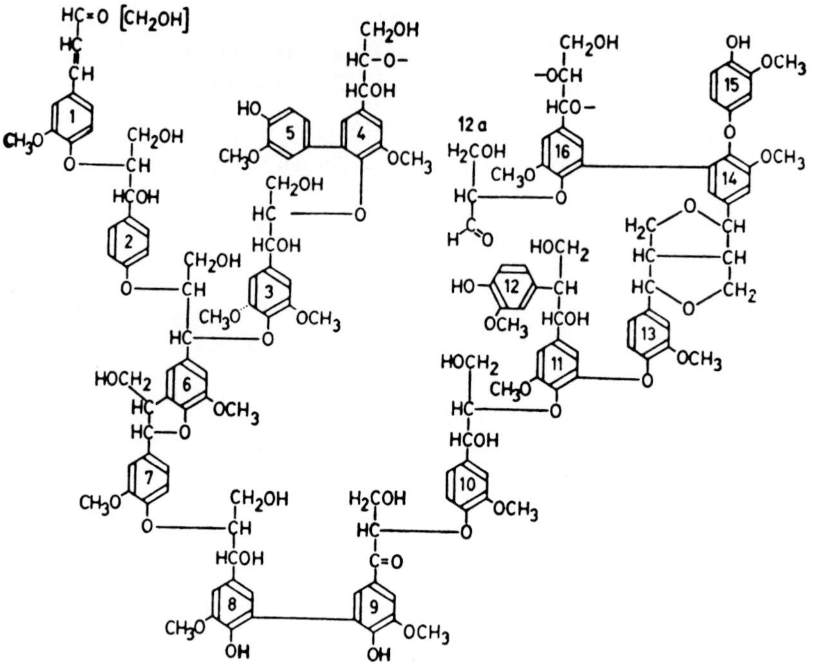

FIGURE 3.   The structure of lignin.

III. COMPOSITION OF AGRI-COMMODITIES

The major components in agricultural residues are the structural cell-wall polysaccharides, primarily cellulose and hemicellulose. The latter two are the most plentiful renewable resource produced by most green plants. These carbohydrates constitute 45 to 70% of the weight of a dried plant, varying according to age and maturity of plant at harvest. Pure cellulose, such as cotton fiber, is rarely found in nature, but rather in combination with other polymers such as lignin, pectin, and hemicellulose. Lignin comprises from 3 to 15% of the dried plant residue. This material is the structural glue that binds filaments of cellulose into fibers for cell integrity and rigidity. Lignin is found in all fibrous plants, and generally increases with age of the plant. Cellulose increases in aging fibrous plants with a decrease in soluble sugars and an increase in lignin. Lignin is a three-dimensional polymer formed by the condensation of cinnamyl alcohol monomers depicted in Figure 3. All possible combinations of the cinnamyl radicals can occur, resulting in various types of bonding. The exact linkage and structure of the lignin-cellulose complex is of considerable debate. There is considerable intermolecular bonding between the uronic acids of hemicellulose and lignin phenolic groups. Lignin apparently forms a three-dimensional net around the cellulose fibers. It is in this fashion that the complex cellulose is rendered unavailable to subsequent enzyme degradation. It is also in this complex area of lignin-cellulose interaction where the ultimate utility of agro-residues has its future. Chemical and/or biological modification of this lignocellulosic complex would result in increased digestibility of the agro-residue, increased hydrolysis rates, and saccharification. Continued research in the area of utilizing lignocellulosics is of paramount importance to the future of these negative value carbohydrate wastes. Table 7 depicts the relative composition of some important U.S. agro-residues.

Table 7
COMPOSITION OF AGRICULTURAL RESIDUES

Plant residue	Carbohydrate (%)						Cellulose	Lignin (%)	Protein (%)
	Arabinose	Xylose	Mannose	Galactose	Glucose	Total			
Cornstalks	1.9	15.5	0.6	1.1	37.7	56.8	29.3	3.1	5.5
Flax straw	2.1	10.6	1.3	2.2	34.7	50.9	34.5	—	7.2
Kenaf stalks	1.5	12.8	1.6	1.3	41.4	58.6	41.9	12.3	4.6
Soybean straw	0.7	13.3	1.7	1.2	43.7	60.6	41.4	—	5.5
Sunflower stalks	1.4	19	1.35	0.05	39.4	43.8	35.1	—	2.1
Sweet clover hay	3.2	7.2	1.2	1.7	31.1	44.4	29.8	—	24.7
Wheat straw	6.2	21.0	0.3	0.6	41.1	69.2	40.0	13.6	3.6
Cattle waste	0.38	0.77	0.73	0.97	24.4	27.2	16.4	6.5	10.1
Swine waste	0.43	0.83	0.98	1.27	25.5	29.8	16.6	1.6	15.1

IV. TECHNOLOGIES FOR UTILIZATION OF RESIDUES

Residue utilization must be considered with optimism due to the large quantities of wastes and by-products available, the need to better utilize existing resources, and the successful processes that have been attained. Successful residue utilization must include the following changes in approach:

1. Residues as resources, not wastes
2. Incentives to change philosophy
3. Evaluation of socioeconomic aspects
4. Use of appropriate technology
5. Beneficial use
6. Proper market
7. Better usage of raw materials

Promising technologies are needed for the utilization of agricultural and agro-industrial residues. Some of the most promising and successful technological processes for the utilization of agro-wastes are described in Table 8.

V. CHEMICALS FROM CARBOHYDRATE RAW MATERIALS

Recent progressive increases in the cost of crude oil have resulted in considerable attention being focused upon fermentation technology. The major production of industrial alcohol and of C_3 and C_4 chemicals is derived from fossil fuels. Alternative process routes for the production of organic chemicals involve fermentation primarily through bioconversion of carbohydrate raw materials to chemicals. Tong[12] has recently described fermentation routes for the production of C_3 and C_4 chemicals from specific available raw materials.

The major organic chemicals that are produced from carbohydrate raw materials by microbial fermentation are identified in Table 9. The main carbohydrate sources for fermentation as follows:

1. Starch grains from corn, wheat, barley, and other cereals
2. Sucrose from beet, cane, and sorghum
3. By-product materials from processing, i.e., fruit and vegetable wastes, starch streams from milling grains, cattle feedlot waste, dairy whey, molasses, and distiller grain

The various chemicals produced via fermentation will be discussed individually in terms of yield, substrate resource, and future opportunities as alternative resource and feedstock chemicals.

VI. CONVERSION OF BIOMASS TO SUGAR

As mentioned previously, the bioconversion of plant biomass to fermentation chemicals depends upon the basic structural composition and integrity of lignocellulose. Most lignocellulosic plant materials require some preliminary biological and/or chemical pretreatment before a direct fermentation to ethanol or other chemicals can be investigated. In general, before a microbial fermentation can be contemplated, the plant polymers, whether lignocellulosics, hemicellulose, or starch, must be hydrolyzed to simple sugars for utilization.

Table 8

TECHNOLOGIES AVAILABLE FOR UTILIZATION OF AGRO-INDUSTRIAL AND AGRICULTURAL WASTES

Residue substrate	Process	Product	Operational[a]	
			Advantages	Disadvantages
Animal waste	Microbial	CH_4, feed supplement	Cheap resource, produces energy, available, reduce pollution	High initial investment
Animal waste	Microbial	Cattle refeeding, single-cell protein		
Sugar cane bagasse	Microbial	Single-cell protein	Surplus availability, technology available	
Dairy whey	Microbial	Single-cell protein, alcohol	Reduce pollution, surplus availability	High salt content, transportation
Cereal process waste	Microbial	Single-cell protein	Reduce BOD and COD	
Cellulosic pulps	Enzymatic (saccharification)	Sugar		
Hemicellulosics (xylans)	Enzymatic	Xylose		Expensive
Starch waste	Microbial	Alcohol	Cheap resource	
Wood pulp sulfite liquor	Microbial	Single-cell protein	Surplus availability	

[a] Items listed on basis of economics, availability, pollutant, and source.

Table 9
CHEMICALS FROM FERMENTATION PROCESSES

Chemicals	Structure	Produced by
Ethanol	CH_3CH_2OH	Species of *Saccharomyces*
n-Butanol	$CH_3CH_2CH_2CH_2OH$	*Clostridium acetobutylicum*
2,3-Butylene glycol	$\underset{\displaystyle CH_3CH-CH-CH_3}{\overset{\displaystyle OH\ \ \ OH}{}}$	Species of *Aerobacter* and bacilli
Glycerol	$\underset{\displaystyle CH_2OH}{\overset{\displaystyle CH_2OH}{HC-OH}}$	Species of *Saccharomyces*
Acetic acid	CH_3COOH	*Clostridium thermoaceticum* *Acetobacter* species
Acetone	$H_3C-\overset{\displaystyle O}{\overset{\|}{C}}-CH_3$	*Clostridium acetobutylicum*
Isopropanol	$CH_3\overset{\displaystyle OH}{\underset{}{CHCH_3}}$	Species of *Clostridium* and bacilli
Fumaric acid	COOH—CH=CH—COOH	Species of *Rhizopus* and mucor
Succinic acid	CH_2COOH—CH_2COOH	Species of mucor, *Rhizopus, Fusarium*
Citric acid	COOH—CH₂—C(OH)(COOH)—CH₂—COOH	*Aspergillus niger, Candida lipolytica*
Lactic acid	CH₃—CH(OH)—COOH	Species of *Rhizopus* and mucor, lactobacilli
Propionic acid	CH_3CH_2COOH	Species of proprionibacterium
Malic acid	COOH—CH(OH)—CH₂—COOH	*A. niger, A. itaconicus, Proteus vulgaris*
Methanol	CH_3OH	Species of *Methylomonas, Pseudomonas, Methylococcus*

Lignin and cellulose crystallinity are the two major deterrents to the effective utilization of lignocellulosic residues for chemical, enzymatic, and microbiological conversion processes to available sugars. The lignin polymer severely restricts enzymatic and microbial access to cellulose. Millet and co-workers[13] have published most comprehensive reviews on specific physical-chemical pretreatments for enhancing cellulose saccharification. These pretreatment steps are applicable to a wide variety of lignocellulosic materials that can be delignified to varying extents depending upon the type of pretreatment methods. Ladisch and co-workers[14] have recently described an organic solvent pretreatment process of cellulosic residues, followed by a cellulase hydrolysis step, to yield quantitative saccharification of the α-cellulose to simple sugars. The process yields 90 to 97% conversion of the cellulose of agri-residues to glucose. Successful application of this type of saccharification technology opens up new horizons to the utilization of biomass as a source of fuels, chemicals, and food.

The other type of pretreatment step of lignocellulosics centers around biological delignification. Kirk[15] has published a most comprehensive review on microorganisms that effect biological lignin degradation. More recent work on the physiological role white-rot fungi play in degradation of lignins draws attention to the synthesis of ^{14}C-labeled lignins and lignocelluloses as specific substrates for microorganisms. Crawford et al.[16,17] and Kirk et al.[18] discuss recent work on the degradation of labeled lignins and lignocelluloses by fungi and actinomycetes to $^{14}CO_2$. These techniques will become invaluable to the study of biodelignification and the role microbes may play in modifying lignocellulosics for subsequent saccharification.

VII. FERMENTATION CHEMICALS: ANAEROBIC AND AEROBIC

The anaerobic products of microbial metabolism consist of various organic solvents, i.e., acetone, ethanol, n-butanol, and isopropanol. Fermentations do not require aeration, and product recovery is accomplished through conventional distillation recovery methods. Fermentation processes to produce these chemicals are not dependent upon pure carbohydrate resources, but can utilize any type of pentose and/or hexose stream generated from biomass feedstocks.

Fumaric acid, glycerol, and 2,3 butylene glycol represent the main chemicals of aerobic fermentation. Tong,[12] in a recent review, discusses the production of various C_3 and C_4 chemicals and the current energy costs of production via fermentation. The aerobic processes are energy-intensive and require cooling, aeration, and agitation since these processes are highly exothermic due to carbohydrate oxidation. A comparison of attained vs. theoretical weight yields on dextrose for the fermentation products mentioned is depicted in Table 10. The major process used exclusively before 1950 was the acetone, n-butanol, and ethanol fermentation. Some improvement has been shown in this process. However, fermentation-derived solvents are presently only a minor factor in North America, although significant quantities are being produced in countries such as South Africa, where inexpensive fermentation resources are available.

In 1976, the total U.S. production of nine C_3 and C_4 chemicals, including ethanol, was near 4 million tons. Only 2% of these chemicals is presently derived via fermentation. Only butanol, acetone, fumaric acid, and ethanol are currently produced from both petroleum and carbohydrate feedstocks. The estimated percentage of organic chemicals produced by fermentation is depicted in Table 11.

Tong[12] indicates that in 1974 the percentage of fermentation-derived industrial alcohol was approximately 10%; however, this had increased to 30% by 1976. This increased industrial grain alcohol production comes largely from integrated grain milling plants where potable and industrial ethanol is produced among other corn products.

Table 10
COMPARISON OF ATTAINED VS. THEORETICAL WEIGHT YIELDS ON DEXTROSE

Fermentation products	% Weight conversion yield attained by fermentation	Theoretical yield % (stoichiometry)	Conversion efficiency, %
Anaerobic processes			
Ethanol	46	51.1	90
n-Butanol, H₂, ace-tone, ethanol	29—35	49.8	58—70
2,3 Butylene glycol	45	50	90
Aerobic processes			
Glycerol	43	76.6	56
Fumaric acid	64	64	100

Modified from Tong, G. E., *Chem. Eng. Prog.*, 74, 70, 1978. With permission.

Table 11
CURRENT STATUS OF C₃ AND C₄ CHEMICALS PRODUCTION

	U.S. 1976 production (million kg)	% Produced by fermentation
Butadiene	1475	0
Acetone	871	5
Isopropanol	780	0
n-Butanol	247	10
Methyl ethyl ketone	237	0
Glycerol	157	0
Maleic anhydride	119	0
Fumaric acid	24	15
2,3 Butylene glycol	—	—
Ethanol total	750	30

Assuming a 34% average weight conversion of carbohydrate to chemicals,[12] the 4 million tons of C_3 and C_4 chemicals can be produced from 12 million tons of starch or other fermentable sugar. The availability of agri-raw materials is not a major problem limiting progress toward fermentation-derived chemicals. This feedstock requirement may be met by expanding the annual cereal grain and sugar crop production by less than 10%, augmented by fermentable sugars in molasses and dairy whey, 0.8 and 0.3 million tons/year, respectively.

A. Ethanol

In view of continuously rising petroleum costs and dependence upon fossil fuel resources in North America, considerable attention has been focused upon alternative energy resources. Primary consideration involves the production of ethyl alcohol from renewable resources and determination of the economic and technical feasibility of using alcohol as an automotive fuel blended with gasoline. Special emphasis has been directed toward fermentation of surplus grains, agricultural residues, and forest waste materials as resources for production of alcohol and other chemicals. Since the major production of industrial alcohol is derived synthetically from ethylene, a major technological breakthrough is required to make the fermentation process competitive with that from ethylene.

The ability to produce ethanol from glucose is widely distributed among different microorganisms; however, the yields vary considerably from almost 2 mol of ethanol per mol glucose fermented, characteristic of yeast, to much smaller amounts produced by numerous bacteria.[19] These variations are attributable to the operation of four different metabolic routes of ethanol formation, three of which involve pyruvic acid as an obligatory intermediate. Pyruvic acid may be produced from glucose by different metabolic sequences, such as, Embden-Meyerhof glycolysis or Entner-Doudoroff cleavage with subsequent conversion to a C_2 unit via decarboxylation to acetaldehyde, or may be a thioclastic reaction to acetyl coenzyme A. Reduction of either C_2 unit yields ethanol.

1. Type I. Glycolysis

(1) glucose + 2NAD + 2ADP + 2Pi $\longrightarrow$ 2pyruvic acid + 2NADH$_2$ + 2ATP

(2) pyruvate $\xrightarrow[\substack{Mg^{++} \\ \text{decarboxylase}}]{TPP}$ acetaldehyde $\xrightarrow[\substack{\text{alcohol} \\ \text{dehydrogenase}}]{NADH_2}$ ethanol
$\quad\quad$ + CO$_2$

The Type I pathway of microbial metabolism of glucose via pyruvate and acetaldehyde leads to essentially quantitative conversion of glucose to ethanol and carbon dioxide. The yeasts are best known for utilizing this pathway, although bacteria are known that possess a yeast-like pathway and ferment glucose almost quantitatively to ethanol.

2. Type II. Thioclastic Reaction

(1) pyruvate $\xrightarrow[\substack{Mg,^{++} CoA}]{\substack{\text{decarboxylase} \\ TPP}}$ acetyl CoA + $\substack{\text{HCOOH} \\ \text{or} \\ H_2 \,\&\, CO_2}$

(2) acetyl CoA $\xrightarrow{NADH_2}$ acetaldehyde $\xrightarrow{NADH_2}$ ethanol
$\quad\quad$ + CoASH

The *Clostridia* and *Enterobacteriaceae* cleave pyruvate to acetyl CoA with subsequent reduction to acetaldehyde and ethanol. For quantitative conversion of glucose to ethanol, H_2 production must be suppressed to provide the reducing power essential for ethanol production.

3. Type III. Entner-Doudoroff Pathway

(1) glucose $\xrightarrow{ATP}$ G—6—P $\xrightarrow{NAD}$ gluconate—6—P

$\rule{2cm}{0pt}$ $\searrow$ $-H_2O$

pyruvate $\longleftarrow$ 2—oxo—3—
$\quad$ + $\quad\quad$ deoxygluconate—6—P
glyceraldehyde—3—P

pyruvate

(2) pyruvate $\longrightarrow$ ethanol

Zymomonas species give similar fermentation balance to yeast, but ethanol derives from C-2, C-3, C-5, and C-6 of glucose with only one half the energy yield.

4. Type IV. Heterolactic Fermentation

$$\text{glucose} \longrightarrow \text{ethanol} + \text{lactic acid} + CO_2$$

Heterolactic microorganisms are capable of glucose fermentation to lactate and ethanol via xylulose $5-PO_4$, which is subsequently cleaved to yield acetyl PO_4 and glyceraldehyde $3-PO_4$. The latter is converted to pyruvate with subsequent reduction to lactic acid. The acetyl PO_4 is reduced to ethanol, utilizing the reducing power generated from the glucose to xylulose $5-PO_4$ conversion.

Conversion of glucose to ethanol by yeast fermentation is well understood in terms of technology and product yield. In defining new possibilities of increasing productivity and reducing distillation costs, very few areas exist in the conventional methods of molasses and starch grains to alcohol. Opportunities exist in strain selection of flocculent yeasts that are tolerant to high sugar concentrations and ferment quickly to around 12% v/v ethanol. The current world production of distilled fermentation alcohol from various substrates is approximately 2.5 million tons/year. It is only in highly industrialized countries that synthetic alcohol from ethylene exists competitively. Trevelyan[20] reported the reverse situation in India where fermentation alcohol is used to produce ethylene. The utility of alcohol as a fuel source begins to reflect various economic factors differently as biomass crops for energy production are taken under consideration.

Ethanol production from plant biomass is being studied extensively by various research laboratories throughout the world. Bellamy[21] and Brooks et al.,[22] at General Electric Company, have pursued the production of both single-cell protein (SCP) and alcohol from agricultural wastes by utilizing various biological conversion processes. The process[22] involves a steam pretreatment to partially delignify wood and enhance cellulose accessibility to microbial utilization. *Clostridium thermocellum* is utilized to ferment the cellulose directly to ethanol and acetic acid. Research has also been conducted upon thermophilic bacteria that produce ethanol from xylose. Mixed culture fermentation of cellulose to ethanol with thermophilic microorganisms has been evaluated by the General Electric group.

Wang and co-workers[23,24] (M.I.T.) continue to investigate the cellulolytic activity of mutants of *C. thermocellum* capable of alcohol tolerance to 5% v/v. These organisms generate after 75 hr growth upon cellulose (10 g/ℓ) some 3 g/ℓ reducing sugars and 2 g/ℓ ethanol. *C. thermocellum* grown on corn cob granules consumes from 8 to 66% of the substrate and produces reducing sugars from 1.38 to 2.95 mg/mℓ.

Research for the past 30 years has mainly concerned the batchwise production of alcohol. However, in recent years considerable work has evolved around continuous fermentation methods. Rosen[25] has recently described various continuous fermentations with starchy material or molasses as substrates. The residence times for continuous molasses fermentations are between 7.5 and 13 hr. By using continuous methods, the conversion is increased and the cubic capacity of fermentor vessels is reduced, but also the instrumentation is simplified.

The large increases in crude oil prices in 1973 have stimulated various research projects for the discovery of new energy sources. The nation of Brazil has developed alcohol processes that utilize numerous raw materials that are plentiful in various regions of the country, i.e., cassava roots, palm trees, and sugar cane. The babassu coconut (23% starch), produced at a rate of 210 million tons, can be utilized to produce a wide variety of products, including charcoal, oil, and alcohol. Considering only the yield of ethanol, theoretically about 8 billion ℓ of ethanol could be produced yearly from the babassu crop in Brazil. This is almost twice the expected production of ethanol in that country, which is estimated at 4.3 billion ℓ by 1980.[26]

Carioca and Scares,[27] experimenting with babassu flour (containing approximately 60% pure starch) as a biomass resource, carried out an alcoholic fermentation. The starch material was gelatinized at 80 to 85°C with subsequent addition of a heat stable α-amylase. Complete saccharification was enhanced by glucoamylase treatment for 40 hr at room temperature. After this hydrolysis procedure, the sugar content measured 9.1%, then pressed yeast and yeast extract nutrients were added. Fermentation was conducted at 28 to 30°C for 42 hr, with subsequent distillation of the mash and redistillation of the initial ethanol product. The yield was 90 mℓ of 92% purity substance from 250 g crude babassu flour in 1 ℓ distilled H_2O. Based upon yields from 60% starch babassu flour (1 kg), the theoretical yield if all starch were fermented to alcohol, would be 568 g ethanol, a relative yield of 76% in their fermentation process.

The cassava plant has commanded considerable attention recently in Brazil as a starch resource for fermentation.[28] Cassava, also known as manioc or tapioca, is characteristically cultivated in many tropical regions of the world for the production of food or animal feed. Cassava, containing 20 to 35% starch and 1 to 2% protein in its roots, is one of the most efficient photosynthesizing plants known. The average crop production in Brazil is 13 tons of roots per hectare. This crop provides a most inexpensive source of starch that is not fully exploited technically for the production of starch products, possibly due to a lack of mechanization in its cultivation and perishability of its roots. The Brazilian Alcohol Program, established in 1975, seeks to utilize 20% ethyl alcohol in gasoline by 1980. To attain this objective, 4.3 billion ℓ of absolute alcohol need to be produced annually by that time. Lindeman and Rocchiccioli[26] have discussed in detail the massive plans of the Brazilian government to produce ethanol from sugar cane into 1981. Productivity factors are evaluated with reference to resources, production, and consumption. In 1978, a new cassava alcohol plant began operations in Brazil with a daily output of 60,000 ℓ of absolute alcohol. The feasibility of alcohol production from starch materials to compete with sugar cane will depend principally upon optimization of the liquefaction and saccharification steps of manufacture. These steps are not a requirement for fermentation of cane juice.

Although cassava starch is readily susceptible to α-amylase, the starch granules are weakly bound; thus, the root fiber creates a barrier to the starch hydrolysis if whole cassava roots are used. Rupture of the lignocellulosic components ensures reduction in the slurry viscosity and less energy in cooking, facilitating starch hydrolysis. This fiber removal can be accomplished through biological pretreatment with the cellulolytic *Thermoactinomyces viridae*. Recent work by Menezes et al.[29] demonstrated that fungal broths of a Basidiomycete and *T. viridae* increased both the rate of sugar formation and degree of solubilization, with subsequent decrease in slurry viscosity.

In discussion of other potentially useful agricultural wastes, the disposal of whey, a by-product of cheese manufacture, has become a serious pollution problem in many areas. In 1974, some 32.5 billion lb of whey were produced, one half of which was disposed of as waste.[30] This biological residue represents some 1.6 million lb of lactose, which can be utilized as a fermentation resource.

O'Leary and co-workers[31,32] have recently reported alcoholic fermentations of a lactase—hydrolyzed acid whey permeate (4.0 to 4.5% lactose) containing 30 to 35% total solids. Fermentations were conducted for 13 days with *Saccharomyces cerevisiae* and *Kluyveromyces fragilis* with maximal yields of 6.5 and 4.5% ethanol, respectively. Although *S. cerevisiae* converted the available glucose present in the lactase—

hydrolyzed whey permeates to alcohol, the galactose generated was not utilized by the organism. More efficient means and/or organisms will be required to utilize the galactose and glucose to alcohol.

Roland and Alm[33] reported that hydrolyzed whey permeate syrups fortified with an N source could be fermented to a 12.5% v/v alcohol beverage with a culture of *S. cerevisiae* var. *ellipsoideus*. Fermentations were conducted, with interval feedings of hydrolyzed whey permeate syrups reaching maximum alcohol in 6 days. Galactose utilization by the yeasts was not measured; however, residual reducing sugars in the wines varied from 0.2 to 4.3%. In summation, a wide variability may exist between the fermentation capacity of *S. cerevisiae* strains to utilize galactose.

The most thoroughly studied process for producing ethanol from biomass is enzymatic conversion of agricultural waste to soluble sugars and subsequent fermentation to ethanol by yeast. Wilke et al.[34-36] and Cysewski and Wilke[37] have provided some preliminary economic evaluations on various principal cost elements. The distribution of costs associated with ethanol production (exclusive of raw material costs) from newsprint and wheat straw by this process is discussed. The major costs of saccharification dominate, because the fermentor capacity required to produce sufficient quantities of fungal cellulase is 30 to 40 times that required to ferment the resulting sugars.

Su and Paulavicius[38] have recently described volumetric production efficiencies for alcohol production by fermentation from newsprint, wheat straw, and molasses. This efficiency in grams per liter per hour is significantly lower than the conventional molasses fermentation by yeast and is reflected in the conversion cost estimates.

Brooks and co-workers,[22] have recently conducted an economic viability study of a process for direct ethanol production from pretreated hardwood chips. These estimates are based upon similarities to the ethanol process from molasses.[39] Cost estimates are based upon a 25 million gal/year-95% ethanol plant from hardwood chips. The first stage involves a high temperature chemical pretreatment followed by a second stage direct fermentation to ethanol. The assumed yield of ethanol was based only on the conversion of the cellulose fraction of the pretreated wood. The conversion of the hemicellulose fraction ($\sim$20% of raw material) to ethanol would enhance the overall conversion yield. Utilization of a continuous fermentation process with cell recycle would provide a means of reducing associated costs.

Any conceptual process for saccharification to produce reducing sugars will require feeding of cellulose at high concentrations. Concentrated cellulose slurries are highly viscous and are difficult to pump and stir in conventional agitated fermentors. The mechanical properties of cellulose have been exploited by Wang and co-workers[24] with a packed-bed fermentor with cellulose as stationary phase. The batch packed-bed fermentor consisted of Solka floc cellulose with *Clostridium thermocellum* with liquid recirculation for 48 hr at 60°C. Cellulose degradation was 67%, with 14 g/ℓ total cells adsorbed onto the cellulose bed, compared to cell concentrations of 1 to 2 g/ℓ in typical stirred tank fermentations. The packed-bed technique may well serve as an excellent cell collector where cell recycle can be achieved and high substrate and product concentrations can be attained. Batch packed-bed fermentation by *C. thermocellum* of Solka floc yielded 8 g/ℓ reducing sugar, 2.2 g/ℓ ethanol, and 2.4 g/ℓ acetic acid.

Recent experiments by Kierstan and Bucke[40] on immobilized cell technology for alcohol production have been reported with two yeasts. Immobilized treated whole cell preparations have been used primarily in single-step reactions, in particular, in isomerization of glucose. Ethanol production from glucose solutions by an immobilized preparation of *S. cerevisiae* was demonstrated over a total of 23 days, with cell half-life of approximately 10 days. The yeast cells were immobilized in calcium alginate gels.

B. Acetone—Butanol—Isopropanol

Clostridium acetobutylicum historically has been the major organism used for the production of acetone and butanol from starch materials. This fermentation became known as the Weizmann process during World War I. Because of the industrial importance of these compounds, it has been studied in greater detail than other clostridial fermentations. The first stage of the fermentation is essentially butyric and acetic acid accumulation, yielding a pH drop to 4.5, with a second stage utilization of the acids to butanol and acetone with concomitant rise in pH. The butanol is formed by the reduction of butyric acid or butyryl—CoA to the alcohol. Minor quantities of ethanol are produced also in this fermentation.

Clostridium butylicum is an isopropanol type fermentation. The products of this type of clostridial fermentation are similar to acetone—butanol fermentation, except that the acetone is reduced to isopropanol. The extra reduction step normally results in a decrease in the amount of H_2 produced during the fermentation. Significant quantities[12] of acetone and butanol have been produced in the last 10 years in countries such as South Africa, where cheap fermentable biomass is available, but not in fossil fuel dependent countries.

Renewed interest in these fermentations has developed in the area of cellulosic waste conversion to butanol and other oil sparing solvents and chemicals. Recent studies[41] on biological production of organic solvents from cellulosics involve conversion of animal feedlot residues to liquid fuels. The process plan involves an alkali pretreatment of cattle feedlot residues followed by addition of a high temperature fungus, *Thermoactinomyces* sp., for cellulase production. The third step involves cellulase hydrolysis of the bulk residue with subsequent fermentation of the sugar syrup by *C. acetobutylicum*. Preliminary economic evaluation indicates that, with present knowledge, butanol can be produced for just over 30c/lb, which is comparable to ethylene based butanol.

Wang and co-workers[23,24] have recently described significant new research data based upon the *C. acetobutylicum* fermentation. Experiments with a corn meal medium with various strains have been initiated and give every indication that there are strains capable of producing mixed solvents near theoretical maximum yields, i.e., 1.05 and 2.26 g/ℓ for acetone and *N*-butanol, respectively.

C. 2,3-Butanediol (2,3-Butylene Glycol)

A number of facultative anaerobes are characterized by their ability to produce 2,3-butanediol (commonly called 2,3-butylene glycol). In general, 2,3-butanediol, produced by species of *Klebsiella, Bacillus,* and *Serratia* is a major fermentation product; however, in the presence of air, the oxidation product acetylmethyl carbinol is formed instead. Butanediol is important industrially as a potential raw material for synthetic rubber and was heavily investigated during World War II.

In the butanediol fermentation, glucose is broken down to pyruvic acid, which is further metabolized to butanediol. Although the major amounts of butanediol are produced by bacteria, yeasts form minor amounts of butanediol.[42] *Bacillus subtilis, Aerobacter aerogenes,* and *Serratia marcescens* produce significant quantities of butanediol from acid hydrolyzed starch,[43] some 35 lb butanediol/100 lb starch. Early investigations by Perlman[44] involved the production of 2,3-butanediol from acid hydrolyzates of hard and soft woods. *Aerobacter aerogenes* fermentations yielded from 24 to 30% butanediol depending upon the type of wood utilized.

D. Propionic Acid

Propionic acid is a major end-product of glucose fermentation in *Propionibacterium* species, occurring also with acetic acid and CO_2. The fermentation involves the reduction of two pyruvic acid molecules to propionic acid, with the oxidation of a third molecule to acetic acid and CO_2.

Recent research has been conducted on the bioconversion of propionic acid to acrylic acid by *Clostridium propionicum* from renewable resources.[23,45] Acrylic acid is a high-volume industrial chemical in high demand (approximately 1 billion lb/year). Two anaerobic organisms, *Peptostreptococcus elsdenii* and *Clostridium propionicum*, accumulate this acid as an intermediate. In *C. propionicum*, lactate is converted to acrylate, then to propionate via activated CoA thio esters. In resting cells of *C. propionicum*, propionate is oxidized to acrylate in the presence of an electron acceptor such as O_2 or methylene blue. Acrylate production is stimulated by sodium lactate. Concentrations of acrylic acid in excess of 4 mM have been achieved with resting cells.

E. Glycerol-Succinic Acid

Oura[42] discusses in detail the formation of glycerol and succinate by yeasts. The formation of glycerol appears to be nonphysiological, and quite useless for the yeast cell that obtains neither energy nor building units from it.[46] During the fermentation of glucose by yeast at pH 6 or below, only small amounts of glycerol are formed. Addition of sulfite to the medium increased glycerol production severalfold. This fermentation is known as the Neuberg 2nd and 3rd forms, in which glycerol accumulates in the fermentation. Two oxidation steps are involved in glycerol formation from glucose, and the redox balance will be achieved by the formation of two units of glycerol. Apparently, there is a direct correlation between redox balance of the cell and the formation of glycerol. When yeasts metabolize glucose under aerobic conditions, no superfluous glycerol is formed. Under these circumstances, the respiratory chain is functioning and transfers electrons to O_2 with no excess of $NADH_2$.

Two mechanisms have been proposed for the formation of succinate in yeast during anaerobic fermentations. One is formation via the normal oxidative mechanism of the TCA cycle, and the other is via a reductive pathway with malate and fumarate as intermediates.[47,48]

The formation of succinate is considerably lower during anaerobic growth than during fermentation, and the physiological state of the cell is different in these two cases. The level of energy-rich nucleotides during growth is low, whereas the energy charge increases strongly during yeast fermentation. The activity of many anabolic ATP dependent enzymes is modified by the energy charge of the cell, such as pyruvate carboxylase.[49] When energy charge is high (during fermentation), pyruvate is matabolized to oxalacetate via an activated pyruvate carboxylase, and the TCA cycle will function actively. The cycle intermediates accumulate as succinate and are excreted into the medium.

Therefore, since fermentation leads to an elevated energy charge in the cell (pyruvate-carboxylase activation), formation of succinate occurs and an excess of reduced respiratory nucleotides. This excessive $NADH_2$ is oxidized in the formation of glycerol, yielding a balance in the redox state of the cell.

F. Acetic Acid

Although numerous organisms are capable of a nonphosphorylative glucose oxidation to acetic acids, recent findings with some anaerobic organisms have stimulated interest in acetic acid production. The anaerobic cellulolytic rumen bacterium *Ruminococcus flavefaciens* normally produces succinic acid as a major fermentation product with acetic and formic acids, H_2 and CO_2. When grown on cellulose and in the presence of the methanogenic rumen bacterium *Methanobacterium ruminantium*, acetate was the major fermentation product. This type of interaction may be of significance in determining the flow of cellulose carbon to the normal rumen fermentation products.

Balch et al.[50] recently described a new genus of fastidiously anaerobic bacteria that produce a homoacetic acid fermentation. The type species, *Acetobacterium woodii*, ferments fructose, glucose, lactate, glycerate, and formate. Hydrogen is oxidized and CO_2 is reduced to acetic acid. Schoberth[51] has demonstrated the formation of acetate by cell extracts of *Acetobacterium woodii*.

Wang and co-workers[24] have recently reported studies on ethanol and acetic acid production by the cellulolytic anaerobe, *Clostridium thermocellum*, on cellulosic biomass. Experiments were conducted in cellulose packed-bed fermentors. Cellulose degradation was 67%, with a yield of 2.4 g/ℓ acetic acid from 110 g/ℓ cellulose.

Brooks and co-workers[22] have described a mixed culture fermentation of cellulose (microcrystalline) at 55°C that yielded acetic acid as the major organic chemical produced, plus ethanol, 2,3-butanediol, and CO_2. This group has also studied a continuous fermentation of a thermophilic *Bacillus* that produced ethanol and acetic acid at various dilution rates.

G. Fumaric Acid

Fumaric acid is produced principally by the fermentation of glucose or molasses with species of the genus *Rhizopus*. Rhodes et al.[52] reported fumaric acid yields of 60 to 70% in 3 to 8 days in shaken flasks containing 10 to 16% glucose or sucrose, or the partially inverted sucrose of molasses. Although fumaric can be produced in high yields by fermentation, it is produced commercially as a by-product in the manufacture of phthalic and malic anhydrides or by isomerization of malic acid with heat and catalyst. A number of chemicals can be produced from fumaric acid, including malic acid, coumaric acid, and maleic anhydride.

H. Citric Acid

The manufacture of citric acid is conducted presently by fermentation of sugar-containing material by microorganisms of the species *Aspergillus niger*. Both surface and submerged fermentation have been utilized for production of 70 kg of citric acid/100 kg of sugar content of raw material (usually molasses).

Usami and Fukutomi[53] recently reported on a citric acid solid fermentation by *A. niger*, sugar cane molasses, and pineapple molasses. After 3 days, 50 to 60% citric acid yield per equivalent sugar was available.

Hang et al.[54] reported upon the production of citric acid by *A. foetidus* from spent grain liquor, a brewery waste. The yields of citric acid varied from 3.5 to 12.3 g/ℓ of the waste fermented. Methanol addition (2 to 4%) markedly increased the formation of citric acid from wastes.

The citric acid-producing fungi can thus be utilized not only for organic chemical production but also for converting the BOD of brewery wastes into fungal protein.

I. Lactic Acid

Wastes from the pulp, paper, and fiberboard industries contain considerable sugar polymers, and thus present a high BOD to receiving waters. Griffith and Compare[55] describe a fixed-film system for continuous lactic acid production from waste waters. Lactic acid yield is in excess of 50%, the carbohydrate is available and readily recovered. The fixed-film unit (2 in. × 6 ft) was seeded with lactobacilli and lactose fermenting yeasts (kefir culture). The wood molasses substrate was pretreated with cellulases, a diastase, and hemicellulases. With a feed rate of 60 g/ℓ wood molasses, 31 to 32 g/ℓ lactic acid yields were obtained.

The production of calcium lactate from molasses by *Lactobacillus delbrueckii* was studied by Tewari and Vyas[56] using different growth factors from moong sprouts and various oil seed cakes. Maximum conversion of molasses plus 5% moong sprouts was achieved within 7 days at 50°C.

J. Malic Acid

Pichia membranaefaciens is capable of converting fumaric acid to L-malic acid. In a recent report, Takao and HoHa[57] describe malic acid yields as high as 80% or more, based on initial glucose when *Rhizopus arrhizus* (fumaric acid production) was grown 2 to 3 days and then associated with *Proteus vulgaris*. Malic acid formation also occurred when *R. arrhizus* was grown in mixed culture.

K. Methanol

Methanol occurs in nature as a breakdown product during microbial decomposition of plant materials and as a metabolite of methane-utilizing bacteria during growth upon methane or natural gas. Foo[58] recently reviewed some of the basic considerations in search of microorganisms with potential for microbial production of methanol. No attempt will be made to discuss the voluminous literature relative to the microbial production of methanol.

Since petroleum feedstocks are no longer cheap (as in the early 1950s), production of liquid fuels via fermentation has gained wide attention, especially alcohol fuels. In recent years, methanol has become a potentially important carbon source for the production of SCP, enzymes, and amino acids.[59] Methanol is also a potential fuel for internal combustion engines, since it possesses cleaner burning properties and produces less pollution than hydrocarbon fuels. A large volume of methanol is used as a solvent and as an intermediate in chemical manufacture.

Methanol can be produced by the destructive distillation of wood; however, most methanol is derived from carbon monoxide with hydrogen reaction.[60] In nature, methanol arises from the breakdown of methyl esters and/or ethers from decomposition of pectin-like plant materials. Very little is known about the microorganisms that produce methanol during decomposition of organic material; however, numerous reviews are available.[61,62]

Methanol inhibition, and the energy and reducing power requirements of methane oxidation present major problems to the excretion of excess methanol by microorganisms. Only small amounts of methanol are excreted by the cell biomass yields of methanol-utilizers in mixed culture studies.[63,64] Greater tolerance is needed to improve yields of methanol and further productivity under possibly elevated pressure. Greater numbers of methane-utilizers will have to be isolated and tested in order to find those more suited to methanol excretion.

REFERENCES

1. Loehr, R. C., An overview—utilization of residues from agriculture and agro-industries, Proc. Symp. on Management of Residue Utilization, United Nations Environment Program, Food and Agriculture Organization of the United Nations, Rome, 1977.
2. Sloneker, J. H., Jones, R. W., Griffin, H. L., Eskins, K., Bucher, B. L., and Inglett, G. E., Processing animal wastes for feed and industrial products, in *Symposium: Processing Agricultural and Municipal Wastes,* Inglett, G. E., Ed., AVI Publishing, Westport, Conn., 1973, 13.
3. Sloneker, J. H., Agricultural residues, including feedlot wastes, *Biotechnol. Bioeng. Symp.*, 6, 235, 1976.

4. **Detroy, R. W. and Hesseltine, C. W.**, Availability and utilization of agricultural and agro-industrial wastes, *Process Biochem.*, 13, 2, 1978.
5. **Lewis, C.**, Energy relationships of fuel from biomass, *Process Biochem.*, 11, 29, 1976.
6. **Ifeadi, C. N. and Brown, J. B., Jr.**, Technologies suitable for recovery of energy from livestock manure, proc. cornell agricultural waste management conference, in *Energy, Agricultural and Waste Management,* Jewell, W. M., Ed., Ann Arbor Science, Ann Arbor, 1975, 373.
7. **Bruns, E. G. and Crowley, J. W.**, Solid manure handling for livestock housing, feeding, and yard facilities in Wisconsin, *Ext. Bull.,* No. A2418, University of Wisconsin, Madison, 1973.
8. **Jewell, W. J.**, Energy from agricultural waste-methane generation, *Agric. Eng. Ext. Bull., 397,* New York State College of Agricultural and Life Sciences, Cornell University, Ithaca, New York, 1974.
9. Agricultural statistics, U.S. Department of Agriculture Washington, D.C., 1973, 617.
10. **Laskin, A. I.**, Single cell protein, *Annu. Rep. Ferment. Processes,* Perlman, D., Ed., Academic Press, New York, 1977.
11. **Srinivasan, V. R. and Han, Y. W.**, Utilization of bagasse, *Adv. Chem. Ser.,* 95, 447, 1969.
12. **Tong, G. E.**, Fermentation routes to C_3 and C_4 chemicals, *Chem. Eng. Prog.,* 74, 70, 1978.
13. **Millet, M. A., Baker, A. J., and Satter, L. D.**, Pretreatments to enhance chemical, enzymatic, and microbiological attack of cellulosic materials, *Biotechnol. Bioeng. Symp.,* 5, 193, 1975.
14. **Ladisch, M. R., Ladisch, C. M., and Tsao, G. T.**, Cellulose to sugars: new path gives quantitative yield, *Science,* 201, 743, 1978.
15. **Kirk, T. K.**, Effects of microorganisms on lignin, *Annu. Rev. Phytopathol.,* 9, 185, 1971.
16. **Crawford, D. L. and Crawford, R. L.**, Microbial degradation of lignocellulose: the lignin component, *Appl. Environ. Microbiol.,* 31, 714, 1976.
17. **Crawford, D. L., Crawford, R. L., and Pometto, A. L., III**, Preparation of specifically labeled ^{14}C-(lignin)- and ^{14}C-(cellulose)-lignocelluloses and their decomposition by the microflora of soil, *Appl. Environ. Microbiol.,* 33, 1247, 1977.
18. **Kirk, T. K., Connors, W. J., and Zeikus, J. G.**, Requirement for a growth substrate during lignin decomposition by two wood-rotting fungi, *Appl. Environ. Microbiol.,* 32, 192, 1976.
19. **Dawes, E. A.**, Comparative aspects of alcohol formation, *J. Gen. Microbiol.,* 32, 151, 1963.
20. **Trevelyan, W. E.**, Renewable fuels: ethanol produced by fermentation, *Trop. Sci.,* 17, 1, 1975.
21. **Bellamy, W. D.**, Cellulose and lignocellulose digestion by thermophilic actinomyces for single-cell protein production, in *Developments in Industrial Microbiology,* Underkofler, L. A., Ed., Society for Industrial Microbiology, Linden, N.J., 1977, 249.
22. **Brooks, R. E., Bellamy, W. D., and Su, T.-M.**, Bioconversion of plant biomass to ethanol, Annual Rep. No. COO-4147-4, National Technical Information Service, Department of Energy, Washington, D.C., 1978, 137.
23. **Wang, D. I. C., Cooney, C. L., Demain, A. L., Gomez, R. F., and Sinskey, A. J.**, Degradation of cellulosic biomass and its subsequent utilization for the production of chemical feed stocks, Research Report No. TID-27977, National Technical Information Service, Department of Energy, Washington, D.C., 1977.
24. **Wang, D. I. C., Cooney, C. L., Demain, A. L., Gomez, R. F., and Sinskey, A. J.**, Degradation of cellulosic biomass and its subsequent utilization for the production of chemical feed stocks, Research Report No. TID-27977, National Technical Information Service, Department of Energy, Washington, D.C., 1978.
25. **Rosen, K.**, Continuous production of alcohol, *Process Biochem.,* 13, 25, 1978.
26. **Lindeman, L. R. and Rocchiccioli, C.**, Ethanol in Brazil: brief summary of the state of the industry in 1977, *Biotech. Bioeng.,* 21, 1107, 1979.
27. **Carioca, J. O. B. and Scares, J. B.**, Production of ethyl alcohol from babassu, *Biotechnol. Bioeng.,* 20, 443, 1978.
28. **Menezes, T. J. B.**, Saccharification of cassava for ethyl alcohol production, *Process Biochem.,* 13, 24, 1978.
29. **Menezes, T. J. B., Arakaki, T., DeLamo, P. R., and Sales, A. M.**, Fungal cellulases as an acid for the saccharification of cassava, *Biotechnol. Bioeng.,* 20, 555, 1978.
30. Production of Manufactured Dairy Products, Da2-1 (75), Statistical Reporting Service, U.S. Department of Agriculture, Washington, D.C., 1975.
31. **O'Leary, V. S., Green, R., Sullivan, B. C., and Holsinger, V. H.**, Alcohol production by selected yeast strains in lactase-hydrolyzed acid whey, *Biotechnol. Bioeng.,* 19, 1019, 1977.
32. **O'Leary, V. S., Sutton, C., Bencivengo, M., Sullivan, B., and Holsinger, U. H.**, Influence of lactose hydrolysis and solids concentration on alcohol production by yeast in acid whey ultra filtrate, *Biotechnol. Bioeng.,* 19, 1689, 1977.
33. **Roland, J. F. and Alm, W. L.**, Wine fermentations using membrane processed hydrolyzed whey, *Biotechnol. Bioeng.,* 17, 1443, 1975.

34. Wilke, C. R., Cellulose as a chemical and energy resource, *Biotechnol. Bioeng. Symp.*, 5, 361, 1975.
35. Wilke, C. R., Yang, R. D., and Von Stockar, U., Preliminary cost analyses for enzymatic hydrolysis of newsprint, Report No. 18, Conf-750992-Z, Lawrence Berkeley Lab., University of California, Berkeley, 44, 1975.
36. Wilke, C. R., Cysewski, G. R., Yang, R. D., and Von Stockar, U., Utilization of cellulosic materials through enzymic hydrolysis. II. Preliminary assessment of an integrated processing scheme, *Biotechnol. Bioeng.*, 18, 1315, 1976.
37. Cysewski, G. R. and Wilke, C. R., Utilization of cellulosic materials through enzymic hydrolysis. I. Fermentation of hydrolyzate to ethanol and single-cell protein, *Biotechnol. Bioeng.*, 18, 1297, 1976.
38. Su, T.-M. and Paulavicius, I., Enzymatic saccharification of cellulose by thermophilic actinomyces, *Appl. Polym. Symp.*, 28, 221, 1975.
39. SRI Report No. 95, Fermentation Process, Stanford Research Institute, California, 1975.
40. Kierstan, M. and Bucke, C., The immobilization of microbial cells, subcellular organelles, and enzymes in calcium alginate gels, *Biotechnol. Bioeng.*, 19, 387, 1977.
41. Pye, E. K., Humphrey, A. E., and Forro, J. R., The biological production of organic solvents from cellulosic wastes, Progress Report No. COO-4070-1, National Technical Information Service, Department of Energy, Washington, D.C., 1977, 40.
42. Oura, E., Reaction products of yeast fermentations, *Process Biochem.*, 12, 19, 1977.
43. Wilkinson, J. F. and Rose, A. H., Fermentation processes, in *Biochemistry of Industrial Microorganisms*, Rainbow, C. and Rose, A. H., Eds., Academic Press, New York, 1963, chap. 11.
44. Perlman, D., Production of 2,3-butylene glycol from wood hydrolyzates, *Ind. Eng. Chem.*, 36, 803, 1944.
45. Dalal, R., Akedo, M., Cooney, C. L., and Sinskey, A. J., Bioconversion of propionate to acrylate acid by resting cells of *Clostridium propionicum*, in *Proceedings of the American Chemical Society*, Pier, L. S., Ed., American Chemical Society, Washington, D.C., September, 1978, No. 38.
46. Sols, A., Gancedo, C., and Delafuente, G., in *The Yeasts*, Vol. 2, Rose, A. H. and Harrison, J. S., Eds., Academic Press, New York, 1971, 271.
47. Chapman, C. and Bartley, W., The kinetics of enzyme changes in yeast under conditions that cause the loss of mitochondria, *Biochem. J.*, 107, 455, 1968.
48. Machado, A., Nunez de Castro, I., and Mayor, F., Isocitrate dehydrogenases and oxoglutarate dehydrogenase activities of baker's yeast grown in a variety of hypoxic conditions, *Mol. Cell. Biochem.*, 6, 93, 1975.
49. Miller, A. L. and Atkinson, D. E., Response of yeast pyruvate carboxylase to the adenylate energy charge and other regulatory parameters, *Arch. Biochem. Biophys.*, 152, 531, 1972.
50. Balch, W. E., Schoberth, S., Tanner, R. S., and Wolfe, R. S., *Acetobacterium*, a new genus of hydrogen-oxidizing, carbon dioxide reducing, anaerobic bacteria, *Int. J. Syst. Bacteriol.*, 27, 355, 1977.
51. Schoberth, S., Acetic acid from H_2 and CO_2: formation of acetate by cell extracts of *Acetobacterium woodii*, *Arch. Microbiol.*, 114, 143, 1977.
52. Rhodes, R. A., Moyer, A. J., Smith, M. L., and Kelly, S. E., Production of fumaric acid by *Rhizopus arrhizus*, *Appl. Microbiol.*, 7, 74, 1959.
53. Usami, S. and Fukutomi, N., Citric acid production by solid fermentation method using sugar cane bagasse and concentrated liquor of pineapple waste, *Hakko Kogaku Zasshi*, 55, 44, 1977.
54. Hang, V. D., Splittstoesser, D. F., Woodams, E. E., and Sherman, R. M., Citric acid fermentation of brewery waste, *J. Food Sci.*, 42, 383, 1977.
55. Griffith, W. L. and Compare, A. L., Continuous lactic acid production using a fixed-film system, *Dev. Ind. Microbiol.*, 18, 723, 1977.
56. Tewari, H. K. and Vyas, S. R., Utilization of agricultural waste materials for the production of calcium lactate by fermentation, *J. Res. Punjab Agric. Univ.*, 8, 460, 1971.
57. Takao, S. and HoHa, K., L-malic acid fermentation by mixed culture of *Rhizopus arrhizus* and *Proteus vulgaris*, *Agric. Biol. Chem.*, 41, 945, 1977.
58. Foo, E. L., Microbial production of methanol, *Process Biochem.*, 13, 23, 1978.
59. Heden, C. G., Microbiological aspects on the methanol economy. *Ann. Microbiol. Enzimol.*, 24, 137, 1974.
60. Waddams, A. L., *Chemicals From Petroleum*, 3rd ed., John Wiley & Sons, New York, 1973.
61. Quayle, J. R., Metabolism of C_1 compounds in autotrophic and heterotrophic microorganisms, *Annu. Rev. Microbiol.*, 15, 119, 1961.
62. Quayle, J. R., Metabolism of one-carbon compounds by microorganisms, *Adv. Microb. Physiol.*, 7, 119, 1972.
63. Wilkinson, T. G., Topiwala, H. H., and Hamer, G., Interactions in a mixed bacterial population growing on methane in continuous culture, *Biotechnol. Bioeng.*, 16, 41, 1974.

64. **Harrison, D. E. F., Drozd, J. W., and Khosrovi, B.,** Proc. 5th Int. Fermentation Symp., The production of single-cell protein from natural gas. Problems associated with the physiology of methylotrophic bacteria, Berlin, Dellweg, H., Ed., Westkreuz-Druckevei and Verlag, Berlin-Bonn, 1976.
65. Development Document for Effluent Limitations Guidelines and New Source Performance Standards for the Grain Processing Segment of the Grain Mills Point Source Category, EPA440/1-74-028a, Environmental Protection Agency, Washington, D.C., 1974.

Chapter 4

GASIFICATION

David L. Brink

TABLE OF CONTENTS

I. DEFINITION

Gasification and pyrolysis are both aspects of thermal processes that are involved in combustion. Gasification may be defined as that regime in which organic materials, including biomass as well as solid fossil fuels, are degraded by thermal reactions in the presence of controlled amounts of oxidizing agents to provide (when carried to completion) a simple gaseous phase comprising hydrogen, water, carbon monoxide, carbon dioxide, methane, small or trace amounts of other components, and residues from contained inorganic matter. Pyrolysis may be defined as that regime in which the same organic matter is degraded by thermal reactions in the absence of added oxidizing agents to provide a complex volatile phase and a carbonaceous char containing inorganic components. Under ambient conditions, the volatile phase (containing an array of organic substances) is condensed to provide an aqueous phase containing soluble organic materials, and an organic phase.

The clear distinction to be made between gasification and pyrolysis is the addition and omission, respectively, of an oxidizing agent, usually oxygen, water, or carbon dioxide, or a mixture of these. When molecular oxygen is the oxidizing agent, the amount used for gasification is substantially below that required for stoichiometric combustion. Furthermore, the system is designed so that the products of combustion (carbon dioxide and water) are passed over the fuel at a temperature in the range required for gasification reactions to proceed. Excluding oxygen, and using water or carbon dioxide as the oxidizing agent, gasification may be carried to completion provided the temperatures required for gasification are achieved by a means other than combustion.

On a practical basis, gasification is normally carried out in a higher temperature region than pyrolysis. To accomplish gasification, it is always necessary to pass through a pyrolytic stage; i.e., pyrolysis is one aspect of gasification. Moreover, in practice, gasification processes are infrequently carried to completion as defined, and both amounts and complexity of volatile products and carbonaceous residues increase as the conditions required for gasification are moderated toward those under which only pyrolysis will occur.

II. GASIFICATION REACTIONS

Gasification of a fuel involves a complex sequence of reactions that are determined and controlled by design of the process. In this discussion, generalized overall reactions are considered.

A. Pyrolytic Reactions

As previously noted, pyrolysis is an aspect of gasification. Since pyrolysis is presented in Chapter 5, only the effect it can have on the modification of a complex fuel is noted here. In their treatment of combustion, Minkoff and Tipper[1] discussed pyrolysis of comparatively simple hydrocarbon fuels and noted the controversy concerning the relative importance of molecular and chain processes in pyrolysis. They pointed out that molecular decompositions will predominate in a lower termperature range. As temperature is increased in a range above about 350 to 500°C, formation of free radicals by homolytic cleavage and by chain reactions increase and predominate. Methane can be pyrolyzed to produce hydrogen, ethane, ethylene, and acetylene. In the pyrolysis of acetylene in a static system, products formed have included low molecular weight polymers at about 350°C, ethylene, benzene, phenanthrene, and more complex condensed aromatic products at about 450 to 500°C, products of thermal cracking above about 600 to 700°C, and acetylenic carbon, hydrogen, and some methane about 1000 to 1500° C. Thus, at high temperatures, there is a system comprised of hydrogen, methane, and carbon.

The pyrolysis of organic fuels, including complex biomass materials and fossil fuels, or simple fuels such as methane and acetylene, will produce volatile phases varying in composition from complex arrays of organic products and carbonaceous residues to systems containing hydrogen, methane, and carbon, depending upon temperature. A carbonaceous residue is always produced. Reactions involved will include decompositions, cracking and free radical formation, and recombinations. Moreover, products such as water and carbon dioxide, formed by thermal degradation of the organic fuels, may participate in gasification reactions as presented in the following discussion.

B. Oxidation-Reduction Reactions

With the introduction of oxidizing agents, gasification is superimposed on pyrolytic degradation as described previously. Oxidizing agents used are oxygen, water, and carbon dioxide.

Fundamental equations of gasification reactions are discussed by Gumz.[2] When oxygen is used as the oxidizing agent, combustion occurs. The exothermic reactions in combustion may be represented by overall Equations 1 and 2.*

$$H_2 + 1/2\ O_2 \rightleftarrows H_2O(1)\ \Delta H^o_{298}\ =\ -68.32\ kcal/gmol \qquad (1)$$

$$C\ (graphite) + O_2 \rightleftarrows CO_2\ \Delta H^o_{298}\ =\ -94.05\ kcal/gmol \qquad (2)$$

The mechanism of oxidation is substantially more complicated than indicated by Equations 1 and 2. It has been shown that oxygen participates in homogenous reactions in the gaseous state with very little or no oxygen diffusing to a solid surface where the heterogenous reactions indicated by Equation 2 can take place. Not only is internal heat provided by combustion, but the formed products (water and carbon dioxide) serve as the oxidizing agents involved in the principle heterogeneous, endothermic gasification Equations 3 and 4.

$$C\ (graphite) + H_2O(1) \rightleftarrows CO + H_2\ \Delta H^o_{298}\ =\ 41.90\ kcal/gmol \qquad (3)$$

$$C\ (graphite) + CO_2 \rightleftarrows 2CO\ \Delta H^o_{298}\ =\ 41.21\ kcal/gmol \qquad (4)$$

The reaction with water is commonly known as the water-gas or steam-char reaction, that with carbon dioxide is the Boudouard or producer gas reaction. A third reaction, Equation 5, involves the formation of

$$C\ (graphite) + 2H_2 \rightleftarrows CH_4\ \Delta H^o_{298} = -17,889\ kcal/gmol \qquad (5)$$

methane, and with Equations 3 and 4, and certain assumptions, provides the basis for deriving the quantitative relationships of most other reactions involved in gasification. The assumptions necessary are that sulfur and other components of the ash have negligible effects upon results obtained. Volatile matter evolved in pyrolysis has an effect and will be discussed later.

The basic relationships among gasification reactions and the three equations noted

---

* Equilibrium reactions given in this discussion are written with the sign for the change in enthalpy for the forward reactions, i.e., left to right.

are provided by a consideration of their equilirium constants. The derivation of equilibrium constants may be found in general textbooks on physical chemistry and thermodynamics, e.g., Lewis and Randall[3] or Moore.[4] A general, reversible gaseous reaction is given by Equation 6.

$$lL + mM \rightleftarrows qQ + rR \qquad (6)$$

At constant temperature and pressure, when the rates of the forward and reverse reactions are equal the system is in equilibrium. Then, for ideal gases, given molar concentrations of products and reactants as partial pressures (P), the condition of equilibrium is expressed by the equilibrium constant, K_P of Equation 7

$$K_P = \frac{P_Q^q \, P_R^r}{P_L^l \, P_M^m} \qquad (7)$$

for the forward reaction of Equation 6. The equilibrium constant for the reverse reaction of Equation 6 is given by the reciprocal of K_P that is represented here by K'_P.

$$K'_P = \frac{P_L^l \, P_M^m}{P_Q^q \, P_R^r} \qquad (8)$$

Expressed in terms of mole fractions where $P_i = X_i P$, Equation 7 becomes:

$$K_P = \frac{X_Q^q \, X_R^r}{X_L^l \, X_M^m} \cdot P^{\Delta\nu} = K_X P^{\Delta\nu} \qquad (9)$$

where P_i = partial pressure of the ith component, P = total pressure in the system, and $\Delta\nu = q + r - l - m$. Similarly, in terms of concentrations where $p_i = \frac{n_i}{v} RT = c_i RT$, Equation 7 becomes:

$$K_P = \frac{c_Q^q \, c_R^r}{c_L^l \, c_M^m} RT^{\Delta\nu} = K_c(RT)^{\Delta\nu} \qquad (10)$$

where c_i = molar concentration of the ith component.

At the temperature involved in gasification and particularly at pressures where ideal gas behavior is exhibited by all of the major gaseous components in the system, Equations 7 to 10 apply. As the behavior of a gas deviates from that of an ideal gas its molar concentration may be expressed in terms of its fugacity, $f_i = \gamma_i P_i$, where γ_i is the fugacity coefficient. Then, the equilibrium constant for Equation 6 becomes:

$$K_f = \frac{f_Q^q \, f_R^r}{f_L^l \, f_M^m} = \frac{\gamma_Q^q \, P_Q^q \cdot \gamma_R^r \, P_R^r}{\gamma_L^l \, P_L^l \cdot \gamma_M^m \, P_M^m} = R_\gamma K_p \qquad (11)$$

where R_γ is the ratio of fugacity coefficients needed to convert partial pressures in K_P into the fugacities in K_f.

For pure liquids and solids at approximately 1 atm and at a given temperature, activities equal one; therefore, terms for such components may be omitted from the equations.

Table 1
EQUILIBRIUM CONSTANTS
FOR GASIFICATION REACTIONS
3, 4, AND 5[a]

	log K_P		
	$K_{PW} = \dfrac{P_{CO}\, P_{H_2}}{P_{H_2O}}$	$K_{PB} = \dfrac{P_{CO}^2}{P_{CO_2}}$	$K_{PM} = \dfrac{P_{CH_4}}{P_{H_2}^2}$
Temperature (K)	log K_{PW}	log K_{PB}	log K_{PM}
700	−2.61853	−3.57358	+0.95261
900	−0.37227	−0.71568	−0.49209
1100	+1.06373	+1.08638	−1.43432
1300	+2.05385	+2.31863	−2.10013
1500	+2.77903	+3.21033	−2.59271

[a] Data from Reference 1.

Returning to gasification Equations 3, 4, and 5, the values of log K_P are given for a range of temperatures from 700 to 1500 K at invervals of 200 K in Table 1. It is noted that K_P for Equations 3 and 4, and K'_P for Equation 5 become significant at about 900 K.

C. Equilibrium of the Components in Gasification Reactions

Given sufficiently high temperatures and sufficient quantities of oxidizing agents, the products of gasification are essentially hydrogen, water, carbon monoxide, carbon dioxide, and methane. The relative amounts of these components are dependent upon temperature, pressure, the composition of the raw material, and the oxidizing agent s) supplied to the process. The equilibrium constants for the forward or for the reverse reactions given by Equations 3, 4, and 5 may be used to express the equilibrium constants for other reactions occurring in gasification.

The so-called water-gas shift reaction is given in Equation 12.

$$CO + H_2O(1) \rightleftarrows CO_2 + H_2 \qquad H^O_{298} = 0.69 \text{ kcal/gmol} \qquad (12)$$

where:

$$K_{PS} = \frac{P_{CO_2}\, P_{H_2}}{P_{CO}\, P_{H_2O}} = \frac{P_{H_2}\, P_{CO}}{P_{H_2O}} \times \frac{P_{CO_2}}{P_{CO}^2} = K_{PW}\, K'_{PB}$$

Thus, the equilibrium constant of the key reaction controlling relative concentrations of the major products of gasification, excepting methane, may be expressed as the product of the equilibrium constants of the water-gas reaction and the reciprocal of the Boudouard reactions.

Reactions concerning the synthesis of methane are given by Equations 13, 14, and 15.

$$CO + 3H_2 \rightleftarrows CH_4 + H_2O \qquad H^O_{298} = -57.79 \text{ kcal/gmol} \quad (13)$$

$$2CO + 2H_2 \rightleftarrows CH_4 + CO_2 \qquad H^O_{298} = -59.10 \text{ kcal/gmol} \quad (14)$$

$$CO_2 + 4H_2 \rightleftarrows CH_4 + 2H_2O \qquad H^O_{298} = -60.48 \text{ kcal/gmol} \quad (15)$$

Again, the equilibrium constants of these reactions can be expressed using the equilibrium constants or the reciprocals of the equilibrium constants of Equations 3, 4, and 5, i.e.,

$$K_{P13} = \frac{P_{CH_4}}{P_{H_2}^2} \cdot \frac{P_{H_2O}}{P_{CO} P_{H_2}} = K_{PM} K'_{PW},$$

$$K_{P14} = \frac{P_{CH_4}}{P_{H_2}^2} \cdot \frac{P_{CO_2}}{P_{CO}^2} = K_{PM} K'_{PB},$$

and

$$K_{P15} = \frac{P_{CH_4}}{P_{H_2}^2} \cdot \frac{P_{H_2O}^2}{P_{CO} P_{H_2}} \cdot \frac{P_{CO}^2}{P_{CO_2}} = K_{PM} (K'_{PW})^2 K_{PB}.$$

Equilibrium constants provide basic information required for calculations of gasifier performance. Experimentally, these are difficult to obtain.

1. Reaction Temperature

The rates of reaction in gasification are highly temperature dependent. Above about 1200 K, the gasification reaction rates are so fast that they can be neglected, and heat and mass transfer are controlling. In the temperature range from 1200 to 1000 K, rates of gasification reactions become controlling, and below about 900 K, rates are so slow that gasification reactions are not of practical importance.

The temperature at which heterogeneous gasification reactions take place cannot be directly measured. The required temperature is not in the gas phase, but rather the temperature at the phase boundary where these reactions are taking place. Given equilibrium conditions over a period of time, the reaction temperature may approach gas phase temperature as the solid particle is heated throughout. Gumz[5] considers that the only accurate procedure for establishing reaction temperature is through calculations based upon known equilibrium constants and heat balance.

2. Equilibrium Constants

To obtain the equilibrium constants, a given experimental unit or piece of equipment must be operated until constant composition of a gas is produced under isothermal and isobaric (i.e., equilibrium) conditions. Two methods have been used,[2,4] static and dynamic. In the static method, operation is carried out keeping the gas phase, appropriately agitated, in contact with the solid phase. Because enthalpy is changing, an isothermal condition throughout the static system is most difficult to achieve. In the dynamic method, removal of heat from, or addition of heat to the system can be solved by adjusting the temperature of the flow of the gasifying agent into the gasifier. However, holding conditions constant until equilibrium has been achieved and demonstrated is again difficult experimentally. These problems, in concert with the problem of true reaction temperature measurement and of obtaining samples of the gas having the composition established at the reaction temperature, illustrate the difficulties encountered in an experimental determination of equilibrium constants. Because of these difficulties, experimental determination of equilibrium constants has been imprecise, and calculations of gasification based on equilibrium constants so determined have varied considerably.

An alternative method for determining equilibrium constants, which gives accurate values, is through calculation based on thermodynamic relationships.[6]

3. Gasification Calculations

Having accurate equilibrium constants, it is then possible to calculate reaction temperature by experimental determination of gas composition obtained under equilibrium conditions as described previously. A considerably more accurate alternative method for determining reaction temperature is to assume a reaction temperature, a total pressure, and ideal gas behavior. Gas composition can then be calculated using the equilibrium constants of Equations 3, 4, and 5, $\Sigma_{P_i} = 1$, and the material balances for carbon, hydrogen, oxygen, and nitrogen. Eight equations can be written in order to calculate eight unknowns, namely, the partial molar volumes of CO, CO_2, H_2, H_2O, CH_4, N_2, and the amounts of gasifying medium required and fuel gasified per unit of product gas. Using this information, the heat balance can be established around a reaction zone of the gasifier including the ash, oxidation, and reduction zones, but not the preheating zone. Then, by a system of iteration, the reaction temperature can be determined.

Gumz[7] describes the technique outlined above and various simplifications that may be used in performing gasifier calculations. The calculations of gas composition when incomplete equilibrium is attained is also discussed. Volatile products, as noted previously, will be produced in pyrolysis and will substantially increase the complexity and uncertainty of gasifier calculations. Volatile organic materials ultimately are decomposed to the most stable hydrocarbon, methane. The reactions involve free radical mechanisms as illustrated in Equation 16.

$$CH_3 \bullet + H \bullet \rightleftarrows CH_4 \quad \Delta H^O_{298} = -104 \text{ kcal/gmol} \qquad (16)$$

Substantially higher concentrations of methane can be produced by free radical reactions than those produced by the gasification reactions discussed at a pressure of about 1 atm. For example, based on the values of log K_P given in Table 1, the values of K_{PM}, K_{P13}, K_{P14}, and K_{P15} calculated at 1 atm and 1100 K, are 3.68×10^{-3}, 3.46×10^{-3}, 3.04×10^{-3}, and 2.26×10^{-5}, respectively. Clearly, the amount of methane formed under the specified conditions (considering that these reactions occur simultaneously and are interacting) will be very low, if not negligible. Then, a relatively high concentration of methane and, clearly, of higher molecular weight hydrocarbons or their derivatives, demonstrate unequivocally that the gasifier has not been operated under, or approaching equilibrium conditions.

D. Effect of Ash Components

The presence of metallic salts has profound effects upon the reactions taking place in the thermal degradation of the fuel, and plays a role in gasification.

1. Metals

Broido[8] showed that yield of levoglucosan decreased sharply and carbonaceous residue increased when cellulose was treated with small amounts of inorganic salts before pyrolysis. Philpot[9] found that certain elements (including phosphorous) in the presence of calcium were especially correlated with a decrease in volatilization rate, a decrease in the threshold temperature at which active pyrolysis begins, and an increase in carbonaceous residues, and that sodium and potassium exhibited lesser effects. Fung et al.[10] showed that all salts (acidic, basic, and neutral) profoundly reduced the yield of levoglucosan, and fire retardant additives accelerated the depolymerization of cellulose in pyrolysis. Thus, in the work cited and in similar studies it has been demonstrated that inorganic ash components decrease volatile materials and increase char in pyrolysis. Similar effects of ash would be expected in gasification thereby providing increased amounts of char. The char is then gasified according to Equations 3, 4, and 5. The vast literature available concerning catalytic effects of various metals, metallic oxides, and salts in coal and petroleum should greatly facilitate studies of the effects of catalysts in biomass gasification.

2. Nonmetallic Elements

As discussed, in an environment where gasification reactions occur, strongly reducing conditions prevail. Even in stoichiometric combustion, the reactions taking place at the solid surfaces of a fuel will involve water and carbon dioxide, and not oxygen as oxidizing agents. Under these conditions, nonmetallic elements such as sulfur, nitrogen, chlorine, phosphorus, selenium, and others may be reduced, i.e., to sulfide, ammonia, chloride, phosphine, and selenide, respectively. All nonmetallic elements, their derivatives produced in gasification, and the states of these derivatives should be thoroughly evaluated in any design to establish possible environmental impacts. In general, the metallic elements of biomass materials, if not already present as such, will react to form inorganic salts when subjected to the conditions of gasification. Any nonmetallic elements that are present will either be retained in the ash as inorganic salts or will be evolved as gasous products.

Sulfur is one of the more widespread nonmetallic elements in fuels, and because of its reactions, is, therefore, of great economic impact when subjected to thermal degradation including either combustion or pyrolysis-gasification. Materials obtained from biomass generally have such a low sulfur content that it is usually insignificant. However, such fuels may be used to augment or be supplemented with high sulfur-containing fossil fuels, either coal or petroleum stocks. Moreover, a very important commercial fuel, spent liquors produced in pulping processes (both kraft and sulfite) have comparatively high sulfur contents in a sodium-based system. Accordingly, this system is used as the basis for the following discussion.

Feuerstein et al.[11] demonstrated that a liquid fuel such as kraft black liquor containing about 3% sulfur, first dried and then pyrolyzed using external heat, could lose as much as 75% of the total sulfur as reduced sulfur-containing gaseous products. The residual sulfur remained as sulfide in a carbonaceous ash. Jones et al.[12] showed that with the continuous introduction of aqueous kraft black liquor into an isothermally heated reaction zone, sulfur was similarly evolved. However, at a temperature above about 900 K, gasification reactions were superimposed on thermal degradation reactions. An array of volatile organic compounds containing reduced sulfur were converted to increasing amounts of hydrogen sulfide as temperature was increased from about 800 to 1000 K. In the range of 1000 to 1300 K, hydrogen sulfide content of the product gas decreased sharply and was accompanied by an equivalent increase of sodium sulfide in the residue. These results can be explained by the general homolytic cleavage of organic sulfur-containing compounds in Equation 17, the recombination of sulfide and hydrosulfide with hydrogen-free radicals to form hydrogen sulfide as illustrated by Equation 18, and the sodium carbonate-hydrogen sulfide reaction in Equation 19.

$$R - SH \rightleftarrows R\bullet + \bullet SH \qquad (17)$$

$$HS\bullet + H\bullet \rightleftarrows H_2S \qquad (18)$$

$$H_2S + Na_2CO_3 \rightleftarrows Na_2S + CO_2 + H_2O \qquad (19)$$

Table 2
H₂S REACTION WITH SODIUM CARBONATE

Temperature				Calculated[a]	
$10^{-3} \times K^{-1}$	K	$\log_{10}K_P$	K_P	$\log_{10}K'_P$	$\log K_P$
1.43	700	0.8800	7.6	—	—
1.111	900	1.5000	32	$\bar{1}.3$	1.3
0.909	1100	1.8950	79	—	—
0.769	1300	2.1450	140	$\bar{2}.2$	2.2
0.667	1500	2.3750	237	—	—

[a] Equilibrium line calculated by Jones[13] using thermochemical data.

The hydrogen sulfide content of a product gas at equilibrium will be controlled by the constant

$$K_{P19} = \frac{(H_2O)(CO_2)}{(H_2S)}.$$

Using this constant, the dry gas composition given by Jones,[13] and the water content of this gas calculated on a hydrogen balance taking into account the formation and consumption of water, the data in Table 2 has been calculated. Data given in Table 2 has been extrapolated by 150 K and 200 K to obtain the points for 700 K and 1500 K, respectively. The lower temperature interval is in a range where pyrolytic thermal decomposition predominates and gasification is almost insignificant. The higher temperature interval represents an upper level, at which many furnaces, including recovery furnaces in the pulping industry, are commonly operated. The kinetic method used in determination of K_P in Table 2 would, at best, provide an approach to equilibrium. Because of possible "short circuiting" as well as the problems raised by Gumz in using this type of procedure, the determined values of K_P could be in considerable error. However, the calculated and experimental values are in relatively good agreement. Equation 19 and its equilibrium constant provide a basis for estimating the hydrogen sulfide content of a gas produced in gasification. Consideration of hydrogen sulfide will further complicate the gasifier calculations presented by Gumz requiring at least nine equations in nine unknowns. Gumz simplified the general procedure for making gasifier calculations by excluding the preheating section of the updraft (countercurrent) gasifier from the boundary used in his heat and material balances. This simplification avoided treatment of volatile materials produced in the preheating section. A substantial portion of sulfur in a fuel will be converted to hydrogen sulfide by reactions illustrated in Equations 17 and 18 in the preheating section and removed in the volatile fraction of a product gas. These volatile products were included in the calculations concerning the concurrent pyrolysis—gasification system presented by Brink et al.[14] However, the latter treatment was confined to atmospheric pressure under conditions whereby both hydrogen sulfide and methane comprised low and relatively constant mole fractions in the product gas. Accordingly, both could be treated as constants leading to a simplified calculation involving four equations and four unknowns.

For other metallic base systems containing sulfur in various states of oxidation that can be reduced to sulfide, the sodium-based system may be used as a model. This model cannot be used, for example, with the magnesium-based system, since magnesium sulfite is not reduced, but is decomposed giving magnesium oxide and sulfur dioxide at temperatures in the gasification range. In those instances where sulfide is found, the amount of hydrogen sulfide liberated will depend upon the metallic cations present and various parameters including the temperature and the pressure of the systems. Fuels containing organically bound sulfur will be decomposed by homolytic reactions followed by recombinations to give volatile organic compounds containing reduced sulfur, and above about 900 K, increasing amounts of hydrogen sulfide. The array of reduced sulfur compounds that can be formed by such a mechanism in the sodium-based system has been described.[15]

Selenium, when present in an organic material, should react in a manner analagous to that described for sulfur. Selenium is known to reach significant levels in biomass grown on certain selenium-rich soils. Because of the high toxicity of selenium, special attention should be given to its presence in any biomass that may be used in a combustion or pyrolysis—gasification processes. The mechanisms determining its distribution in gaseous and residual products should be elucidated, and processing should be developed to ensure its safe handling.

Nitrogen is present in biomass in various forms and especially in proteinaceous materials. It may be anticipated that oxidized nitrogen will ultimately be reduced to either molecular nitrogen or to ammonia in pyrolysis-gasification. Intermediate nitrogen-containing compounds that may be formed should undergo cleavage, yielding the more stable forms of nitrogen noted. At intermediate temperatures, amines and other nitrogen-containing compounds can survive conditions of pyrolysis-gasification and will, therefore, be present in product gases. The toxicity of nitrogen compounds formed should be established by appropriate studies similar to those currently being made on coal gasification products.[16]

Chlorine and other halides in oxidized states or present as substituents in organic compounds will form hydrogen chloride (halide) under the strongly reduced conditions of gasification. The presence of small amounts of these products can lead to serious problems of corrosion. In the presence of metallic carbonates normally formed in gasification, the hydrogen halide will react to form the corresponding metallic halide. Some of these salts, e.g., sodium chloride, have significant vapor pressures under gasification temperatures and are partially volatilized. Volatilized salts condense on cooling to form aerosols that must be removed from the product gas or its combustion products. These salts may also be the cause of extensive corrosion problems.

Phosphorous present in phosphates can be reduced to phosphine, PH_3, a highly toxic compound. If the product gas is burned directly, phosphine would be oxidized to water and oxides of phosphorous. The latter could be removed by scrubbing the combustion gases. If the product gas is not used directly in combustion, phosphine should be removed by scrubbing.

3. Environmental Considerations

The discussion of the fate of both metallic and nonmetallic elements illustrates the necessity for considering the specific reactions of each element under the conditions of gasification used. Subsequent treatment of a gas stream or an ash will depend upon the properties of each element involved in gasification, its distribution between gaseous and solid phases, and the toxicity of each compound that is formed. Current studies in this area of investigation[17-19] indicate its pertinence to the field of pyrolysis-gasification, as well as to combustion. Though the amount of a given element may be trivial in the overall process of pyrolysis-gasification, its presence can have serious impact environmentally. In this respect, the pathways involved in the formation of each product should be elucidated, then processes can be designed to eliminate potential environmental problems.

III. PROCESSES OF GASIFICATION

Gasification of solid fuels in an old technology concerned with energy production, in which even research has been comparatively dormant over a period of 4 decades. At the beginning of this period, solid fuels gasification was almost totally superceded by the use of natural gas and petroleum fuels. Currently, it is being thrust back into consideration because natural gas and petroleum can no longer satisfy our energy needs. This reconsideration of solids gasification is based upon the widespread abundance of coal as a fossil fuel and a limited, but renewable solar resource in biomass. Together, these can provide the raw material for one of several alternate energy technologies that can serve over a long term. During this time, energy production must inevitably be completely converted to renewable solar energy resources and an unpredictable geothermal contribution as fossil fuels production declines. With total stocking and utilization of the land resource, biomass production can be greatly expanded. In fact, of the several raw materials that may be used in gasification, biomass is the one that can be managed to enhance our environment and watershed. This is in sharp contrast to utilization of fossil fuels, either coal or shale oil, which inevitably degrade the environment. Because of the vast economic base enjoyed by natural gas, petroleum, and indeed, coal and shale oil, it will probably be essential to protect and probably subsidize any substantial development of biomass utilization in energy production until it becomes competitive.

In assessing the feasibility of a commercial process in the past, it has been a necessary and sufficient condition that technical feasibility should be accompanied by economic feasibility. These two are no longer sufficient conditions. Over the past decade, a third criterion, environmental feasibility, has evolved from an optional to a mandatory basis. Specifications to be met are established on the basis of national welfare, rather than management decision. Any technology, new or revised, that is to be practiced by an emerging gasification industry will be subject to close scrutiny and tight environmental regulation. This will be true without regard to scale of operation. Because of the vast range in scale being contemplated in gasification of biomass, on the order of 10,000-fold, as discussed the next section, it is pertinent to consider those design features that will control environmental acceptibility of a process as well as technical and economic feasibility.

In solids gasification, two or three process streams are produced depending upon conditions, and whether the mainstream, i.e., the product gas, is cooled before being used. The other process streams are an ash and a liquid phase that is condensed when the product gas is cooled. In practice, these three phases are not clearly separable. It is here that environmental aspects can control acceptibility of a process or add significantly to processing complexity and costs. Accordingly, a consideration of gasification design should include aspects of separation and processing of the three process streams. Specifically, the composition of vapors and gases in the product gas can be varied over an extreme range depending upon the basic design of the gasification process. This feature alone calls for widely different process designs. A classification of designs used in solid fuels and petroleum fuels gasification is given in an excellent review by Elliott and Linden.[20] Categories given in their classification include:

1.   Heat transfer: internal heating—autothermic or direct; external heating—heat exchange or indirect
2.   Reactant contact: fixed bed; fluidized bed; suspension of solid fuel in reaction medium
3.   Reactant flow: cocurrent; countercurrent
4.   Gasifying medium: steam with oxygen, oxygen enriched air; air and hydrogen
5.   Residue removal: dry ash; slagging

The classification used in their review[20] is primarily based upon product being gasified, i.e., solid fuels and petroleum fuels, reactant contact, and the gasifying medium. Specific designs of coal gasifiers are given for each category. Advantages and disadvantages of various designs and specific gasifiers are presented.

The design criteria presented by Elliott and Linden are essentially all applicable to gasification of biomass materials. Other pertinent articles concerned with reactor design are found in more general literature dealing with gasification, including symposia preprints[21] and symposia proceedings.[22-24] A preliminary publication presents single page summaries of 18 biomass gasifiers.[25] These may be grouped, according to the above classification, into nine countercurrent, fixed-bed types, five cocurrent fixed-bed types, three fluidized-bed types, and one cocurrent suspension type. Two countercurrent fixed-bed designs using air, a third using oxygen, and one fluidized-bed design for gasification of biomass are reviewed in differing detail by Katzen et al.[26] The process they considered most advanced for gasification of biomass was one of the countercurrent fixed-bed gasifiers using air as the gasifying medium. Bliss and Blake[27] reviewed the same processes discussed by Katzen et al. Brink et al[28] present a process involving a cocurrent-suspension: fixed-bed-slagging type pryolysis-gasification system that has been studied for the gasification of residual pulping liquors. In addition to the gasifiers noted, several designs using rotary kilns are being studied for biomass gasification.

Review of various gasifier designs and technology peculiar to these designs is beyond the scope of this chapter. Some of the literature pertinent to this subject has been cited. It is most pertinent here to consider those aspects of technology in a general way that can have a profound effect on implementation of the technology being developed.

Reactant flow is considered to be the single design criterion that has the greatest and, indeed, most profound effect upon the technology that must be developed. This is attributable to the composition of the main product gas that is produced and subsequent design required to give an environmentally acceptable process. To correctly assess the efficiency of a process, the entire process must be considered and compared to an alternative process design. This systems approach was taken, for example, by Brink et al.[29] in comparing production of electrical power by a conventional hogged fuel boiler vs. a pyrolysis-gasification combustion process using either air or oxygen. For purposes of this chapter, it is possible to present an approach that takes into consideration the overall process in assessing process feasibility as defined herein.

For current purposes, reactant flow is considered as presented on the basis of countercurrent and cocurrent flow. Fluidized-bed technology often gives results intermediate between these two technologies.

A. Countercurrent Design

A fixed bed, which in reality moves by plug flow in a shaft type furnace employing an updraft gas stream, i.e., countercurrent flow, is recognized for several characteristics that are cited[20,30] as attributes. Some of these are that this design leads to maximum heat recovery, low temperatures of exit gases, drying and devolatilization of green fuels, high carbon conversion, and minimal contamination of product gas with solids.

In the countercurrent design, heat recovery is frequently calculated as the cold gas efficiency or simply gas efficiency. This is the volume of gas produced (e.g., in standard cubic meters, SCM) per unit of fuel (kilogram fuel) multiplied by the gross heating value per unit volume of the product gas (kcal/SCM), and divided by the gross heating value of the solid fuel (kcal/kg of fuel) expressed as a percentage. This number is used to compare the efficiencies of gas producers in conversion of a fuel into gas as the single product. Used in this restricted sense, this is a useful number. It does not account for waste heat recovery or represent the overall efficiency of the process. Cold gas efficiency reaches a maximum as a function of the water vapor used in the blast of a countercurrent shaft type gasifier. Gumz states that gasifiers of this design can be operated using a high partial pressure of water vapor in the blast, thereby producing a higher cold gas efficiency as long as cheap steam is available from within the process. However, steam from outside the process should be minimized since its use is usually uneconomical, but is needed to control temperature below the slagging point. He further states that steam represents a significant element of cost in the operation of gasifiers and that it is economical to utilize all available sources of by-product steam.

The low temperature of product gas is achieved in the countercurrent gasifier designs by direct heat exchange between the hot product gas and the descending colder fuel. The production of volatiles, i.e., by evaporation of water and pyrolysis, cools the product gas that picks up the gases and vapors produced. This process can be carried close to the dew point of water vapor. In considering the operation of the gasifier as an isolated unit, efficient heat exchange and maximum heat recovery are realized. However, a major loss of heat is still represented by the latent heats of vaporization of the components present in the product gas, that are normally liquids at the reference conditions, i.e., 25°C and 1.0 atm pressure. Since water vapor is commonly used as the gasifying agent, this must be provided as steam to the gasifier. Moreover, the steam is raised to gasification temperature by burning a part of the fuel. Once again, it is necessary to consider the total process with respect to heat economy and not just the gasifier per se.

The advantages of low temperatures of the product gases from the gasifier, and the drying and devolatilization of green fuels should be considered together. Because the temperature of the gases passing upward in the shaft furnace is reduced, condensation of vapors and tars occurs in the colder fuel bed. This is an advantage in that the condensed materials are carried down into a higher temperature zone where further degradation can take place. However, it is a characteristic of pyrolysis gases that on cooling, aerosols are produced from the high boiling vapors. A portion of the aerosols will be broken by impingement on the fuel bed, but another portion will remain suspended in the product gas. Because of the presence of uncondensed vapors and of aerosols in the product gases, countercurrent product gases issuing from the gasifier have an appreciable content of tars and higher boiling organic compounds, as well as lower boiling organic products. These products, i.e., the so-called devolatilization products, are condensed from the product gas. In some instances, products can be separated from these condensates that have a commercial value, e.g., phenols from coal tar. In future years, gasification of solid fuels can again be a significant source of organic chemicals that can replace petrochemicals. Economics will determine the relative advantage of such processing. Any condensates that cannot be converted into commercial products and that are recovered in an organic phase separating from an aqueous phase can be returned to the gasifier for reprocessing. Using a cyclical process of this kind, the water insoluble organic phase can be reprocessed to extinction. However, such recycling is done at the cost of additional ancillary equipment and its operation, added requirements for energy in the process and heat losses attendant to such processing.

The aqueous phase separating from the condensates contains high concentrations of water-soluble organics. Some of these can be removed by additional processing if this is attractive commercially. Alternatively, the aqueous stream requires treatment to eliminate the dissolved organic materials before the water can be discarded to a receiving water.

Aerosols that persist through normal condensation processes create an additional problem. This material must be removed in any product gas that is not used directly in combustion. Even when gasifiers and combustion units are close coupled, aerosols present in product gas can break and foul gas pipes, conduits, and burners. Removal of the aerosol to provide a clean product gas can be costly and requires sophisticated equipment and processing. Thus, drying and devolatilization of green fuels and low product gas temperature can be serious deficiencies rather than benefits when the total system is evaluated relative to economic and environmental feasibility.

In his treatment of the countercurrent gasifier, Gumz[30] not only discusses the role of water vapor in the gasifying medium, but also the effects of preheating the blast, recycling off-gas, and external heat losses. These effects are especially important in consideration of the gasification of biomass having a high moisture content.

B. Cocurrent Design

A fixed bed employing a down-draft gas stream, i.e., cocurrent flow, has been described by Gumz[30] as requiring more heat, and producing a product gas having a lower heating value and a substantially higher temperature than are typical in countercurrent flow. When calculations are carried out around the gasifier alone, and especially when volatiles and moisture content of the fuel are eliminated from the calculations, these relationships are, indeed, valid. However, in a systems approach, it will be found there are compensating factors that minimize or may even reverse the advantages claimed for the countercurrent systems. Use of the high temperature product gas to generate steam and to preheat incoming fuel in the cocurrent design will reduce the advantage attributed to heat economy, but not the efficiency of heat exchange in the countercurrent design. By heating to a sufficiently high temperature and maintaining the temperature for a sufficient time, organic products are degraded to carbon monoxide, carbon dioxide, and methane. Thus, all processing and ancillary equipment required for gas cleanup is avoided. This distinct advantage of a cocurrent system is in sharp contrast to the complications resulting from having to process a substantial amount of complex organics present in the product gas in the countercurrent design. Once evaporated, the moisture in biomass that is normally relatively high serves as the gasifying agent in the cocurrent gasification design. Steam is not required in the gasifying medium. By decreasing the moisture content of wood and supplying external heat and by preheating the gasifying agent, the gasifying efficiency in down-draft gasifiers is substantially increased.[30]

The removal of particulate and gaseous contaminants, e.g., hydrogen sulfide, from cocurrent-generated product gas is possible after heat transfer has reduced temperature to levels that permit efficient gas scrubbing. For example, the removal of particulate by the phenomenon of nucleation may be realized and efficient scrubbing of hydrogen sulfide may be efficiently carried out by lowering the product gas temperature below the dew point. Simultaneously, recovery of low level heat is possible as the water vapor in the product gas is condensed.

The flexibility in design of either cocurrent or countercurrent systems permits adjustment of product gas temperatures. However, as countercurrent product gas temperature is increased, the attributes of this system due to gas temperature are diminished. Likewise, as cocurrent product gas temperature is decreased, an increasingly complex product gas composition is obtained. This problem can be overcome, in part, by close coupling of a gasifier and a combustion unit.[31]

C. Fluidized-Bed Design

A fluidized-bed design is obtained using a countercurrent flow of gas and particulate material. This is done using a fluidizing medium to control particle size and to provide a condition that allows for the expansion of the bed, i.e., the particulate is not suspended. Using this design, a uniform temperature is maintained throughout the bed and particles are broken by mechanical attrition, yielding high rates of gasification at temperatures that are lower than required in fixed-bed processes. However, it suffers from the problem of high loss of sensible heat in the product gas, a requirement for relatively uniform particle size and density, and carry-over of both fuel and ash in the product gas. As in cocurrent technology, the loss of sensible heat can be compensated by heat exchange. Although tar and high molecular weight organic components can be virtually eliminated, a considerable problem is created by the carry-over of inorganic ash. Carry-over of fuel introduces an element of cocurrent suspension design. Energy required for fluidizing is an appreciable factor.

D. Cocurrent Suspension Design

By suspending the particulate material in the gasifying medium, the particles are conveyed in that medium. Several processes have been designed for use with coal,[20] having the distinct advantage that all grades of coal, including caking grades, can be processed. Ash fusion temperature is not important except in determining whether a slagging or nonslagging design is used. Product gases contain no tars and very low methane contents; therefore, these gases are particularly well suited for use in chemical synthesis. The high rates of gasification obtained provide for minimum reactor volume per unit of fuel processed. High conversion of fuel can be realized by recycling carbonaceous residues. High thermal losses are compensated, in part, by provision of adequate heat exchange capacity as in the cocurrent fixed-bed design. This design requires a finely divided feed, preferably with a low moisture content. The low moisture content results in production of a product gas having a high content of carbon monoxide and a lower content of hydrogen in the product gas.[32] High temperature is required to minimize the production of tars and lower temperature can be used for the production of organics,[33] but with the problems attending the production of a complex array of organic products, including tars.

E. Selection of Gasification Technology

The question of selecting a given gasifier design should be based upon a total systems approach. The design of the gasifier unit provides one important input in an overall system. Selection of the system design should be based on as complete an assessment as possible of the technical, economic, and environmental information that is available on the subject.

The basic flows used in gaisifiers have been used to illustrate some of the issues involved and questions to be asked. No answers have been provided. Reactor design is currently in a state of high flux. Each type of application will be governed by a set of parameters. Based upon these parameters the thorough analysis suggested above should be made. In the next decade, numerous variations of the basic designs will be proposed. The challenge will be to advance the technology of gasification by selecting designs for given applications that will be competitive in a volatile energy era.

IV. APPLICATION OF GASIFICATION TECHNOLOGY

A very broad range is currently being considered in the application of gasification technology. This may be illustrated by consideration of commercial installations capable of supplying boiler energy at the rate of 2.5×10^8 kcal/hr ($\sim 1 \times 10^9$ Btu/hr) at the upper end to the retrofit of gas or oil burners in single family dwellings that will deliver under 2.5×10^4 kcal/hr ($\sim 1 \times 10^5$ Btu/hr) at the lower end of the scale. Facilities that are capable of satisfying this range can be designed, built, and operated. Sites required for facilities at the upper end of the scale, requiring on the order of 1500 oven-dry tons of lignocellulosic material per day, are very limited. As the rated capacity of gasification facilities is decreased, sites become increasingly more numerous. With application of this technology, it will be generally possible to improve land management practices and environmental quality.

A most important area of application is in the mid-range of the scale given where various industries concerned with silvicultural and agricultural activities have quantities of residues available. Gasification technology can be provided to retrofit and/or expand existing steam generating facilities. Many of these industries can, in fact, supply their energy needs based on this technology. Arrangements can be made to supply neighboring industries or communities with excess product gas, steam, or electricity. The basis for this type of energy supply was highly developed in some areas of the U.S. as late as the 1930s. It is still used in some countries and no doubt will be on the increase from this time forward.

Application of gasification technology to mobile power units was practiced in the crisis periods of 1914 to 1918 and 1940 to 1945. A part of this technology has been reported recently.[34] Application to diesel power units and farm machinery is currently under consideration.[25,31]

Also, gasification of biomass is being considered for the production of product gas to be used in the synthesis of chemicals such as methanol and ammonia[26,27,35] as well as the hydrocarbons obtainable by the Fischer—Tropsch reaction. Methanol can be further processed to formaldehyde.

For these purposes, large quantities of biomass must be assured over extensive periods of time. Thus, locations for such plants will be a limiting factor. It is particularly with respect to such applications that national policy must be established, first with respect to the improvement of forestry practices noted above, and then to the commitment of resources. Production of chemicals from syn-gas is highly sensitive to economies of scale. Use of biomass for this type of production will be in direct competition with coal. Although coal is available in vast amounts at given locations, its use entails high costs to maintain environmental quality, if this is possible. In sharp contrast, growing and harvesting plants, including trees, by modern silvicultural and agricultural techniques can result in definite overall environmental benefits.

REFERENCES

1. Minkoff, G. J. and Tipper, C. F. H., *Chemistry of Combustion Reactions,* Butterworths, Reading, Mass., 1962, 237.
2. Gumz, W., *Gas Producers and Blast Furnaces,* John Wiley & Sons, New York, 1950, 3.
3. Lewis, G. N. and Randall, M., revised by Pitzer, K. S. and Brewer, L., *Thermodynamics,* 2nd ed., McGraw-Hill, New York, 1961, 169.
4. Moore, W. J., *Physical Chemistry,* 3rd ed., Prentice-Hall, Englewood Cliffs, N.J., 1962, 167.
5. Gumz, W., *Gas Producers and Blast Furnaces,* John Wiley & Sons, New York, 1950, 28.
6. Wagman, D. D., Kilpatrick, J. E., Taylor, W. J., Pitzer, K. S., and Rossini, F. D., Heats, free energies, and equilibrium constants of some reactions involving O_2, H_2, H_2O, C, CO, CO_2, and CH_4, *J. Res. Natl. Bur. Stand.,* 34, 143, 1945.
7. Gumz, W., *Gas Producers and Blast Furnaces,* John Wiley & Sons, New York, 1950, 32.
8. Broido, A., Thermogravimetric and differential thermal analysis of potassium bicarbonate — contaminated cellulose, *Pyrodynamics,* 4(3), 243, 51, 1966.

9. **Philpot, C. W.**, Influence of mineral content on the pyrolysis of plant materials, *For. Sci.*, 16(4), 461, 1970.
10. **Fung, D. P. C., Tsuchiya, Y., and Sumi, K.**, Thermal degradation of cellulose and levoglucosan — the effect of inorganic acid, *Wood Science*, 5(1), 38, 1972.
11. **Feuerstein, D. F., Thomas, J. F., and Brink, D. L.**, Malodorous products from the combustion of kraft black liquor. II. Analytical aspects, *Tappi*, 50(6), 276, 1967.
12. **Jones, K. F., Thomas, J. F., and Brink, D. L.**, Control of maloders from kraft recovery operations by pyrolysis, *J. Air Pollut. Control Assoc.*, 19(7), 801, 1969.
13. **Jones, K. T.**, The control of malodors from kraft recovery operations by pyrolysis, Ph.D. thesis, University of California, Berkeley, 1968.
14. **Brink, D. L., Faltico, G. W., Thomas, J. F.**, Pyrolysis, gasification, combustion. Feasibility in pulping recovery systems. The first stage — second stage reactors as a production unit, in *TAPPI Conference Papers, Alkaline Pulping and Testing*, TAPPI, Atlanta, Ga., 1976, 169.
15. **Brink, D. L., Pohlman, A. A., and Thomas, J. F.**, Analysis of sulfur-containing and sulfur-free organic products formed in kraft back liquor pyrolysis, *Tappi*, 54(5), 714, 1971.
16. **Ho, C.-h., Clark, B. R., Guerin, M. R., Ma, C. Y., and Rao, T. K.**, Aromatic nitrogen compounds in fossil fuels — a potential hazard, *Am. Chem. Soc. Div. Fuel Chem. Prepr.*, 24(1), 281, 1979.
17. **Koppenaal, D. W., Schulz, H., Lett, R. G., Brown, F. R., Booker, H. B., and Hattman, E. A.**, Trace element distributions in local gasification products, *Am. Chem. Soc. Div. Fuel Chem. Prepr.*, 24(1), 299, 1979.
18. **Kuhn, J. K., Cahill, R. A., Shiley, R. H., Dickinson, D. R., Kruse, W.**, The effect of trace elements associations and mineral phases in the pyrolysis products of coal, *Am. Chem. Soc. Div. Fuel Chem. Prepr.*, 24(1), 307, 1979.
19. **Dreher, G. B., Russell, S. J., and Ruch, R. R.**, Chemical and minerological characterization of coal liquification residues, *Am. Chem. Soc. Div. Fuel Chem. Prepr.*, 24(1), 299, 1979.
20. **Elliott, M. A. and Linden, H. R.**, Gas, manufactured, in *Kirk-Othmer: Encyclopedia of Chemical Technology*, 2nd ed., Vol. 10, Mark, H. F., McKetta, J. J., Jr., and Othmer, D. F., Eds., John Wiley & Sons, New York, 1966, 353.
21. **Division of Fuel Chemistry, American Chemical Society**, Preprints to symposia presented at National Meetings, Volumes 1—24, Washington, D.C
22. **Jones, J. L. and Radding, S. B., Eds.**, *Solid Wastes and Residues, Conversion by Advanced Thermal Process*, ACS Symp. Ser. 76, American Chemical Society, Washington, D.C., 1978.
23. **Massey, L. G., Ed.**, *Coal Gasification*, Adv. Chem. Ser. 131, American Chemical Society, Washington, D.C., 1974.
24. **Schora, F. C., Jr.**, *Fuel Gasification*, Adv. Chem. Ser. 69, American Chemical Society, Washington, D.C., 1967.
25. **Reed, T. B. and Jantzen, D. E.**, *Directory of Air Biomass Gasifiers in the U.S. and Canada*, Solar Energy Research Institute, Golden, Colo., 1979.
26. **Katzen, R. and Associates**, Chemicals from Wood, Forest Products Laboratory, U.S. Department of Agriculture, Forest Service, Madison, Wis., 1975.
27. **Bliss, C. and Blake, D. O.**, Silviculture biomass farms, Vol. 5, *Conversion Processes and Costs*, Division of Solar Energy, Energy Research and Development Administration, Washington, D.C., 1977.
28. **Brink, D. L., Faltico, G. W., and Thomas, J. F.**, The University of California pyrolysis-gasification-combustion process, in *Forum of Kraft Recovery Alternatives*, Howells, T. A., Ed., The Institute of Paper Chemistry, Appleton, Wis., 1976, 148.
29. **Brink, D. L., Faltico, G. W., and Thomas, J. F.**, The pyrolysis-gasification-combustion process: energy effectiveness using oxygen vs air with wood-fueled systems, in *Fuels and Energy from Renewable Resources*, Tillman, D. A., Sarkanen, K. V., and Anderson, L. L., Eds., Academic Press, New York, 1977, 141.
30. **Gumz, W.**, *Gas Producers and Blast Furnaces*, John Wiley & Sons, New York, 1950, 109.
31. **Williams, R. O., Goss, J. R., Mehlschau, J. J., Jenkins, B., and Ramming, J.**, Development of pilot plant gasification system for the conversion of crop and wood residues to thermal and electrical energy, in *Solid Wastes and Residues, Conversion by Advanced Thermal Processes*, Jones, J. L. and Radding, S. B., Eds., ACS Symp. Ser. 76, American Chemical Society, Washington, D.C., 1978, 142.
32. **Brink, D. L.**, Pyrolysis-gasification-combustion: a process for utilization of plant material, in *Applied Polymer Symposium*, Timell, T. E., Ed., John Wiley & Sons, New York, 1976, 1377.
33. **Brink, D. L. and Massoudi, M. S.**, A flow reactor technique for the study of wood pyrolysis. I. Experimental, *J. Fire Flammability*, 9(2), 176, 1978.

34. **Reed, T. B. and Jantzen, D. E.,** Generator gas, the Swedish experience from 1939—1945, SP-33-140, Solar Energy Research Institute, Golden, Colo., 1979.
35. **Anon.,** Economic pre-feasibility study: large scale methanol fuel production from surplus Canadian forest biomass, Part 1, Summary report, Intergroup Consulting Economists, for Policy and Program Development Directorate, Winnipeg, Canada, 1976.

Chapter 5

PYROLYSIS

Ed J. Soltes and Thomas J. Elder

TABLE OF CONTENTS

I. DEFINITION OF PYROLYSIS

The pyrolysis of carbonaceous materials refers to incomplete thermal degradation resulting in char, condensible liquid or tar, and gaseous products. In its strictest definition, pyrolysis is carried out in the absence of air. However, pyrolysis has more recently found broader definition in describing any chemical changes brought about by the application of heat, even with air or other additives.

Some solids, liquids, and gases are produced in every thermal degradation process, including gasification. However, pyrolysis differs from gasification in that the products of interest are the char and liquids, which as a result of the incomplete nature of the process, retain much of the structure, complexity, and signature of the raw material undergoing pyrolysis. In reference to wood, pyrolysis processes are alternately referred to as carbonization, wood distillation, or destructive distillation processes. As generally accepted, carbonization refers to processes in which the char is the principal product of interest (wood distillation, the liquid; and destructive distillation, both char and liquid).

II. OLDER WOOD PYROLYSIS PROCESSES

A. Early Growth of the Wood Pyrolysis Industry

Wood is the world's oldest raw material, and possibly still its most widely used. The pyrolysis of wood was probably man's first chemical process. Destructive distillation is known to have been practiced by the ancient Chinese. The Egyptians, Greeks, and Romans made charcoal by wood carbonization, and collected the condensible volatiles for embalming purposes and for filling joints in wooden ships.[1] Today, wood pyrolysis processes are still important, particularly in developing nations and in Russia. In the U.S. and more generally the Western Hemisphere, competition from alternate feedstocks has forced out all but a few wood distillation and carbonization plants.

Until the late 1800s, wood carbonization was the major pyrolysis process, and supplied the increasing amounts of charcoal that were required for iron ore smelting in a growing America. Although coke from coal was introduced in 1760, U.S. iron furnaces virtually used charcoal exclusively until about 1840. After 1869, the share of the iron market produced with charcoal was increasingly surpassed by that of coal and coke iron.[2] The charcoal industry had almost exhausted surrounding woodlands in the East, and soon concentrated in Michigan and the South, where supplies of wood were still adequate. It is interesting to note that iron foundries followed the industry to these regions, because charcoal iron was the best available. Steel made with charcoal can resist rust and hold an edge better than steel made with coke.[3] Chestnut was the preferred wood for carbonization, and was relatively plentiful before the blight of 1904.

In the mid-1800s, charcoal found many uses. Charcoal vendors in cities sold charcoal by the peck and bushel from push carts to heat fashionable Victorian homes. Charcoal was a better fuel than wood in that its Btu content was 50% higher than dry wood. Charcoal was also used to pack meat, and make printer's ink, black paint, and medicines. The general populace brushed their teeth and sweetened their breath with it, and ingested it for upset stomach.[2,3]

At the same time, increasing amounts of methanol (wood alcohol) were needed in a growing intermediate dye and synthetic organic chemicals industry. Soon methanol was being distilled from hardwoods in processes that recovered the volatile fractions of wood pyrolysis. In the period from 1870 to 1900, marked advances were made in crude liquor recovery and refining. The wood spirits that contained methanol also contained acetic acid, and in 1870, a process was developed to produce acetic acid and acetone via calcium acetate. By 1900, destructive distillation of wood was practiced widely on a commercial scale by heating wood in closed retorts. The vapors produced were condensed to give a tar (wood tar or sedimentation tar) and an aqueous layer (pyroligneous acid). A fraction of the vapors were noncondensible (wood gas), and solid charcoal was left in the retort. Further processing of pyroligneous acid produced methanol and acetic acid. Wood gas could be scrubbed, then burned as fuel.

By the turn of the century, wood was the only source for bulk quantities of methanol, acetic acid, and acetone. With the advent of this wood chemical industry that utilized the by-products of carbonization, charcoal became the by-product and was available in considerable quantities for low cost.

B. Competition for the Industry

The reign of wood as the principal raw material for the organic chemicals industry was short-lived. Alternate processes for the production of acetone, acetic acid, and methanol were soon developed and these depressed the wood distillates market.[4] Even these newer processes were soon competing with, and losing to, a growing and more efficient petrochemicals industry.

During World War I, acetone became available as a starch fermentation product, and was later made from propylene. More recently, acetone has been produced in the cumene phenol synthesis and by the oxidation/dehydrogenation of isopropanol. Acetic acid was soon made from calcium carbide, then synthetically through the oxidation of acetaldehyde or butane. It is produced today from methanol and synthesis gas. Wood alcohol suffered its first blow with the availability of synthetic methanol from Germany, and is now readily produced from synthesis gas, as well as in other processes.

During the period in which wood-derived chemicals were being supplanted with synthetic chemicals, wood distillers attempted to compete with process improvements. Wood waste, instead of marketable timber, was used as a raw material, the drying time for wood was shortened, retort operation was made continuous, and processes were introduced for obtaining acetic acid directly, but to no avail. Charcoal marketing was however extended, and many U.S. plants formerly operated for wood distillation were converted exclusively to wood carbonization.[1,4]

In the period from 1950 to 1965, acetic acid was produced from wood distillation in the U.S. at a rate of about 20 million lb/year.[5] In 1972, the last plant producing acetic acid via this route was closed due to economic pressures. Total synthetic acetic acid production in the U.S. today is over 2 billion lb. Acetic acid is the fastest growing chemical derivative of methanol, and by 1981 will be the second largest use for methanol in the U.S.

During the period from 1950 to 1965, about 13 to 15 million lb of methanol was produced annually in the U.S. by wood distillation. About 10 million lb were produced yearly from 1966 to 1976.[6] This "natural methanol", as it is called, was produced primarily for alcohol denaturization, and until recently, the U.S. Treasury Department required that the denaturant had to be from wood distillation. Natural methanol was sold for three to four times the market price of synthetic methanol.

Where it is applied commercially today, the basic process of wood destructive distillation is essentially unchanged, save for process improvements aimed at better yields of more refined products. Hardwoods are generally preferred to softwoods because they produce higher yields of acetic acid and methanol, and a denser charcoal. Softwoods, however, additionally produce turpentine and various oils in destructive distillation, and pine was used extensively in a southern U.S. wood distillation industry.

Under present conditions, the production of acetone, acetic acid, and methanol from wood distillates is rarely economically attractive in the Western Hemisphere. Russia, and other Soviet Bloc Countries, however, continue to use wood for chemicals, both through distillation and hydrolysis processes. Much of the rest of the world still relies on wood carbonization to yield charcoal for heating, cooking, and industrial purposes, while chemicals are only occasionally produced as by-products. In many developing nations, relatively crude carbonization kilns are used to produce charcoal in much the same way as the U.S. did some 2 centuries ago.

As a chemical, there is no true synthetic counterpart for charcoal. Its use in iron ore smelting has declined, except in countries such as Brazil,[7] while in the U.S., charcoal, although having many potential industrial applications, is now most widely used as a specialty item for domestic and campers' fuel for outdoor cooking.

Is there a future for wood pyrolysis? Recent economic developments have increased the cost of fossil fuel raw materials to the point that they can no longer be considered relatively inexpensive fuels and chemical raw materials. In addition, recent developments in thermal processing of wood, materials of construction technology, and increased availability of logging and other lignocellulosic waste materials suggest that routes to chemicals and fuels via thermal degradation of wood may once again become economically attractive.

C. Hardwood Carbonization and Distillate Recovery Processes
1. Carbonization Via Kilns

Charcoal can be readily produced from wood with no capital investment in equipment through the use of charcoal piles, earth kilns, or pit kilns. As the names of these processes suggest, hardwood (2 to 50 cords) is carefully stacked in a mound or pit around a central air channel, then covered with dirt, humus, moss, clay, or sod. After providing openings at the bottom of the pile to admit and regulate air, and on the sides of the pile to allow for the release of smoke and volatiles, the wood pile is lit in the center, and carbonization proceeds. As carbonization is slow, the pile subject to collapse, and hot charcoal subject to spontaneous combustion necessitating a long cool-down period, charcoal tenders were required to stay with the pile until carbonization was complete, sometimes as much as 6 weeks later. As described, the production of charcoal in earth kilns is an art, and experience was required to achieve yields of 20% based on dry wood weight.[3,8]

Variations in the art that require some investment in facilities and equipment make use of permanent (20 to 90 cords) kilns. Here, brick or masonry are used to cover and contain wood during carbonization, and doors are provided for charging wood and removing product. Holes and chimneys may be optimally located to regulate air flow and discharge flue gases. Higher yields of charcoal can be obtained in these kilns than in earth kilns, with less hazard or dependence on operator skills. In the U.S., particularly in Michigan, when charcoal was produced for iron ore smelting, dome-shaped beehive kilns were extensively used. In larger plants, several smaller kilns were usually operated in series to even out production. Smaller portable metal kilns (one half to four cords) were also developed, and could be set up in the forest and moved about as sources of wood supply in the surrounding areas became exhausted.

A fundamental problem in the use of charcoal piles or beehive-type kilns is that carbonization is accomplished by heat generated from combustion of part of the wood charge and charcoal product. The resulting low yields, coupled with undesirably long cycle times, and the impracticability in not being able to recover by-products resulted in the development of retort furnaces for carbonization. It is important to recognize, however, that with the exception of more highly industrialized societies, most of the world still uses the simpler processes for wood carbonization.

2. Retort Furnaces

Retort furnaces have been more recently used to produce charcoal on a commercial scale. Although various retort processes differ somewhat in design, all make better use of the heat involved in wood pyrolysis to effect higher charcoal yields.[9] Retort furnaces typically yield 30 to 40% charcoal from hardwood, as opposed to some 20 to 25% attainable in kilns.

Retort furnaces also have provision for by-product recovery. At the higher temperatures prevalent in the pyrolysis zone of these furnaces, volatile by-products can crack, decompose, and condense. Short residence time at these undesirable higher temperatures is generally accomplished by extracting volatile products with externally circulated gases. The volatile components in the exit gas stream are then subjected to a condensation train to remove condensibles before the gases are returned to the reactor. These condensibles are the tars and aqueous pyroligneous acid referred to previously. The circulated gases can be composed of the wood gas product of pyrolysis, flue gas from some other operation, or both. After condensing the desired volatile products, gases can be heated before recirculation to dry the wood feed, or supply energy for the pyrolysis process. Alternately, the stripped gas can be used countercurrently to cool product charcoal before entering the pyrolysis zone to recover volatiles.

Several texts already provide detailed descriptions and process flow sheets of important retort processes.[8,9] These will be discussed only briefly here. The Stafford and Seaman batch processes developed in the U.S. were primarily suited to the pyrolysis of small pieces of wood obtained as wood waste and sawdust from wood processing industries. Wood is dried and preheated before entering the retort in which a fire starts the carbonization process. As carbonization proceeds, the exothermic pyrolysis reaction is used to complete pyrolysis without the application of external heat. Wood waste and sawdust become densely packed in a stationary pyrolysis reactor (Stafford vertical retort), and result in poor heat conduction. In the Seaman process variation, an inclined, rotating retort was used to mix the feed and improve heat transfer.

The Badger-Stafford process is a continuous process version of the Stafford process.[8] Wood chips are predried in sloping rotary driers before being admitted to the top of cylindrical vertical retorts. Continuous pyrolysis is accomplished by carefully regulating wood chip flow to effect wood distillation by the heat liberated in the exothermic pyrolysis reaction. Gases and vapors are led through a condenser on the side of the retort, and charcoal is discharged on the bottom.

Three continuous pyrolysis processes are described by Wenzl.[9] The Reichert process used in Germany is a continuous reactor that uses short blocks of wood to overcome the heat transfer problem. In this process, rate of carbonization is controlled by the amount and temperature of a circulating gas stream. The Lambiotte process is a three-stage continuous process favored in several European countries. Here, hot gases move upward and countercurrently through a vertical retort, releasing heat into wood moving downward. Charcoal product is discharged at the bottom of the reactor after first being cooled by the externally circulated gas.

Wood must be dried before undergoing pyrolysis in continuous retorts. Hot gases from pyrolysis generally supply some of this energy, although most processes rely on waste gases from power and steam plants. If these are not available, part of the wood feed to pyrolysis can be burned to provide hot flue gases for drying and pyrolysis energy purposes. Such a provision is incorporated into a Swedish continuous softwood pyrolysis operation described by Wenzl.[9] This operation has provision for recovery of turpentine.

3. Distillate Recovery Processes
a. General Observations

All hardwood distillation plants using retort furnaces produce about the same quantity of the major products, charcoal and raw pyroligneous liquor. Each 2000 lb of dry hardwood (northern U.S.) will yield some 650 to 785 lb of charcoal and 145 to 200 gal of raw pyroligneous liquor.[8]

The major products recovered from this raw pyroligneous liquor are about 5 to 6 gal of refined methanol and 75 to 90 lb of glacial acetic acid. Minor products include allyl alcohol, methyl acetone (a mixture of methanol, methyl acetate, and acetone), wood tar and oils, pitch, and formic and propionic acids. Not all of these products were recovered in every plant, and some plants produced others not listed. Many additional unit operations beyond the pyrolysis process are required to separate and purify the desired distillate products, and it is these down-stream requirements that were subject to improvement and modernization.

b. Refining Raw Pyroligneous Acid

The condensed distillates of wood pyrolysis are generally referred to as raw pyroligneous liquor or crude wood vinegar (see Figure 1). This liquor is first settled and decanted to separate sedimentation tar (A-tar) from the aqueous raw pyroligneous acid (crude wood acid).[9,10] Sedimentation tar can be fractionated into tar oils and tar pitch. Tar oils are further fractionated into light, heavy, and high-boiling fractions.

Raw pyroligneous acid contains methanol, acetic acid and soluble tar (B-tar). Soluble tar is easily decomposed to an insoluble pitch by heating, and can thus be removed by distilling raw pyroligneous acid to yield low-tar boiled liquor. Methanol can be distilled continuously from boiled liquor, and crude acetic acid can be recovered through multiple-effect evaporation.

There exist several variations to this process scheme.[8-10] B-tar can be recovered in good yields by scrubbing retort gases. Methanol can be distilled directly from raw pyroligneous acid. Many plants neutralized the acids in boiled liquor with lime before distilling off weak alcohol, and evaporated the remaining acetate liquor to yield grey calcium acetate. Weak alcohol was fractionated to yield a 92 to 95% methanol, or condensed to a crude 82% wood alcohol. The grey calcium acetate was acidified and refined to produce acetic and other acids. Alternately, distilling acetic acid was taken up in calcium hydroxide and recovered as the acetate for further processing. Acidification of the acetate with sulfuric acid, followed by distillation, yielded an 82% acetic acid. This was further rectified to produce glacial acetic acid. Calcium acetate could also be processed to produce acetic acid and acetone. Other refinements in these processes existed, especially if some of the minor components in raw pyroligneous acid were marketed.

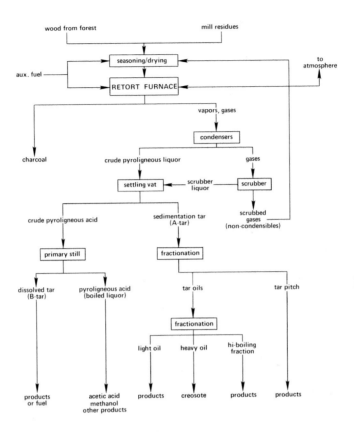

FIGURE 1.   Typical flow sheet for hardwood distillation.

c. Acetic Acid Recovery Processes

Most of the technically more interesting developments in distillate refining processes were in the methods for concentration and recovery of acetic acid. Three different methods are particularly worthy of mention: the Coahran liquid-liquid extraction process, the Suida scrubbing process, and the Othmer azeotropic distillation process.

The liquid-liquid extraction process, developed by Coahran,[4,11] was a single continuous cold treatment of liquor with either ethyl ether or ethyl acetate. Extraction was followed by a distillation with an oxidizing and purifying treatment using 2 to 4 lb of sodium dichromate per ton of acid. This process was favored in the U.S., and still used extensively in the U.S.S.R. with several modifications, since essentially quantitative extraction is possible with over 95% of the water content flowing to waste without the necessity of evaporating it. Recent studies in Russia indicate a preference for ethyl acetate or binary mixtures for acetic acid extraction.[12,13]

In the Suida process,[14] raw pyroligneous acid was vaporized and scrubbed countercurrently with a wood oil obtained by vacuum distillation of sedimentation tar. The oil and acetic acid were passed through a dehydrating column, acetic acid was boiled off in a vacuum stripping column, then further rectified. Vapors from the scrubbing column could be concentrated to yield methanol, allyl alcohol, acetone, and weak alcohol. The Suida process was highly efficient, but required substantial capital investment. Additionally, it was not popular in the U.S. because of comparatively excessive steam costs.

The Othmer process produced crude methanol in a demethanolizing distillation column.[1,15-18] The dealcoholized liquor was metered into an evaporator that separated

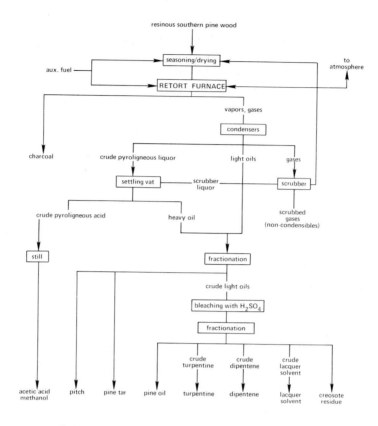

FIGURE 2.   Typical flow sheet for pine distillation.

any residual tar and distilled off a dilute acetic acid. An azeotropic dehydrating column, using butyl acetate, ethylene chloride, or ethyl tertiary amyl ether as the azeotropic withdrawing liquid, stripped water from the acetic acid. The crude acetic acid
so derived contained some propionic and butyric acids, tar, and 0.5% water. This was
further rectified with a sodium dichromate treatment as described earlier. Steam consumption was relatively high for the Othmer process, but the process employed a minimum of equipment. The Othmer process was the last to survive in the U.S.

D. Destructive Distillation of Pine

Destructive distillation of U.S. southern pine wood was once of great economic importance in the manufacture of naval stores.[19] Naval stores included turpentine, as
well as pitch, tar, and resin, which were indispensable in shipyards for protecting surfaces and caulking seams of wooden ships. The principal raw material for this industry
was resinous heartwood from fallen trees, logs, and stumps that had so rotted away
as to contain no sapwood.[8]

The destructive distillation process for pine was similar to that used for hardwoods,
except that the products were charcoal, turpentine, pine oil, dipentene, pine tar, tar
oils, and pitch (see Figure 2). Pine yielded less methanol and acetic acid, and these
were rarely recovered. It was a true distillation industry, with charcoal produced as a
by-product residue.

Vapors and gases from pine distillation in batch retorts were passed through con-
densers, with noncondensible gases used for retort fuel. The crude distillate was settled
to remove a weak acid-water solution, and the oily fraction, consisting of terpenes and

Table 1
YIELDS OF PYROLYSIS PRODUCTS
FROM 21 U.S. HARDWOOD SPECIES (%
OF OVEN DRY WOOD)

Product	Average yield	Yield range
Charcoal	42.7	37.0—50.4
Tar	9.4	4.2—11.7
Pyroligneous acid	33.7	30.7—36.5
Methanol	1.7	0.9—2.2
Total acid as acetic acid	5.7	4.3—6.8

Adapted from Hawley, L. F. and Wise, L. E., *The Chemistry of Wood*, Chemical Catalog, New York, 1926, 200.

resin decomposition products, was distilled to produce pine tars, pine tar oils, and crude light ends. Subsequent refining of the light ends in a solvent still and fractionation column yielded turpentine, dipentene, and pine oil. Pitch was obtained as a residue in distillations, and directly from the retort towards the end of distillation. Approximate yields of products from 2000 lb pine stump wood were 400 to 450 lb charcoal and 30 to 50 gal of total oils, including 4 to 6 gal wood turpentine, about 1 gal pine oil, and 20 to 30 gal of pine tar.[8]

The development of a more efficient solvent-steam distillation process for the extraction of naval stores and the eventual depletion of highly resinous old-growth stumpwood, forced the demise of this industry.

III. PRODUCTS AND THEIR USES

A. General Observations

The major products of wood pyrolysis are char, tar, pyroligneous acid, and wood gas. Commercial retort processes using the same hardwood species differ little in yields of these products. Yields are somewhat more variant as a function of hardwood species used. Average yield and yield ranges of charcoal, tar, and pyroligneous acid from 21 U.S. hardwood species are given in Table 1.[19]

There are major differences in yields of products when softwoods are pyrolyzed. The yields of pyrolysis products from two hardwood and two softwood species are compared in Table 2.[9,20] For softwoods, char yield is generally higher, but of lower density. The yields of acetic acid and methanol are higher for softwoods.[20,21]

B. Hardwood Pyrolysis Products
1. Char

At the usual carbonization temperature of about 400°C, char represents the largest component in wood pyrolysis products. Commercial char produced under these conditions contains about 80% carbon, 1 to 3% ash, and 12 to 15% volatile materials.[9]

The elemental composition of char, its yield in pyrolysis, and its properties, are dependent on final carbonization temperatures. As shown in Table 3, chars can be produced that are essentially all carbon, or alternately with oxygen contents of over 40%. Table 4 indicates how charcoal properties are affected by carbonization temperature,[9] suggesting that end product uses can dictate carbonization temperatures employed. The wide range of properties of charcoal result in many commercial applications (Table 5).

Table 2
YIELDS OF PYROLYSIS PRODUCTS FROM SOFTWOODS VS. HARDWOODS (% OF OVEN DRY WOOD)

Product	Softwoods		Hardwoods	
	Pine	Spruce	Birch	Beech
Charcoal	37.8	37.8	31.8	35.0
Tar	11.8	8.1	7.9	8.1
Gases	14.7	14.9	14.0	15.8
Acetic acid	3.5	3.2	7.1	6.0
Methanol	1.0	0.9	1.6	2.1
Water	22.3	25.7	27.8	26.7

Adapted from Klason, P., Heidenstam, G., and Norlin, E., *Z. Angew. Chem.*, 23, 1252, 1910.

Table 3
ELEMENTARY COMPOSITION AND YIELD OF CHARCOAL AS A FUNCTION OF CARBONIZATION TEMPERATURE

Carbonization temperature °C	Elementary Composition			Yield %
	C	H	O	
200	52.3	6.3	41.4	91.8
300	73.2	4.9	21.9	51.4
400	82.7	3.8	13.5	37.8
500	89.6	3.1	6.7	33.0
600	92.6	2.6	5.2	31.0
800	95.8	1.0	3.3	26.7
1000	96.6	0.5	2.9	26.5

From Wenzl, H. F. J., *The Chemical Technology of Wood*, Academic Press, New York, 1970, 267. With permission.

Table 4
PROPERTIES OF CHARCOAL AS A FUNCTION OF CARBONIZATION TEMPERATURE

Increases with increasing carbonization temperature	Decreases with increasing carbonization temperature	Goes through maximum with increasing carbonization temperature
C content	H content	Micropore volume (800—850°C)
Ash content	O content	
True density	Apparent density	Intermediate pore volume (500—550°C)
Content of nonvolatiles	Content of volatiles	Absorbency (450-650°C)
Shrinkage	Charcoal yield	
Ignition temperature	Water sorption	
CO in oxidation products	CO + H$_2$O in oxidation products	
	Electrical conductivity	

The microscopic structure of wood charcoal reflects the anatomical structure of the wood from which it was formed, and it is possible to identify the wood of origin from the microstructure of its char. Further, the charring of wood produces a conducting surface, thus eliminating the need for coating in scanning electron microscopy.[22,23]

2. Pyroligneous Acid

Settled pyroligneous acid, i.e., the aqueous acid layer decanted from sedimentation tar, contains many components (see Table 6). Due to the earlier importance of the hardwood distillation industry, the composition of hardwood pyroligneous acid was well known over 50 years ago.[24]

The major components of commercial interest were methanol and acetic acid. As discussed previously, at the turn of the century, hardwood was the only source of bulk quantities of methanol, acetic acid, and acetone (derived from acetic acid). Other com-

Table 5
COMMERCIAL APPLICATION OF CHARCOAL

Domestic and specialized fuels	Metallurgical	Chemical
Recreational	Copper	Carbon disulfide
Curing tobacco	Brass	Calcium carbide
Cooking in dining cars	Pig iron	Silicon carbide
Heating in shipyards and citrus groves	Steel	Sodium cyanide
	Nickel	Potassium cyanide
	Aluminum	Carbon monoxide
	Electro-manganese	Activated carbon
	Armor plate	Carbon black
	Foundry molds	Fireworks
		Gas adsorbent
		Water purifier
		Crayons
		Soil conditioner
		Pharmaceuticals
		Poultry and animal feed

From Wenzl, H. F. J., Ed., *The Chemical Technology of Wood,* Academic Press, New York, 1970, 270. With permission.

ponents of commercial significance were allyl alcohol, methyl acetone, and formic and propionic acids. All of these chemicals are still important commodities (with the exception of methyl acetone), but can be produced more economically (in the Western Hemisphere) from petrochemical sources.

3. Wood Tars

The composition of wood tar is an awesome mixture of phenolic and nonphenolic compounds that has taxed the capabilities of several analytical laboratories for over a century. Characterization continues today.[25,26] Table 6 is a new compilation of molecular species identified in hardwood and softwood wood tars and pyroligneous acids. Such tables have been published every so often as new information on tar and pyroligneous acid composition was obtained.[9,19,24,26,27] Because lignin by-products nomenclature is confusing, several names for the same molecular species are given in the phenolics section of Table 6.

The most valuable chemical products from wood pyrolysis today may be the phenols that are derived from the lignin component of wood. Sedimentation tar, because of its high phenolic content (up to 60%), has received much attention in the Russian literature. Laboratory attempts to isolate single reactive phenols have been numerous, and although technical feasibility has been generally established in many cases, economic feasibility has not. Most products from sedimentation tar rely on the use of mixtures or fractions with similar chemical or physical properties.[9,28-30]

Sedimentation tar is first fractionated into tar oils and tar pitch. The pitch can be used as a cement and embedding agent,[9] and the tar oils are further fractionated into light, heavy, and high-boiling fractions. Wood creosote is prepared by extracting phenols from heavy oil (bp 190 to 240°C) with alkali. Alternately, the heavy oil may be further separated into 4-methylguaiacol, cresol, and guaiacol. The high boiling oil is used as a flotation and impregnation medium.[31-33] Phenols in the 240 to 300°C fraction have been shown to be effective as an antioxidant or cracking inhibitor for gasoline.[34-37]

B-tar is the soluble tar still left in the aqueous pyroligneous acid after sedimentation

Table 6
COMPOUNDS IDENTIFIED IN THE PYROLIGNEOUS ACID AND TAR PRODUCTS OF SOFTWOOD AND HARDWOOD PYROLYSIS

Phenols

	Compound	Hardwood	Softwood	Tar	Pyroligneous acid	Ref.
C_6H_6O	Phenol	X	X	X		9, 24, 26, 27
$C_6H_6O_2$	Catechol 1,2-Dihydroxybenzene	X		X		24, 27
$C_6H_6O_3$	Pyrogallol 1,2,3-Trihydroxybenzene					27
C_7H_8O	o-Cresol 2-Hydroxytoluene	X	X	X		9, 24, 26, 27
C_7H_8O	m-Cresol 3-Hydroxytoluene	X	X	X		9, 24, 26, 27
C_7H_8O	p-Cresol 4-hydroxytoluene	X	X	X		9, 24, 26, 27
$C_7H_8O_2$	Guaiacol methylcatechol o-hydroxyanisole 2-methoxyphenol Pyrocatechol monomethyl ether	X	X	X		9, 24, 26, 27
$C_7H_8O_3$	1,2-Dihydroxy-3-methoxybenzene Pyrogallol monomethyl ether					27
$C_7H_8O_3$	5-Methyl pyrogallol 3,4,5-Trihydroxytoluene					
$C_8H_{10}O$	o-Ethylphenol 2-Ethylphenol Phlorol	X		X		9, 24, 27
$C_8H_{10}O$	2,3-Xylenol 2,3-Dimethylphenol					9, 27
$C_8H_{10}O$	2,4-Xylenol 2,4-Dimethylphenol					9, 27
$C_8H_{10}O$	3,5-Xylenol 3,5-Dimethylphenol	X	X			9, 24, 27
$C_8H_{10}O$	2,5-Xylenol 2,5-Dimethylphenol					9, 27
$C_8H_{10}O$	3,4-Xylenol 3,4-Dimethylphenol	X		X		9, 24, 27
$C_8H_{10}O$	Xylenols		X	X		26
$C_8H_{10}O_2$	4-Methylguaiacol 2-Methoxy-4-methylphenol Homocatechol dimethyl ether Creosol	X	X	X		9, 24, 26, 27
$C_8H_{10}O_2$	6-Methylguaiacol 2-Methoxyl-6-methylphenol					27
$C_8H_{10}O_2$	4-Ethyl resorcinol 1,3-Dihydroxy-4-ethylbenzene		X	X		26
$C_8H_{10}O_2$	1,2-Dimethoxybenzene Dimethyl pyrocatechol Pyrocatechol dimethylether Veratrole					9
$C_8H_{10}O_3$	2,6-Dimethoxyphenol Pyrogallol-1,3-dimethyl ether	X		X		9, 24, 27
$C_8H_{10}O_3$	Methoxy-3,4-dihydroxytolvene Methoxy-4-homocatechol					27
$C_9H_{10}O_2$	4-Vinyl guaiacol 2-Methoxy-4-vinyl phenol					27

Table 6 (continued)
COMPOUNDS IDENTIFIED IN THE PYROLIGNEOUS ACID AND TAR PRODUCTS OF SOFTWOOD AND HARDWOOD PYROLYSIS

Phenols

	Compound	Hardwood	Softwood	Tar	Pyroligneous acid	Ref.
$C_9H_{10}O_3$	4-Acetyl guaiacol 4-Hydroxy-3-methoxyacetophenone					26
$C_9H_{12}O_2$	4-Ethylguaiacol 4-Ethyl-2-methoxyphenol		X	X		9, 24, 26, 27
$C_9H_{10}O_2$	3,5-Dimethylguaiacol 2-Methoxy-3,5-dimethylphenol					27
$C_9H_{12}O_2$	3,4-Dimethoxytoluene Homoveratrole					27
$C_9H_{12}O_3$	2,6-Dimethoxy-4-methylphenol					27
$C_9H_{12}O_3$	5-Propyl pyrogallol 1,2,3-Trihydroxy-5-propylbenzene					27
$C_9H_{12}O_3$	2,6-Dimethoxy-4-methylphenol Methyl pyrogallol dimethyl ether	X		X		24
$C_{10}H_{12}O$	Estragole 1-Allyl-4-methoxybenzene					27
$C_{10}H_{12}O_2$	Eugenol 4-Allylguaiacol 4-Allyl-2-methoxyphenol		X	X		26, 27
$C_{10}H_{12}O_2$	Isoeugenol 4-(1-Propenyl)guaiacol		X	X		26, 27
$C_{10}H_{12}O_3$	1-Guaiacyl-propan-2-one		X	X		26
$C_{10}H_{14}O$	Thymol 2-Isopropyl-5-methylphenol					27
$C_{10}H_{14}O$	2-Methyl-4-n-propylphenol		X	X		26
$C_{10}H_{14}O$	n-Propylanisole		X	X		26
$C_{10}H_{14}O_2$	4-Propyl guaiacol 2-Methoxy-4-propylphenol		X	X		24, 26, 27
$C_{10}H_{14}O_3$	2,6-Dimethoxy-4-ethylphenol					27

$C_{10}H_{14}O_3$	1,2-Dihydroxy-3-methoxy-5-propyl-
	  benzene
	5-Propyl pyrogallol monomethyl
	  ether
						27

$C_{11}H_{16}O_3$	2,6-Dimethoxy-4-propylphenol Propyl pyrogallol dimethyl ether	X		X		24, 27

Acids and Lactones

	Compound	Hardwood	Softwood	Tar	Pyroligneous acid	Ref.
CH_2O_2	Formic acid	X	X	X	X	9, 24, 26, 27
$C_2H_4O_2$	Acetic acid	X	X	X	X	9, 24, 26, 27
$C_2H_4O_3$	Glycolic acid Hydroxyacetic acid					27
$C_3H_4O_2$	Acrylic acid Propenoic acid					27
$C_3H_6O_2$	Propionic acid	X	X	X	X	24, 26, 27
$C_4H_2O_3$	Succinic anhydride					27
$C_4H_6O_2$	Butyrolactone					27
$C_4H_6O_2$	Crotonic acid 2-Butenoic acid	X			X	24, 27
$C_4H_6O_2$	3-Butenoic acid	X			X	24, 27
$C_4H_6O_2$	Methacrylic acid 2-Methylpropenoic acid					27

Table 6 (continued)
COMPOUNDS IDENTIFIED IN THE PYROLIGNEOUS ACID AND TAR PRODUCTS OF SOFTWOOD AND HARDWOOD PYROLYSIS

Acids and Lactones

	Compound	Hardwood	Softwood	Tar	Pyroligneous acid	Ref.
$C_4H_8O_2$	Butyric acid	X	X	X	X	24, 26, 27
$C_4H_8O_2$	Isobutyric acid					27
$C_5H_4O_2$	2-Furoic acid					27
	Pyromucic acid					
$C_5H_8O_2$	Valerolactone	X			X	24
	Pentanoic acid lactone					
$C_5H_8O_2$	γ-Hydroxyvalerolactone					27
	4-Hydroxypentanoic acid lactone					
$C_5H_8O_2$	δ-Hydroxyvalerolactone					27
	5-Hydroxypentanoic acid lactone					
$C_5H_8O_2$	Angelic acid	X			X	24, 27
	cis-2-methyl-2-butenoic acid					
	α-methylisocrotonic acid					
$C_5H_8O_2$	Tiglic acid					27
	trans-2-Methyl-2-butenoic acid					
	α-Methylcrotonic acid					
$C_5H_8O_2$	2-Pentenoic acid					27
$C_5H_8O_2$	3-Pentenoic acid					27
$C_5H_8O_3$	Levulinic acid					27
	4-Oxopentanoic acid					
	γ-Ketovaleric acid					
$C_5H_8O_3$	α,γ-Dihydroxyvalerolactone					27
	2,4-Dihydroxypentanoic acid lactone					
$C_5H_{10}O_2$	2-Methyl butyric acid	X	X		X	24, 27
	2-Methyl butenoic acid					
	Valeric acid					
$C_5H_{10}O_2$	3-Methyl butyric acid		X	X		26, 27
	3-Methyl butenoic acid					
	Isovaleric acid					
$C_6H_6O_3$	4-Methyl-2-furoic acid					27
$C_6H_{12}O_2$	Pyrocinchonic anhydride					27
	Dimethylmaleic anhydride					
$C_6H_{12}O_2$	Caproic acid	X	X		X	24, 27
	Hexanoic acid					
$C_6H_{12}O_2$	α-Methyl valeric acid		X		X	24, 27
	2-Methyl pentanoic acid					
$C_6H_{12}O_2$	Isocaproic acid					27
	4-Methyl pentanoic acid					
$C_7H_{12}O_2$	2-Heptenoic acid					27
$C_7H_{14}O_2$	Enanthic acid		X			24, 27
	Heptanoic acid					
$C_8H_{16}O_2$	Caprylic acid		X			24, 27
	Octanoic acid					
$C_8H_{16}O_4$	Dihydroxycaprylic acid	X			X	24, 27
	Dihydroxyoctanoic acid					
$C_9H_{18}O$	Pelargonic acid					27
	Nonanoic acid					
$C_{10}H_{20}O_2$	Capric acid					27
	Decanoic acid					
$C_{16}H_{32}O_2$	Palmitic acid		X			24, 27
	Hexadecanoic acid					
$C_{18}H_{34}O_2$	Oleic acid		X			24, 27
	cis-9-Octadecenoic acid					

Table 6 (continued)
COMPOUNDS IDENTIFIED IN THE PYROLIGNEOUS ACID AND TAR PRODUCTS OF SOFTWOOD AND HARDWOOD PYROLYSIS

Acids and Lactones

	Compound	Hardwood	Softwood	Tar	Pyroligneous acid	Ref.
$C_{18}H_{36}O_2$	Stearic acid					27
	Octadecanoic acid					
$C_{20}H_{30}O_2$	Abietic acid		X			24, 27
$C_{20}H_{30}O_2$	Pimaric acid		X			24, 27
$C_{20}H_{40}O_2$	Arachidic acid		X			24, 27
	Eicosanoic acid					
$C_{22}H_{44}O_2$	Behenic acid	X		X		24, 27
	Docosanoic acid					
$C_{24}H_{48}O_2$	Lignoceric acid	X		X		24, 27

Aldehydes

	Compound	Hardwood	Softwood	Tar	Pyroligneous acid	Ref.
CH_2O	Formaldehyde	X			X	24, 27
$C_2H_2O_2$	Glyoxal					27
	Ethanedial					
C_2H_4O	Acetaldehyde	X	X		X	24, 26, 27
	Ethanal					
$C_2H_4O_2$	Glycolaldehyde					27
	Hydroxyethanal					
C_3H_6O	Propionaldehyde	X	X	X		24, 27
$C_3H_8O_2$	Methylal	X			X	24, 27
	Dimethylacetal					
C_4H_6O	Crotonaldehyde					27
	2-Butenal					
$C_5H_4O_2$	Furfural	X	X	X	X	26, 27
	2-Furaldehyde					
C_5H_8O	Tiglaldehyde					27
	2-Methyl-2-butenal					
$C_5H_{10}O$	Pivalaldehyde					24, 27
	2,2-Dimethyl propanol					
$C_5H_{10}O$	Valeraldehyde	X		X		24, 27
	2-Methyl butenal					
$C_5H_{10}O$	Isovaleraldehyde	X			X	24, 27
	3-Methyl butenal					
$C_6H_6O_2$	5-Methyl furfural	X			X	24, 27
$C_6H_6O_3$	5-Hydroxymethyl furfural					27
$C_6H_{10}O$	2-Methyl-3-ethyl acrolein					27
	2-Methyl-2-penten-1-al					
$C_7H_{12}O$	Cyclohexane carboxaldehyde					27
$C_7H_{14}O$	Enanthaldehyde					27
	Heptanal					
$C_9H_{10}O_3$	Veratraldehyde		X	X		26
	3,4-Dimethoxybenzaldehyde					
$C_9H_{16}O$	Cyclohexane Propionaldehyde					27

Ketones

	Compound	Hardwood	Softwood	Tar	Pyroligneous acid	Ref.
C_3H_6O	Acetone	X	X		X	24, 27
$C_4H_6O_2$	Hydroxypropanone					27
	Hydroxyacetone					

Table 6 (continued)
COMPOUNDS IDENTIFIED IN THE PYROLIGNEOUS ACID AND TAR PRODUCTS OF SOFTWOOD AND HARDWOOD PYROLYSIS

Ketones

	Compound	Hardwood	Softwood	Tar	Pyroligneous acid	Ref.
$C_4H_6O_2$	2,3-Butanedione	X	X		X	24, 27
	Diacetyl ketone					
C_4H_8O	2-Butanone	X		X	X	24, 27
	Methyl ethyl ketone					
C_5H_6O	Cyclopent-2-en-1-one					27
C_5H_8O	Cyclopentanone	X		X	X	24, 27
	Adipic ketone					
$C_5H_8O_2$	2,3-Pentanedione					27
$C_5H_8O_3$	Acetoxypropanone					27
$C_5H_{10}O$	3-Methyl-2-butanone					27
	Methyl isopropyl ketone					
$C_5H_{10}O$	2-Pentanone	X		X	X	24, 27
	Methyl propyl ketone					
$C_5H_{10}O$	3-Pentanone diethyl ketone	X		X	X	24, 27
$C_5H_{10}O_2$	4-Hydroxy-2-pentanone					27
$C_6H_6O_2$	2-furyl methyl ketone		X	X		26, 27
$C_6H_6O_3$	Maltol					27
	3-Hydroxy-2-methyl-4-pyrone					
C_6H_8O	2-Methyl-2-cyclopenten-1-one					24, 27
$C_6H_8O_2$	4-Methyl-2-cyclopenten-1-one					27
$C_6H_8O_2$	2-Hydroxy-3-methyl-2-cyclopenten-1-one		X	X		26, 27
$C_6H_8O_2$	1,3-Cyclohexanedione					27
$C_6H_{10}O$	1,4-Cyclohexanedione					27
$C_6H_{10}O$	Cyclohexanone	X		X		24, 27
	Pimelic ketone					
$C_6H_{10}O$	3-Hexen-2-one	X			X	24, 27
$C_6H_{10}O$	4-Methyl-3-penten-2-one					27
	Mesityl oxide					
$C_6H_{10}O$	Methylcyclopentanone	X		X		27
$C_6H_{10}O_2$	2,3-Hexanedione					27
$C_6H_{12}O$	2-Hexanone	X		X		24, 27
	Methyl butyl ketone					
$C_6H_{12}O$	2-Methyl-3-pentanone					27
	Ethyl isopropyl ketone					
$C_7H_{10}O$	1-Methyl-2-cyclohexen-5-one					27
$C_7H_{12}O$	Methyl cyclohexonone	X		X		24
$C_7H_{12}O$	Dimethyl cyclopentanone					27
$C_7H_{14}O$	4-Heptanone Dipropyl ketone Butyrone					27
$C_8H_{14}O$	Dimethylcyclohexanone	X		X		24
$C_8H_{12}O$	3-Isopropyl-2-cyclopenten-1-one					27
$C_8H_{12}O$	2,4-Dimethyl-4-cyclohexen-1-one					27
$C_8H_{14}O$	Trimethylcyclopentanone					27
$C_8H_{14}O$	3,6-Octanedione					27
$C_9H_{14}O$	3,5,5-Trimethyl-2-cyclohexen-1-one					27
	Isophorone					
$C_9H_{16}O$	2,4,4-Trimethylcyclohexanone					27
$C_9H_{16}O$	3,3,5-Trimethylcyclohexanone					27
$C_{10}H_{16}O$	Camphor					27
$C_{11}H_{16}O$	1,3,3-Trimethylbicyclo[2.2.2]-5-octen-2-one					27
$C_{15}H_{12}O_3$	2,5-Difuryl-1-cyclopentanone					27

Table 6 (continued)
COMPOUNDS IDENTIFIED IN THE PYROLIGNEOUS ACID AND TAR PRODUCTS OF SOFTWOOD AND HARDWOOD PYROLYSIS

Alcohols

	Compound	Hardwood	Softwood	Tar	Pyroligneous acid	Ref.
CH_4O	Methanol	X	X		X	24, 27
C_2H_6O	Ethanol					27
C_3H_6O	Allyl alcohol	X	X		X	24, 27
C_3H_8O	Propyl alcohol					27
C_3H_8O	Isopropyl alcohol					27
C_4H_8O	1-Buten-3-ol	X			X	24
C_4H_8O	2-Buten-1-ol					27
$C_4H_{10}O$	Isobutyl alcohol	X		X		24, 27
C_5H_6O	Furfuryl alcohol					27
$C_5H_{10}O_2$	1,2-Cyclopentanediol					27
$C_5H_{12}O$	Isoamyl alcohol	X		X		24, 27
$C_6H_{12}O$	Cyclohexanol					27
$C_{10}H_{18}O$	Borneol					27
$C_{10}H_{18}O$	Cineole					27
$C_{10}H_{18}O$	Fenchyl alcohol					27
$C_{10}H_{18}O$	Isofenchyl alcohol					27
$C_{10}H_{18}O$	α-Terpineol					27

Esters

	Compound	Hardwood	Softwood	Tar	Pyroligneous acid	Ref.
$C_2H_4O_2$	Methyl formate	X		X		24, 27
$C_3H_6O_2$	Methyl acetate	X		X	X	24, 27
$C_4H_8O_2$	Methyl propionate	X		X		24, 27
$C_5H_8O_2$	Methyl crotonate					27
	2-Butenoic acid methyl ester					
$C_5H_{10}O_2$	Methyl butyrate	X		X		24, 27
$C_5H_{10}O_2$	Methyl isobutyrate		X			24, 27
$C_6H_{12}O_2$	Methyl-2-furoate					27
$C_6H_{12}O_2$	Methyl valerate	X		X		24, 27
	Methyl pentanoate					
$C_7H_{14}O_2$	Methyl caproate					27
	Methyl hexanoate					
$C_8H_{16}O_2$	Methyl heptanoate					27

Ethers

	Compound	Hardwood	Softwood	Tar	Pyroligneous acid	Ref.
$C_5H_{12}O_2$	Propionaldehyde dimethyl acetal					27
$C_6H_{10}O_5$	Levoglucosan		X	X		26, 27
$C_4H_{10}O_2$	Acetaldehyde dimethyl acetal					27
$C_4H_{10}O_2$	1,2-Dimethoxyethane					27

Hydrocarbons

	Compound	Hardwood	Softwood	Tar	Pyroligneous acid	Ref.
C_5H_{10}	Pentene					27
C_5H_{12}	Pentane					27
C_6H_6	Benzene		X		X	24, 27
C_7H_8	Toluene	X	X		X	24, 27
C_7H_{12}	1-Heptyne					27
C_7H_{16}	Heptane					27

Table 6 (continued)
COMPOUNDS IDENTIFIED IN THE PYROLIGNEOUS ACID AND TAR PRODUCTS OF SOFTWOOD AND HARDWOOD PYROLYSIS

Hydrocarbons

	Compound	Hardwood	Softwood	Tar	Pyroligneous acid	Ref.
C_8H_{10}	m-Xylene	X	X		X	24, 27
C_9H_{12}	Cumene Isopropylbenzene	X			X	24, 27
C_9H_{12}	Pseudocumene 1,2,4-Trimethylbenzene					27
$C_{10}H_8$	Naphthalene					27
$C_{10}H_{14}$	Cymene 1-Isopropyl-4-methylbenzene					27
$C_{10}H_{14}$	Durene 1,2,4,5-Tetramethylbenzene					27
$C_{10}H_{16}$	Camphene					27
$C_{10}H_{16}$	Limonene					27
$C_{10}H_{16}$	α-Pinene					27
$C_{10}H_{16}$	β-Pinene					27
$C_{10}H_{16}$	Sylvestrene					27
$C_{10}H_{16}$	γ-Terpinene					27
$C_{10}H_{16}$	Terpinolene					27
$C_{15}H_{24}$	Cadinene					27
$C_{15}H_{32}$	Pentadecane					27
$C_{17}H_{36}$	Heptadecane					27
$C_{18}H_{12}$	Chrysene					27
$C_{18}H_{18}$	Retene		X	X		24, 27
$C_{18}H_{38}$	Octadecane					27
$C_{19}H_{40}$	Nonadecane					27
$C_{20}H_{12}$	Benzopyrene					9
$C_{20}H_{42}$	Eicosane					27
$C_{21}H_{44}$	Heneicosane					27
$C_{22}H_{46}$	Docosane					27
$C_{23}H_{48}$	Tricosane					27
$C_{30}H_{60}$	Melene					27

Furans

	Compound	Hardwood	Softwood	Tar	Pyroligneous acid	Ref.
C_4H_4O	Furan		X			24, 27
C_5H_6O	2-Methylfuran	X	X	X		24, 27
C_5H_6	3-Methylfuran					27
C_6H_8O	2,5-Dimethylfuran	X	X	X		24, 27
$C_6H_{12}O$	2,2-Dimethyltetrahydrofuran	X			X	24
$C_6H_{12}O$	2,5-Dimethyltetrahydrofuran					27
$C_7H_{10}O$	Propylfuran					27
$C_7H_{10}O$	2,3,5-Trimethylfuran	X		X		24, 27
$C_7H_{12}O$	2,2-Ethyl-5-methyl-4,5-dihydro- furan	X			X	24, 27
C_8H_6O	Benzofuran					27
$C_8H_8O_3$	2,3-Diacetylfuran	X		X		24
$C_9H_{14}O$	Amylfuran					27

Bases

	Compound	Hardwood	Softwood	Tar	Pyroligneous acid	Ref.
NH_3	Ammonia	X			X	24, 27
C_5H_5N	Methylamine					27

Table 6 (continued)
COMPOUNDS IDENTIFIED IN THE PYROLIGNEOUS ACID AND TAR PRODUCTS OF SOFTWOOD AND HARDWOOD PYROLYSIS

Bases

Compound		Hardwood	Softwood	Tar	Pyroligneous acid	Ref.
C_2H_7N	Demethylamine					27
C_3H_9N	Trimethylamine					27
C_5H_5N	Pyridine	X		X	X	24, 27
C_6H_7N	2-Picolene					27
C_7H_9N	Lutidine	X		X		24, 27

Note: Where indicated in these previous tabulations, especially those by Schorger[24] and Lin et al.[26]; origin is shown. Otherwise, it can be assumed that most of the other compounds were derived from hardwood distillation. Compounds identified as being of softwood origin may also be found in the pyrolysis of hardwoods.

of A-tar. This soluble tar is the main product of decomposition of wood carbohydrates. Although much work has been reported in the chemical characterization of soluble tar,[38-42] little promise for commercialization has been noted except for particle board binders,[43] and foundry mold binders after further processing.[41,44,45] Hydroxyacids, anhydrosugars, and simple phenols predominate. Levoglucosan recovery has been suggested.[46]

The phenolic composition of wood tar is very complex. As with phenols derived from hydrolysis lignin, commercialization of individual components or products derived from further processing awaits further research.

Some work has been initiated in hydrocracking of sedimentation tar fractions.[47-49] Distillate yields are reported to be increased by prior treatment of light oils with aqueous alkali solutions.[50] Phenolic levels in a pyrolytic oil produced in the U.S. have been found to be about 30%.[26]

4. Wood Gas

Wood gas is the noncondensible gaseous fraction of wood pyrolysis, evolved during various stages of wood decomposition. Its primary use in retort furnace distillation is as a vehicle to remove volatile products and/or as a supplemental fuel.

Air used in pyrolysis results in concentrations of carbon dioxide, carbon monoxide, and nitrogen in wood gas. Methane, hydrogen, ethane, and several unsaturated hydrocarbons are also found in wood gas,[8] although carbon dioxide predominates. The percentage of combustible gases produced is a function of the inside wood temperature, and will exceed carbon dioxide concentration at temperatures over 350°C (excluding nitrogen). Pyrolysis and gasification are part of the same continuum in wood thermal degradation as temperatures and oxygen feeds increase, and a higher percentage of wood is decomposed into gaseous products (see Chapter 4).

C. Softwood Pyrolysis Products

The major softwood pyrolysis process was the destructive distillation of southern U.S. pine, described earlier. The more important products of this industry (turpentine, dipentene, pine oil, and pine tars) are still produced today, but primarily as by-products of the southern pine kraft pulping industry (see Chapter 9).

The relatively harsh conditions employed in pine pyrolysis vs. those encountered in kraft pulping do not permit the recovery of fatty or resin acids that are decomposed during pyrolysis. Examination of the chemical composition of the nonphenolics of softwood pyrolysis reflects this fact (see Table 6). The low methanol and acetic acid yields in softwood pyrolysis, relative to hardwood pyrolysis, provided little economic incentive to find applications for the pyroligneous acid fraction.

There is much similarity between phenolic constituents found in softwood tars and those found in hardwood tars. Differences are primarily due to the structure of softwood guaiacyl lignin vs. that of hardwood guaiacylsyringyl lignin (see Chapter 8). Interest in the chemical composition of tars of softwood pyrolysis has been renewed recently, since these are produced as by-products of various pyrolysis and gasification processes aimed at recovering usable energy and chemical values from lignocellulosic materials. Since the major wood species in the production of paper and solid wood products in the U.S. is pine, municipal wastes that typically contain 50% paper, contain many pine-derived materials. In addition, several processes have been developed to make better fuel and chemical use of the residues of our large pine utilization industry, through thermal degradation. Knowledge of the composition of the products of these pyrolysis processes is a prerequisite to their utilization as chemical feedstocks.

IV. PYROLYSIS REACTIONS

All organic materials decompose upon heating. Upon being subjected to pyrolytic conditions, wood behaves much like a mixture of its three major components: cellulose, hemicellulose, and lignin.[51] We can therefore, consider the major constituents of wood singly, and finally wood as an aggregate, in discussing the pyrolysis reactions of wood.

A. Pyrolysis of Cellulose

The thermal degradation of cellulose has been extensively studied by Shafizadeh,[52-55] and it has been reported to occur in two steps. The first takes place at temperatures below 300°C. Degradation at this level is indicated by an increase in brittleness and a decrease in strength of the material. Chemical reactions that may occur in this temperature range are oxidation, depolymerization, dehydration, and decarboxylation. Products that may be evolved include carbon monoxide, water, and other low molecular weight compounds that contain carboxyl, carbonyl, and hydroxyl groups.[52,54]

Degradation under 300°C is markedly increased in the presence of oxygen, water, acid, or alkali.[56-62] The presence of oxygen will increase the rate of depolymerization.[52] Although X-ray diffraction patterns of cellulose before and after heating to temperatures below 300°C are identical, chromatographic data indicates the presence of acids, aldehydes, glucose, and phenols.[62] It has been postulated that oxygen attacks the amorphous regions of cellulose, leaving the crystalline regions that are responsible for the X-ray diffraction patterns, intact.[63,64]

Kilzer has made a study of the possible mechanisms involved in the formation of carbon monoxide and water.[65] Water is lost first through an inter-ring mechanism to produce dehydrocellulose, which decomposes to form carbon monoxide and additional water.

Heating of cellulose above 300°C will result in char, tar, and gaseous products. Investigations into the tar fraction have revealed that the major component is levoglucosan.[27,51,52] Yields of 38 to 50% levoglucosan in the tar have been reported,[27,66] depending on the nature of the starting material and heating conditions. Heating under vacuum will increase yields to 78% levoglucosan, 18% light products and gases, and 6 to 9% char.[52] Levoglucosan is heat sensitive, but also volatile so that it will distill

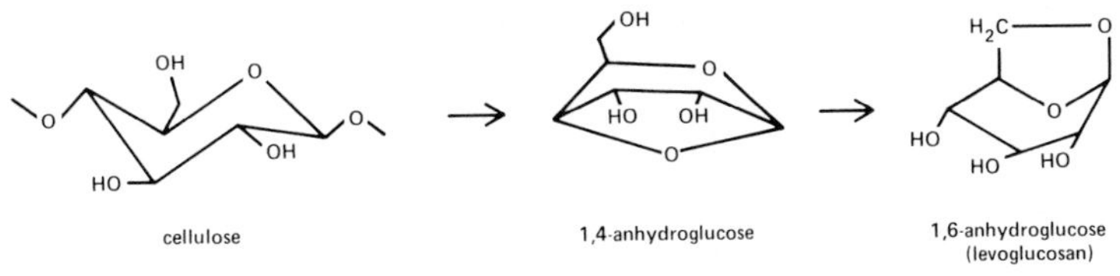

FIGURE 3.   Formation of levoglucosan according to Kilzer.

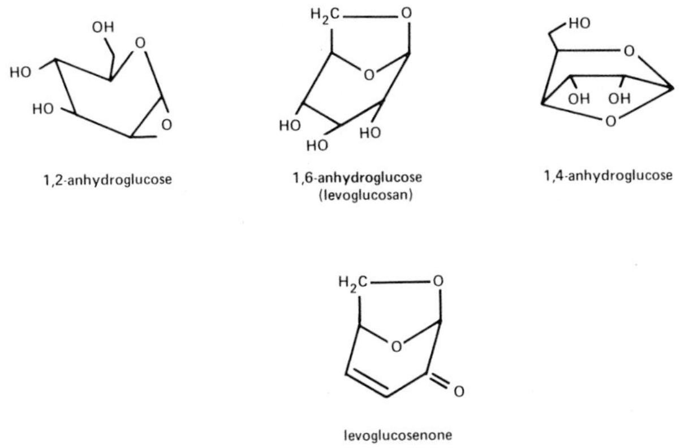

FIGURE 4.   Anhydroglucose and levoglucosenone structures.

out of the reaction zone under vacuum conditions before it can undergo secondary reactions.[66-68]

The mechanism by which levoglucosan is formed has been the object of considerable work. Russian researchers (who are particularly prolific in the field of pyrolysis) have proposed a mechanism in which levoglucosan formation is the major reaction in areas of high packing density. Dehydration and hydrolysis reactions are said to be of secondary importance in these regions. In parts of the molecule with different physical characteristics, the predominance of reactions may be reversed, with dehydrations occurring to a large extent.[69,70] Levoglucosan can be formed through the migration of the hydrogen on the hydroxyl group of carbon number six, to the oxygen of carbon number four. The free radical center thus formed, is then attached to carbon number one. The one-six linkage is preferred because the distance from one to six is less than that between any other neighboring carbon atoms.[69]

Kilzer states that the formation of levoglucosan proceeds through a depolymerization that involves the displacement of the C-1 oxygen by an S_N2 reaction with the C-4 oxygen within the same unit.[65] The C-6 hydroxyl could then attack carbon number one, displacing the one-four bridge and forming a one-six bridge (see Figure 3). Shafizadeh states that the depolymerization of cellulose occurs by intramolecular substitution of the one-four link in cellulose by one of the free hydroxyl groups to produce the first three anhydro sugars shown in Figure 4.[53]

Potential industrial applications of thermally degraded cellulose involve the hydrolysis of the tarry fraction. Hydrolysis gives a 50% yield of glucose based on cellulose. Sugars derived from pyrolysis of cellulose could then be used either as glucose or anhydro sugars.[71] Levoglucosenone (1,6-anhydro-3, 4 dideoxy-β-D-*Glycero*-hex-3-enopyranos-2-ulose; see Figure 4) that may be formed from levoglucosan is a polyfunctional substance that could possibly be used as an industrial intermediate.[59]

Some minor lighter products are always formed in cellulose pyrolysis. Formation of these products begins at temperatures lower than those required for levoglucosan formation, thus indicating that they do not arise from secondary reactions of levoglucosan.[52] The volatile fraction is, however, qualitatively similar whether it comes from cellulose or levoglucosan.[54]

B. Pyrolysis of Lignin

When subjected to pyrolytic conditions, lignin produces aromatic compounds,[72-75] and more char than cellulose.[27,74,76,77] Further, lignin produces no single predominating product analogous to levoglucosan from cellulose. The products of lignin pyrolysis reflect the complexity of lignin molecules in which many repeating units are present with a variety of possible linkages.

The tarry residue from lignin pyrolysis accounts for about 15% of the lignin. This fraction has been investigated by various authors who found a mixture of phenolic compounds that are homologs of phenol, guaiacol, or 2,6-dimethoxyphenol.[78] Substituent groups are usually para to the phenolic hydroxyl, and are usually three carbons or less. Substituents that have been reported are methyl, ethyl, propyl, isopropyl, vinyl, allyl, propenyl, carboxyl, carboxymethyl, or carboxyaldehyde.[78] Fletcher has reported the finding of phenol, o-cresol, p-cresol, 2,4-xylenol, guaiacol, 4-ethyl guaiacol, and 4-propyl guaiacol,[73,74] all of which are logical, based on the structure of lignin. Phenols found in the pyrolysis of wood (see Table 6) are essentially similar to those reported for the pyrolysis of lignin.

The solid residue from the pyrolysis of lignin may account for 55% of the total yield, but has been largely ignored by research workers. The only substantial work has been performed by Gillet and Urlings, who found a resemblance between the char and a type of lignite coal.[79]

An aqueous distillate (pyroligneous acid) has also been derived from thermally degraded lignin. Yields of 20% are not uncommon, and substances that have been found include water, methanol, acetic acid, and acetone.[78]

The gaseous products from lignin pyrolysis may account for 12% of the products and has been found to contain carbon monoxide, methane, and ethane.

C. Pyrolysis of Hemicelluloses

Of the three major components of wood, the hemicelluloses are the most temperature sensitive, decomposing within the range of 200 to 260°C.[51] It is thought that hemicellulose degradation occurs in two steps: decomposition of the polymer into water soluble fragments, followed by conversion to very short or monomeric units that, in turn, decompose into volatiles.[80] In comparison with cellulose, hemicelluloses produce more gas, less tar, and about the same amount of aqueous distillate, but no levoglucosan.

It might be expected, based on the large number of five carbon sugars associated with the hemicelluloses, especially in hardwoods, that many furan derivatives would be obtained. However, low yields of furfural are found relative to acetic acid.[81] These results may be due to the reactivity of furfural, which, under pyrolytic conditions, can undergo many secondary reactions.[9,27]

The five carbon sugars, when pyrolyzed, have been shown to be a major source of acetic acid,[27,62,72] and Brown has proposed that xylose units split to form two units of acetic acid and one formaldehyde (see Figure 5).[51] The hemicelluloses, with their associated acetyl groups, may also be a source of acetic acid.[9,27]

The effect of temperature on hemicelluloses has been studied by heating wood and then extracting with water.[80-82] Wood heated to temperatures greater than 100°C

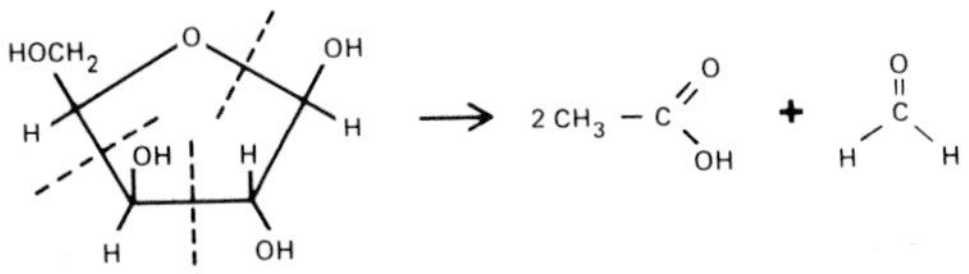

FIGURE 5. Formation of acetic acid and formaldehyde from xylose.

showed increases in hot water extractables, while cold water extractables began to increase only after the wood was heated to 120°C. Both extracts were maximized between 150 and 180°C.

D. Pyrolysis of Wood and Related Materials

If wood is completely pyrolyzed, resulting products are about what would be expected by pyrolyzing the three major components separately.[51] The thermal decomposition of wood does not progress at an even pace, but rather in a step-wise manner, with the hemicelluloses breaking down first, at temperatures of 200 to 260°C. Cellulose follows in the temperature range 240 to 350°C, with lignin being the last component to pyrolyze at temperatures of 280 to 500°C. [51,75,83-86]

Pyrolysis of wood has been studied as a zonal process with zone A occurring at temperatures up to 200°C.[51] The surface of the wood becomes dehydrated at this temperature, and along with water vapor, carbon dioxide, formic acid, acetic acid, and glyoxal are given off.[79,87] When temperatures of 200 to 280°C are attained, the wood is said to be in zone B and is evolving water vapor, carbon dioxide, formic acid, acetic acid, glyoxal, and some carbon monoxide.[18,81] The reactions to this point are mostly endothermic, the products are largely noncondensible, and the wood is becoming charred.[88-90] Pyrolysis actually begins between 280 and 500°C, which is called zone C. The reactions are exothermic, and unless heat is dissipated, the temperature will rise rapidly. Combustible gases such as carbon monoxide, methane, formaldehyde, formic acid, acetic acid, methanol, and hydrogen are being liberated and charcoal is being formed. The primary products are beginning to react with each other before they can escape the reaction zone.[92-94] If the temperature continues to rise above 500°C, a layer of charcoal will be formed that is the site of vigorous secondary reactions and is classified as zone D. Carbonization is said to be complete at temperatures of 400 to 600°C.[27,88]

Three points have been defined in the exothermic reaction of wood.[94] The first is the flame point at 225 to 260°C. At this temperature, gases are evolved from wood decomposition and will burn if a pilot flame is present. The second, or burning, point at 260 to 290°C is represented by a steady flame. The third, or flash, point at 330 to 470°C is a region of spontaneous combustion.

After water is removed from wood at temperatures above 140°C, four classes of compounds are produced.[95] These are noncondensible gases (200 to 450°C, maximum at 350 to 400°C, carbon monoxide, carbon dioxide, hydrogen, and methane), condensible pyroligneous products (maximized at 250 to 300°C, ceases at 350°C, contains more than 50% moisture), condensible tar (300 to 400°C, moisture-free), and charcoal. The yields of the major products found in various species have been examined, and it is shown that the char yield is lower in hardwoods, while the amounts of acetic acid, methanol, and acetone are higher (see Table 2).

Many pyrolysis and gasification processes have been studied and developed in recent years that use municipal waste or agricultural residues as feed. Municipal waste contains much wood-derived paper. Agricultural residues are similar to wood residues in

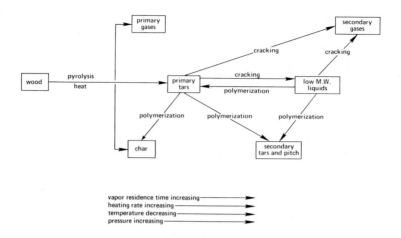

FIGURE 6.   Effect of vapor residence time, heating rate, temperature, and pressure on the formation of char, tars, and gases in the pyrolysis of wood.

chemical composition. As with wood, these materials will behave in pyrolysis as a mixture of these components, and further, because of their chemical similarity to wood, produce the same types of products. Pyrolysis can then be considered a leveling process,[96] in which the many physical forms of lignocellulosic waste materials can be transformed into similar gases, liquids, and solids.

E. Products as Functions of Pyrolysis Reaction Conditions

As indicated earlier, carbonization, pyrolysis, and gasification can be considered as part of the same continuum in wood thermal degradation. Yields of gaseous, liquid, and solid products will be functions of pyrolysis reaction conditions, such as heating rate, gas residence time, temperature, and pressure. In general, the effects of such parameters on product composition in the absence of oxygen are shown in Figure 6. Air or oxygen will result in increased amounts of CO and CO_2 as products. At higher temperatures, CO, CO_2, and N_2 concentrations in the gas product will reduce its higher heating value. As can be seen, pyrolysis is a combination of simple, sequential, competing, and complex reactions. There is little difference in this respect between the pyrolysis of wood or coal. Some of the more important parameters affecting product yields and compositions can now be discussed.[97]

The yields of volatile products (gases and liquids) will increase as the heating rate increases. For high heating rates, the residence time of tars at intermediate temperatures that favor repolymerization is low, and most of the reactions are taking place at higher temperatures that favor cracking. The two major factors affecting heating rates are the type of pyrolysis equipment used, and the particle size of the feed material. For a given reactor, as particle size increases, the rate of thermal diffusion within a particle decreases, resulting in a lower heating rate. Therefore, liquid products are favored when pyrolyzing smaller particles at high heating rates and char is favored by subjecting larger particles to slower heating rates.

As is the case for most chemical reactions, temperature is a most critical variable. As temperature is increased, the yield of volatile products increases. However, the nature of these volatile products will also be a function of pyrolysis temperature. As the temperature increases, average molecular weight of the tar will decrease, due mainly to cracking of the tar to low molecular weight liquids and secondary gases. Therefore, increasing pyrolysis temperature will decrease char yield, decrease the heavy tar yield, and increase the production of low molecular weight liquids and gases.

As the residence time of vapor products at high temperatures increases, yield of total liquid product will decrease due to cracking. Low molecular weight gas yield will therefore increase. A dramatic illustration of this effect is reported by Choi,[98,99] who heated cellulose sheets for very short residence times using heated rollers, and achieved yields of 80% liquid pyrolysis tar.

Three major observations can then be made as a result of this discussion:

1. If the purpose is to maximize the yield of liquid products, a low temperature, high heating rate, short gas residence time process would be required.
2. For a high char production, a low temperature, low heating rate process would be chosen.
3. If the purpose is to maximize the yield of fuel gas resulting from pyrolysis, a high temperature, low heating rate, long gas residence time process would be preferred.

V. PROSPECTS FOR PYROLYSIS

A. Newer Biomass Pyrolysis Processes

In the past decade, there has been considerable effort in the development of solid waste pyrolysis processes. Although many were originally designed for municipal waste, most have also been applied to wood residues. With few exceptions, products were aimed at fuel applications. Nevertheless, several processes resulted in significant yields of liquid products (see Table 7), and it is these processes that can now be considered in the identification of future prospects for wood pyrolysis.[71,96,100-102]

A logical technique for distinguishing between pyrolysis processes is by reactor type. As will be seen, reactor type will dictate, to a large extent, the conditions of pyrolysis (heating rate, vapor residence time, etc.), and thus, the pyrolysis products and composition.

1. Ovens and Kilns

Batch ovens and kilns, by their very nature, operate at relatively low heating rates and long vapor residence times. These conditions result in very low yields of liquid products. Early work by the Bureau of Mines confirmed such observations in the pyrolysis of solid waste materials in an electrically heated, 18 in diameter, batch-type oven.[103] Monsanto's Langard System used a standard off-the-shelf horizontal rotary kiln to dispose of all types of solid waste to produce a low Btu gas and a char product.[97] The 35 ton/day kiln was operated at 1800°F and fired with fuel oil and air. The high operating temperature contributed to low liquid product yield.

2. Gravitating-Bed Processes

Gravitating-bed pyrolysis processes generally result in high gas yields, primarily because of long vapor residence times. Gas product yields are enhanced when these reactors are used to pyrolyze municipal wastes of relatively large particle size. Processes such as the Union Carbide Purox and Battelle-Northwest systems are municipal waste gasifiers designed around a gravitating-bed pyrolysis reactor.[97,104]

In the Union Carbide Purox System,[102] solid waste is fed into the top of a vertical shaft furnace, and gravitates downwards. Hot combustion gases, formed at the bottom of the reactor by the combustion of char with pure oxygen, move upward. The hot gases pyrolyze the solid waste and are thereby cooled. Because of heat-transfer limitations in gas-solid contact, large temperature gradients exist in this type of reactor. Char combustion temperatures in the bottom combustion zone are in excess of 2000°F, while

Table 7
COMPARISON OF PYROLYSIS PROCESSES

Process	Reactor type	Temperature (°F)	Gas yield	Gas HHV Btu/SCF	Oil yield	Oil HHV Btu/lb	Char yield	Scale
Bureau of Mines	Laboratory furnace (Batch)	900—1650	18—40%	450—570	0.6—3.6%	10,000—17,000	8—40%	18 in diameter retort
Monsanto Landgard	Horizontal kiln	1800	Gas is combusted to generate steam	125	—	—	N.A.[a]	35 tons/day
Union Carbide Purox	Vertical shaft furnace below a gravitating bed	High	70%	300	Some oil in gas phase, but it is not separated	—	small	5 tons/day pilot plant, 200 tons/day demonstration plant
Battelle-Northwest oil yield	Gravitating bed	1300—1800	100%	100—220	Some oil in gas phase, but it is not separated	—	small	3 ft diameter pilot reactor
Tech-Air	Gravitating bed	100—1900	14—30%	200	15—26%	12,000—13,500	28—40%	50 tons/day
University of West Virginia	Fluidized-sand bed	1500	85%	360—420	7%	—	10%	15 in diameter reactor laboratory unit
Thermax, Alberta Industrial Developments Ltd.	Fluidized-sand bed	—	—	low	—	—	N.A.[a]	75 tons/day second unit under construction
Garrett Flash Pyrolysis	Entrained-flow transport reactor	900 for oil production 1400° for gas production	27% (This gas is used to supply process heat)	550	40%	10,500	20%	4 tons/day pilot plant, 200 tons/day unit being constructed for county of San Diego

ᵃ   Not available.

exit pyrolysis gas can be below 1000°F. Low temperatures in the pyrolysis zone might be conducive to liquid formation, were it not for the low heating rates and long gas residence times. Any liquid condensed in a gas cleaning train is recycled for combustion. It has been claimed that an alternate processing system could provide a liquid fuel oil by-product,[102] suggesting potential for a significant liquid product yield.

The Battelle-Northwest process is basically similar except that the char is gasified with steam and air, instead of oxygen. The air injection results in a lower-Btu fuel product.

There is, however, a gravitating-bed pyrolysis system that produces a high liquid yield. The Tech-Air pyrolysis operation is similar to the Union Carbide Purox and Battelle-Northwest systems, in that waste is fed into the top of a vertical furnace, gravitates downward, and is pyrolyzed by heat generated in gasifying char in the lower section of the reactor with air.[105-107] A similar temperature profile to the Union-Carbide Purox process has been reported. The Tech-Air unit, however, yields about 23% char and 25% oil from a mixed pine sawdust and bark feed.

Due to the fixed-bed nature of the gasifier and the high temperatures in some sections of the reactor, it would be anticipated that the oil yield should be low. Actual, atypically high liquid yields could be explained a number of ways. The feed material contained a large fraction of pine sawdust, i.e., extremely small particles that would result in a high heating rate compared with the heating rates of the very large chunks fed to the Union Carbide and Battelle reactors. High heating rates favor high oil yields. Secondly, air is introduced not only at the bottom of the bed, but also at some locations up the side of the reactor. It could therefore be postulated that the air introduced near the top of the bed, not only quenches the pyrolysis gases (low temperatures inhibit cracking of the oils), but also sweeps them rapidly out of the top of the reactor (short gas residence times lead to higher liquid yields).

All of the proposed explanations lead to one important conclusion. For this particular type of gasifier, it would appear that the yield of oils will be particularly sensitive to the geometry of the reactor and the particle size of the feed material.

3. Fluidized-Bed Systems

Fluidized-bed systems, with high heat transfer rates between sand and feed particles, exhibit relatively high heating rates conducive to high liquid yields. Liquids, however, can be cracked by high pyrolysis temperatures and low fluidization velocities (high residence time). Some 7% liquid yield was reported for the West Virginia University process, designed to produce gas from solid waste, sawdust, and coal.[108] This 1/2 ton/ day system was unique in that it consisted of two fluidized beds capable of separating char combustion gases from pyrolysis gas.

The Thermax process, another fluidized-bed pyrolysis system operated by Alberta Industrial Developments, Ltd. and B.C. Research utilizes wet wood residues (75 tons/ day) to produce charcoal.[102] All off-gas from the pyrolysis reaction is burned and used for fluidization. No liquid yields have been reported. A second unit that has been redesigned markedly, is under construction.[102]

4. Entrained-Flow Systems

This type of system is best illustrated by the Garrett Flash Pyrolysis/Occidental Research process. A 4 tons/day pilot plant was constructed in 1971 by Garrett to produce liquid and solid fuels from solid waste or coal. The system was later operated by Occidental, which was to construct a 200-ton/day unit for the county of San Diego.[102,109,110]

The Garrett design encompasses all of the features required to maximize liquid product. In general, the process involves feeding a finely divided ground solid material through a hot entrained-flow transport reactor. The unit under construction is a vertical, stainless-steel pipe, through which organics are pneumatically blown. In the reactor, heating rates are even higher than is possible in a fluidized bed. Temperature is maintained at 900°F and vapor product residence time is extremely short.

The low temperature, high heating rates and short vapor residence times do produce the expected distribution of products: 40% oil from municipal solid waste and 35 to 50% oil from Douglas-fir bark. Char yield is some 20% and gas 25% (used for process heat requirements).

5. Conclusions

If our objective is, in general, chemicals production through wood pyrolysis, the only processes that should be considered are those that result in significant yields of liquid products. We've discussed the effects of various reaction parameters on product yields and composition, and confirmed these observations in the examination of several existing pyrolysis processes. Processes of interest are those typified by the Tech-Air and Garrett approaches, although fluidized-bed processes also offer promise at lower process temperatures and higher fluidization rates.

B. Liquefaction Processes

The liquefaction of cellulosic materials has been studied on a laboratory level for over 50 years. In general, these efforts were based on heating in the presence of various catalysts, alkali, hydrogen, or carbon monoxide.[111] Although these are not true pyrolysis processes, and although emphasis is often on the fuel value of the liquid products, they deserve mention here as an alternate route to high liquid yields through thermal degradation.

1. Effect of Various Catalysts

It was reported as early as 1922 that cotton heated at 250°C in the presence of hydrogen iodide and red phosphorus produces some liquid hydrocarbons.[112] More recent studies have also reported that cellulose heated in the presence of hydrogen iodide and iodine produces some oil.[113]

In 1931, Lindblad investigated the effect of various catalysts on wood destructive distillation,[114] and found that cobalt sulfide, cupric hydroxide, molybdic acid, nickel sulfide, and zinc chloride assisted in the production of tars and oils from wood. Similar observations were made on the effects of copper, iron, and cobalt hydroxides, nickel formate, ammonium molybdate, and ammonia chromate on sawdust pyrolysis.[115] The effects of some of these and other catalysts on wood and clothing flammability have received much attention. Generally, such catalysts assist in the evolution of volatile products at lower pyrolysis temperatures.

2. Effect of Alkali

Over 50 years ago, pressure oxidation of cellulose in alkaline medium at elevated temperatures was shown to produce high yields of volatile acidic products.[116-118] Cotton, straw, grass, and foliage treated with alkali yield a tarry liquid upon heating,[119-121] especially at higher pressures. So-called bitumens, produced by the action of alkali and heat on cellulose, bagasse, and bamboo have been hydrogenated to yield gasoline, middle oils, and lubricating oils.[9,122-124] Apparently the low boiling gasoline-like fraction is produced from cellulose, and the heavier fractions from lignin.[9] It is surprising that this route to liquid products from lignocellulosic materials has not received recent attention.

3. Effect of Hydrogen and Carbon Monoxide

Hydrogenation of cellulose-containing materials results in relatively high yields of liquid products. Early work has demonstrated that cellulose, upon hydrogenation with nickel catalysts, could be completely converted into liquids and gases.[125,127] Berl studied the conversion of agricultural residue into substances that could be hydrogenated into petroleum-like products.[128-130] Others reported similar results, with liquid yields as high as 85% in the destructive hydrogenation of cellulose and of wood.[131]

Most recent work with thermal degradation in the presence of hydrogen or carbon monoxide was carried out by the U.S. Bureau of Mines.[132,133] High liquid yields were reported using cellulose, municipal waste, newsprint, and sewage sludge as feeds.

The only biomass liquefaction process under active development today is the U.S. Bureau of Mines/Bechtel process.[102,134] Bechtel, in a Department of Energy (DOE) sponsored study, is looking into the feasibility of producing synthetic crude oil from biomass in a 3 tons/day pilot plant constructed at Albany, Oregon. The objective is to convert wood and agricultural residues into a liquid fuel and chemical feedstock at a price competitive with probable future crude oil costs. The system is based on the earlier experimental work of the U.S. Bureau of Mines, in which hydrogen or carbon monoxide were used to remove oxygen content while maintaining favorable carbon/hydrogen ratios in the products.

Pyrolysis may be considered a process in which water is removed from carbohydrates to produce carbon. Weiss postulates that the only way to form a liquid product from cellulose is for oxygen (rather than water) to be stripped from the molecule.[111] Liquefaction is then the process of removing oxygen from an oxygenated compound to produce hydrocarbons.

C. Pyrolysis Fuels and Chemicals

1. Role of Pyrolysis

The industrial and affluent growth of the U.S. can be linked to the past availability and costs of petroleum that were used to produce inexpensive energy and petrochemical products. Today, the U.S. uses more energy per capita than any other nation. However, in less than a decade, the U.S. has seen the advent of higher petroleum costs, has experienced interruptions in supplies on local levels, and continues to become more and more dependent on imports. It is not surprising that renewable domestic biomass is receiving much attention today, as an underutilized resource with potential for the production of bulk chemicals and energy products.

What role will biomass play? There is a growing realization that even our renewable resources have production limits, that biomass cannot be grown in sufficient quantities to displace our national energy needs, and, that basic human needs in the form of food, clothing, or shelter products are largely derived from biomass materials, and may take precedence over chemical and energy needs. However, associated with the increasing volumes of biomass products grown and harvested for our basic needs, are similarly increasing volumes of residue streams that have found little or no product value. With rising costs of all raw materials, even residues are now being processed, and the nature and availability of residues is changing. Nonetheless, it is a basic truism that biomass utilization in any form always produces some residues. Even residues beget residues! It is here, in the disposition of various lignocellulosic residues (be they production residues, process residues, or municipal waste discarded residues) that we will find our biomass raw materials for chemical and energy products. This is not to say that biomass from plantations will not also serve, but our growing needs for forestry and agricultural land will push these plantations into rangeland and the sea.

What role will pyrolysis play? As we've seen, pyrolysis, in the forms of hardwood and pine destructive distillation, was the only process for the production of chemicals to feed a growing organic chemicals industry at the turn of the century. For a number of reasons, pyrolysis cannot again assume this major role.

Pyrolysis is a nonspecific process in that the relatively drastic decomposition conditions acting on multicomponent lignocellulosic materials result in an awesome mixture of products. We must realize as well, that as opposed to the original pyrolysis industry that used clean wood chips from trees or residues acquired for the purpose, the raw materials for pyrolysis in the future will be relatively dirty. As we develop better separation processes for utilization of our current residues for higher-valued purposes, residues from these operations will be become more heterogeneous.

Pyrolysis has a future in the processing of these dirty residues into clean products. The various processes described can accept, e.g., municipal waste, manure, or a 50/50 mixture of sand and bark, to yield clean volatile gaseous and liquid products. If inorganic content is low, a char can be produced. If high, the char can be pyrolyzed to yield more volatile products and a slag.

Pyrolysis is also a leveling process.[96] Even though the physical forms of the myriad types of forestry and agricultural residues differ, they are all basically very similar in chemical composition (cellulose, various hemicelluloses and lignin, or phenolic lignin-like materials). This observation suggests that a process suitable for producing oils, chars, and gases from wood residues would not only be suitable for agricultural residues, but should also produce products with similar chemical and physical characteristics as well.

What products can, should, pyrolysis produce? If we accept the thought that essentially all utilization of biomass residue materials will also produce some residues, and the fact that all biomass materials in any altered or degraded form (save CO_2) have fuel value, then pyrolysis will produce a mixture of chemical and fuel products that will depend on the residue used, the type of reactor used, the conditions employed in the reactor, and the marketability of the respective products. Pyrolysis then is flexible, and lends itself to a solution approach that has not yet received enough attention: consider wood, agricultural, and municipal waste residue streams as energy raw materials first and transform them, via pyrolysis, into liquid, solid, and gaseous fuel products.

If such fuels can be developed economically, further exploitation of the chemical or potential chemical values of these fuels could only add to venture profitabilities. Here, however, we would want to produce only those chemicals of economic importance, and in quantities geared to market demand. The rest would be relegated to the already established economic fuel status. It is surprising that this approach (energy before chemicals) receives little support today, even though it has historical precedence in our use of petroleum, coal, natural gas, and even wood!

Let's now turn our discussion into the nature of fuels and chemicals that can be produced via biomass pyrolysis.

2. Pyrolysis Fuels

What is our current energy problem? It can be argued that the heart of our energy problem is not our current dependence on petroleum itself, but on petroleum technology in energy and chemicals production. As a result of the past ready availability and relatively inexpensive nature of petroleum, our best understood, our most efficient, and usually our least expensive technologies for chemicals and energy production revolve around the use of petroleum, to the point where this desired technology dictates the use of petroleum. It should be obvious that our thinking about new energy materials should include fuel preparation processes for biomass that would permit the use of existing high efficiency energy installations, and yet, little attention has been paid to this requirement. We must address ourselves to the transformation of low Btu materials into higher Btu materials, and in physical forms that can be utilized in existing natural gas, oil, and coal burning equipment, with little or no additional investment in modifications, or losses in energy production efficiencies.

By definition, pyrolysis produces gaseous, liquid, and solid products, and again, is flexible in the relative yields of these products, and can be used on a variety of biomass raw materials. Further, the oils and chars are high Btu materials that can be, and have been, used in oil and coal furnaces. The gaseous product can be, has been, burned for on-site energy needs.

Charcoal is a superior fuel to wet wood in that half the caloric content of wood (8000 to 8600 Btu/lb dry hardwood) can be processed into a high-grade fuel (up to 14000 Btu/lb) approaching 30 to 40% of dry wood weight. It has been suggested that, in energy usage, biomass be first carbonized into charcoal because of the inherent advantages of using charcoal vs. wet wood as fuel.[135]

Pyrolysis oils have been compared with No. 6 fuel oil. Table 8 gives typical properties of No. 6 fuel oil vs. Garrett/Occidental's pyrolysis oil from municipal waste,[102] and Tech-Air's pyrolysis oil from pine sawdust and bark.[136] There is a striking similarity in the physical properties of the two pyrolysis oils produced from different sources, by different equipment. It should be noted that, although the Btu/lb for pyrolysis oil is considerably lower than the Btu/lb for No. 6 pyrolysis oil, the higher density of pyrolysis oil makes its Btu/gal value much closer to fuel oil. Both Garrett/Occidental's and Tech-Air's oils have been fired in oil burners.

Some corrosivity has been experienced with this oil,[25,26,136] but recent work has identified the nature of this corrosive action[25] and processes are being developed to remove it. One fault with Occidental's pyrolysis oil is its high ash-producing capabilities that may make it unfeasible to use with conventional fuel oil.[102]

The use of wood gas as a fuel, especially in providing heat for drying wood and the pyrolysis process itself, has already been discussed. The relatively low Btu content of most wood gases (200 to 500 Btu/scf [standard cubic foot]) suggests on-site utilization.

The higher densities and higher energy densities of oil and char also suggest lower storage and transportation costs in the utilization of residues.[137] It is these costs, along with those of harvesting, that have been identified as being largely responsible for the usual unfavorable economic outlook in residue utilization.

Further, seasonal energy needs in agriculture (for irrigation, for example) do not coincide with the seasonal availability of harvest residues. Processing of agricultural residues to produce fuels, instead of direct combustion to produce energy, permits the more economical storage of homogeneous high density fuels instead of heterogeneous, low density, spoilable residues.

3. Pyrolysis Chemicals

Many past attempts at developing chemical products from forest products industry residues have been less than successful, either because processes were aimed at recovering only a small percentage of a chemical product from a complex mix of reaction products, the chemicals isolated had limited market potentials, or processing interfered with a major wood utilization activity.[96] Many process schemes suggested today again rely on prior separation and/or use of only one of wood's cellulose, hemicellulose, or lignin components, with minimal attention paid to the utility of the other components. Processes must be identified that use all of the wood as a primary raw material and transform all of the wood into marketable commodity products.

Pyrolysis offers a rapid route to cracking all of the complex polymeric structures

Table 8

TYPICAL PROPERTIES OF NO. 6 FUEL OIL AND TWO
PYROLYTIC OILS

	No. 6 fuel oil[a]	Occidental pyrolytic oil[a]	Tech-Air pyrolytic oil[b]
C, wt%	85.7	57.0	59.5[c]
H	10.5	7.7	7.0[c]
S	0.7—3.5	0.2	<0.01
Cl	—	0.3	NR
Ash	0.05	0.5	<0.1
N	—	1.1	0.9[c]
O	2.0	33.2	NR
Density, g/mℓ	0.98	1.30	1.14 at 14% H_2O, 1.35 at 4% H_2O
Btu/lb	18,200	10,600	10,600[c]
Btu/gal	148,800	114,900	119,000[c]
Pour point, °C	18—29	32[d]	27[d]
Flash point, °C	66	56[d]	111[d]
Viscosity			
SSU at 190°F	340	1150[d]	—
cP at 25°C, Brookfield LV, 60 rpm	2260	—	225[d]
Pumping temperature, °C	46	71	NR
Atomization temperature, °C	105	116	NR

[a]   Occidental pyrolytic oil and No. 6 fuel oil as reported by Levelton.[102]
[b]   Tech-Air pyrolytic oil as reported by Knight.[136]
[c]   C, H, N, Btu/lb and Btu/gal corrected for moisture content for direct comparison with Occidental pyrolytic oil.
[d]   Pyrolytic oil containing 14% moisture.

found in lignocellulosic residue materials. As opposed to gasification, in which all structural identity is lost in the formation of simple molecules, pyrolysis offers high yields of liquid products that retain some of the integrity of the monomeric units present in the original polymers. As such, these liquids offer promise as feedstocks for chemical opportunities.

The complexity of the starting material and the drastic nature of the pyrolysis process, however, yield many product components. There are two general routes to developing utility for this mixture: pyrolysis products must either be fractionated into simpler mixtures of components with similar chemical and/or physical properties, or pyrolysis products must be reformed into more useful chemicals and chemical intermediates.

We have seen how hardwood sedimentation tars in Russia can be fractionated into useful products. Further utility can be envisaged in the use of refined tars as diesel oil substitutes or in phenol-formaldehyde condensation resins.

We have also seen how chars can be produced with different properties for many chemical applications.

An alternate route to chemical products from wood is hydrolysis of carbohydrate polymers followed by catalytic hydrocracking of lignin. This concept in providing for chemical utilization of wood is still being researched and refined today.[138] If phenols, however, are the more valuable chemicals that can be derived from lignocellulosic residue materials, then pyrolysis offers the more direct route.

Both lignin cracking and pyrolysis result in a mixture of phenolics that have to be reprocessed. Fortunately, these phenolic mixtures bear some resemblance to fossil fuel process sidestreams.

Coal and petrochemical processes, particularly those using catalytic conversion technology, have been developed to transform coal and petroleum sidestreams into useful energy and chemical products. Further work is needed in identifying those processes developed for fossil stream processing that have applicability to phenolics, oils, and chars from wood.

Ironically, the future of biomass pyrolysis may rest in the similarities of its products to fossil fuels and fossil fuel side streams, either as direct replacements for natural gas, oil, and coal in power generation, or as chemical feedstocks amenable to upgrading through coal and petrochemical conversion processes. This is not necessarily surprising. After all, our fossil fuels were derived from biomass.

REFERENCES

1. Shreve, R. N., Wood chemicals, in *Selected Process Industries,* McGraw-Hill, New York, 1950, chap. 30.
2. Youngquist, W. G. and Fleischer, H. O., *Wood In American Life 1776—2076,* Forest Products Research Society, Madison, Wis., 1977, 78.
3. Sloane, E., *A Reverence for Wood,* Ballantine, New York, 1965.
4. Shreve, R. N., Wood chemicals, in *Chemical Process Industries,* 3rd ed., McGraw-Hill, New York, 1967, chap. 32.
5. Klapproth, E. M., *Acetic Acid,* CEH Product Review, Stanford Research Institute, 1977.
6. Blackford, J. L., *Methanol,* CEH Product Review, Stanford Research Institute, 1977.
7. Glasser, W. G., Lima, A. F., and Assumpcao, R. M. V., Brazil — How to harness the biomass giant, paper presented at TAPPI conference, Atlanta, 1977.
8. Panshin, A. J., Harrar, E. S., Bethel, J. S., and Baker, W. J., Carbonization and destructive distillation of wood, in *Forest Products: Their Sources, Production and Utilization,* 2nd ed., McGraw-Hill, New York, 1962, chap. 19.
9. Wenzl, H. F. J., Ed., Further destructive processing of wood, in *The Chemical Technology of Wood,* (transl.), Academic Press, New York, 1970, chap. 5.
10. Bunbury, H. M., *Destructive Distillation of Wood,* Van Nostrand, New York, 1926.
11. Coahran, J. M., U.S. Patents 1,680,452, 1928; 1,784,270, 1930; 1,845,128, 1932; 1,845,129, 1932; 1,865,887, 1932; 1,870,839, 1932; 2,197,069, 1940.
12. Glukhareva, M. I. and Chashchin, A. M., Effect of alcoholic products on the extraction of HOAc from pyroligneous distillate of wood, *Sb. Tr. Tsentr. Nauchno Issled. Proektn. Inst. Lesokhim. Promsti.,* 23, 55, 1973; *Chem. Abstr.,* 83:81587k.
13. Glukhareva, M. I., Taravkova, E. N., Chashchin, A. M., Kushner, T. M., and Serafinov, L. A., Planning of an active experiment to determine the optimum composition of a complex extractant in acetic acid production, *Gidroliz. Lesokhim. Promst.,* 8, 18, 1975; *Chem. Abstr.,* 84: 76027w.
14. Poste, J. R., Suida process for acetic acid recovery, *Ind. Eng. Chem.,* 24, 722, 1932.
15. Othmer, D. F., Dehydrating aqueous solutions of acetic acid, *Chem. Metall. Eng.,* 40, 631, 1933.
16. Othmer, D. F., U.S. Patent 2,186,617, 1940.
17. Othmer, D. F., Acetic acid and a profit from wood distillation, *Chem. Metall. Eng.,* 42, 356, 1935.
18. Othmer, D. F. and Schurig, W. F., Destructive distillation of maple wood, *Ind. Eng. Chem.,* 33, 188, 1941.
19. Hawley, L. F., The thermal decomposition of wood, in *Wood Chemistry,* Wise, L. E., Ed., Reinhold, New York, 1944, chap. 21.
20. Klason, P., von Heidenstam, G., and Norlin, E., *Z. Angew. Chem.,* 23, 1252, 1910.
21. Bagrova, R. K. and Kozlov, V. N., Birch, pine, and spruce wood pyrolysis on heating to various temperatures, *Tr. Inst. Khim. Uralskago Filiala AN SSSR,* 1, 97, 1958.

22. McGinnes, E. A., Jr., Harlow, C. A., and Beall, F. C., Use of scanning electron microscopy and image processing in wood charcoal studies, *Scanning Electron Microscopy/IITRI/II,* 1976, 543.
23. Elder, T. J., Murphey, W. K., and Blankenhorn, P. R., Thermally induced changes of intervessel pits in black cherry, unpublished manuscript, Department of Forest Science, Texas A & M University, College Station, 1978.
24. Schorger, A. W., The distillation of cellulose and wood, in *The Chemistry of Cellulose and Wood,* McGraw-Hill, New York, 1926, chap. 13.
25. Lin, S. -C., Volatile acid constituents in a wood pyrolysis oil, M.S. thesis, Texas A & M University, College Station, 1978.
26. Lin, S. -C., Wolfhagen, J. L., and Soltes, E. J., Constituents of a commercial wood pyrolysis oil, paper presented at American Chemical Society National Meeting, Miami Beach, September, 1978.
27. Goos, A. W., The thermal decomposition of wood, in *Wood Chemistry,* Wise, L. E. and Jahn, E. C., Eds., Reinhold, New York, 1952, chap. 20.
28. Kiprianov, A. I., Foliadova, Z. I., and Soitonen, G. P., Some physical properties of wood-tar oils, *Izv. Vyssh. Uchebn. Zaved. Lesn. Zh.,* 8, 145, 1965; *Chem. Abstr.,* 64: 9943d.
29. Kiprianov, A. I., Faliadova, Z. I., and Bystrova, O. N., Some physical properties of wood tar oils, *Izv. Vyssh. Uchebn. Zaved. Lesn. Zh.,* 8, 148, 1965; *Chem. Abstr.,* 64: 3832g.
30. Klanduch, J., Kosik, M., Rendos, F., and Domansky, R., Production of furfural by catalytic pyrolysis of wood at different temperatures, *Holzforsch. Holzverwert.,* 17, 1, 1965; *Chem. Abstr.,* 63: 5885a.
31. Gerasimova, N. and Klasen, V. I., Use of neutral compounds from beech-wood tar as flotation agents, *Vuglishta (Sofia),* 19, 16, 1964; *Chem. Abstr.,* 61: 15705g.
32. Pikkat-Ordynskii, G. A., Federova, A. M., Golovskaya, T. G., Uvanov, I. P., Vinogradov, L. N., and Shilnikova, R. A., Use of neutral oils from the wood chemical industry for the flotation of coal, *Koks Khim.,* 4, 8, 1968; *Chem. Abstr.,* 69: 20978c.
33. Krylova, E. P., Ponomarev, M. A., Uvanov, I. P., Shilnikova, R. A., and Vinogradov, L. N., Flotation of Kursk Magnetic Anomaly iron ores by wood tar pyrolysis products, *Porysh. Kach. Rud. KMA,* 143, 1969; *Chem. Abstr.,* 72: 14790f.
34. Kravchenko, M. I. and Kiprianov, A. I., Preparation of a gasoline cracking inhibitor in the process of wood tar distillation in a tubular furnace, *Nauchn. Tr., Leningr. Lesotekh. Akad.,* 135, 56, 1970; *Chem. Abstr.,* 74: 143528s.
35. Vodzinskii, Y. V. and Kosyukova, L. V., Antioxidative effectiveness of phenols from wood tar pyrolyzate, *Sb. Tr. Tsentr. Nauchno Issled. Procktn. Inst. Lesokhim. Promsti.,* 23, 105, 1973; *Chem. Abstr.,* 83: 81665.
36. Czechowski, Z., Hulisz, S., and Dudzinska, K., Use of wood tars as inhibitors stabilizing cracked gasoline, *Rocz. Akad. Roln. Poznaniu,* 62, 47, 1973; *Chem. Abstr.,* 81: 65904g.
37. Foliadova, Z. I., Korotova, O. N., and Kiprianov, A. I., Potential antioxidant level in wood-tar oils of the Ashinsk complex, *Gidroliz. Lesokhim. Promst.,* 7, 1974; *Chem. Abstr.,* 81: 107670t.
38. Fedorishchev, T. I., Soluble tar from dry distillation as a source of phenolic raw material, *Sb. Tr. Tsentr. Nauchno Issled. Procktn. Inst. Lesokhim Promsti.,* 39, 1961; *Chem. Abstr.,* 57: 16942b.
39. Kromina, L. V. and Tishchenko, D. V., Chemical composition of the hydroxy-acid-hydroxylactone part of soluble wood tar, *Gidroliz. Lesokhim. Promst.,* 23, 17, 1970; *Chem. Abstr.,* 74: 8884t.
40. Ivanov, N. A. and Piyalkin, U. N., Composition of hydroxyacids of soluble tar studied by gas chromatography, *Izv. Vyssh. Uchebn. Zaved., Lesn. Zh.,* 13, 101, 1970; *Chem. Abstr.,* 74: 143526g.
41. Toporkova, G. V., Kalashev, V. A., and Sorokin, I. S., Mild oxidation of the carbohydrate-lactone part of the soluble resin of wood thermolysis, *Izv. Vyssh. Uchebn. Zaved. Lesn. Zh.,* 17, 108, 1974; *Chem. Abstr.,* 82: 100477k.
42. Shirikova, M. N., Krutov, S. M., Sorokin, I. S., and Kostina, S. P., Study of the hydroxyacid-lactone part of soluble wood tar by reduction with lithium aluminum hydride, *Khim. Drev.,* 112, 1976; *Chem. Abstr.,* 85: 22956y.
43. Vinogradov, L. N., Ulzutueva, E. 6., Goldshmidt, Y. M., and Uvanov, I. P., Wood-particle board binder based on wood-derived phenols, *Gidroliz. Lesokhim. Promst.,* 18(5), 7, 1965; *Chem. Abstr.,* 63: 11859g.
44. Chetverikov, D. I. and Plekhanova, E. A., The technology of processing soluble wood tar, *Khim. Pererab. Drev. Nauchn. Tekhn. Sb.,* 30, 7, 1964; *Chem. Abstr.,* 63: 8603g.
45. Lyass, A. M., Borsuk, P. A., Usabov, Z. G., Kuznetsov, U. G., Kagan, N. Y., Razumeev, Y. A., Bortnik, U. M., Kurenblyum, I. V., and Dinitrieva, V. A., Self-hardening mixture for foundry molds and cores, British Patent 1,358,641, 1974; *Chem. Abstr.,* 82: P89372n.
46. Peniston, Q. P., Separating levoglucosan and carbohydrate-derived acids from aqueous mixtures by treatment of metal compounds, U.S. Patent 3,374,222, 1969.
47. Tishchenko, L. V. and Toporkova, G. V., Hydrogenation of neutral oils from wood pyrolysis, *Nauchn. Tr. Leningr. Lesotekh. Akad.,* 82, 1969; *Chem. Abstr.,* 73: 132177e.

48. Rachinskii, A. V., Levin, E. D., and Skusova, T. D., Hydrocracking of a broad resin fraction obtained by pyrolyzing wood, *Khim. Khim. Tekhnol. Drev.*, 1, 101, 1974; *Chem. Abstr.*, 80: 61231z.

49. Sultanov, A. S., Abidova, M. F., Repyakh, S. M., and Rachinskii, A. V., Hydrocracking of the resin products of wood chemical production on a modified aluminum-cobalt-molybdenum catalyst, *Khim. Khim. Tekhnol. Drev.*, 3, 83, 1975; *Chem. Abstr.*, 85: 126137y.

50. Federov, E. V. and Maltsev, M. P., Wood Tar, U.S.S.R. Patent 379,605, 1973; *Chem. Abstr.*, 79: P68037d.

51. Browne, F. L., Theories of Combustion of Wood and its Control, U.S. Forest Service, U.S. Department of Agriculture, Forest Products Laboratory, Rep. 2136, 1958.

52. Shafizadeh, F., Pyrolysis and combustion of cellulosic materials, *Adv. Carbohydr. Chem.*, 23, 419, 1968.

53. Shafizadeh, F. and Fu, Y. L., Pyrolysis of Cellulose, *Adv. Carbohydr. Chem.*, 29, 113, 1973.

54. Shafizadeh, F., Industrial pyrolysis of cellulosic materials, *Appl. Polym. Symp.*, 28, 153, 1975.

55. Shafizadeh, F., Utilization of biomass by pyrolytic methods, paper presented at TAPPI Forest Biology/Wood Chemistry conference, Madison, Wis., 1977.

56. Weigerink, J. G., Effects of drying conditions on properties of textile yarns, *Text. Res. J.*, 10, 493, 1940.

57. Weigerink, J. G., Moisture relations of textile fibers at elevated temperatures, *J. Res. Natl. Bur. Stand.*, 24, 645, 1940.

58. Heuser, E., *The Chemistry of Cellulose*, John Wiley & Sons, New York, 1944.

59. Waller, R. C., Bass, K. C., and Roseveare, W. E., Degradation of rayon tire yarn at elevated temperatures, *Ind. Eng. Chem.*, 40, 138, 1948.

60. Conrad, C. M., Tripp, V. W., and Mares, T., Thermal degradation in tire cords. I. Effects on strength, elongation and degree of polymerization, *Text. Res. J.*, 21, 726, 1951.

61. Tripp, V. W., Mares, T., and Conrad, C. M., Thermal degradation in tire cords. II. Effects on modulus, toughness and degree of resilience, *Text. Res. J.*, 21, 840, 1951.

62. Nikitin, N. I., Thermal decomposition of wood, in *The Chemistry of Cellulose and Wood*, Israel Program for Scientific Translations, Jerusalem, 1966, chap. 25.

63. Davidson, G. F., The progressive oxidation of cotton cellulose by periodic acid and meta periodate over a wide range of oxygen consumption, *J. Text. Inst.*, 32, T 109, 1941.

64. Nevell, T. P., A qualitative X-ray study of the oxidation of cotton cellulose by nitrogen dioxide, *J. Text. Inst.*, 42, T 130, 1951.

65. Kilzer, F. J. and Broido, A., Speculation on the nature of cellulose pyrolysis, *Pyrodynamics*, 2, 151, 1965.

66. Venn, H. J. P., Yield of β-glucosane obtained from low pressure distillation of cellulose, *J. Text. Inst.*, 15, T 414, 1924.

67. Madorsky, S. L., Hart, V. E., and Straus, S., Pyrolysis of cellulose in a vacuum, *J. Res. Natl. Bur. Stand.*, 56, 343, 1956.

68. Madorsky, S. L., Hart, V. E., and Straus, S., Thermal degradation of cellulosic materials, *J. Res. Natl. Bur. Stand.*, 60, 343, 1958.

69. Golova, O. P. and Krylova, R. G., Thermal decomposition of cellulose and its structure, *Dokl. Akad. Nauk SSSR*, 115, 419, 1957; *Chem. Abstr.*, 52: 4979e.

70. Golova, O. P., New data on the bonds of polysaccharide (cellulose) structure and the chemical reactions occurring during thermal decomposition, *Dokl. Akad. Nauk SSSR*, 116, 61, 1958.

71. Shafizadeh, F., Development of pyrolysis as a new method to meet the increasing demands for food, chemicals, and fuel, special paper, 8th World Forestry Congress, Jakarta, Indonesia, 1978.

72. Katzen, R., Muller, R. E., and Othmer, D. F., Destructive distillation of lignocellulose, *Ind. Eng. Chem.*, 35, 302, 1943.

73. Fletcher, T. L. and Harris, E. E., Destructive distillation of Douglas-fir lignin, *J. Am. Chem. Soc.*, 69, 3144, 1947.

74. Fletcher, T. L. and Harris, E. E., Products from the destructive distillation of Douglas-fir lignin, *Tappi*, 35, 536, 1952.

75. Kuriyama, A., Studies on carbonization phenomena in wood. VIII. Pyrolysis of cellulose and lignin in wood, *J. Jpn. Wood Res. Soc.*, 4, 30, 1958.

76. Bobrov, P. A., The thermal decomposition of wood with superheated steam, *Tr. Tsentr. Nauchno Issled. Lesokhim. Int. Narkomlessa SSR*, 5, 3, 1934; *Chem. Abstr.*, 29: 1616.[4]

77. Bobrov, P. A., Thermal decomposition of wood with superheated steam, *Lesokhim. Promst.*, 4, 9, 1935; *Chem. Abstr.*, 29: 8310.[8]

78. Allan, G. G. and Mattila, T., High energy degradation, in *Lignins — Their Occurence, Formation, Structure and Reactions*, Sarkanen, K. V. and Ludwig, C. H., Eds., Wiley-Interscience, New York, 1971, chap. 14.

79. Gillet, A. and Urlings, J., Comparative pyrolysis of wood, cellulose, lignin and coals. I. Systematically graduated pyrolysis of wood, *Chemie et Ind. (Paris)*, 67, 909, 1952.

79a. Gillet, A. and Urlings, J., Comparative pyrolysis of wood, cellulose, lignin and coals. II. From coal to wood, *Chemie et Ind. (Paris),* 68, 55, 1953.

80. Fengel, D., On the changes of the wood and its components within the temperature range up to 200°C. I. Hot and cold water extracts of thermally treated spruce wood, *Holz Roh Werkst.,* 24, 9, 1966.

81. Merritt, R. W. and White, A. H., Partial hydrolysis of wood, *Ind. Eng. Chem.,* 35, 297, 1943.

82. Beall, F. C. and Eickner, H. W., Thermal degradation of wood components: a review of the literature, U.S. Forest Service, U.S. Department of Agriculture, Forest Products Laboratory, Research Paper 130, 1970.

83. Hawley, L. F. and Wiertelak, J., Effect of mild heat treatment on the chemical composition of wood, *Ind. Eng. Chem.,* 23, 184, 1931.

84. Hawley, L. F., Wiertelak, J., and Harris, E. E., Synthetic lignin, *Ind. Eng. Chem.,* 24, 873, 1932.

85. Mitchell, R. L., Seborg, R. M., and Millett, M. A., Effect of heat on the properties and chemical composition of Douglas-fir wood and its major components, *J. For. Prod. Res. Soc.,* 3, 38, 1953.

86. Kudo, K. and Yoshida, E., The decomposition process of wood constituents in the course of carbonization. I. The decomposition of carbohydrate and lignin in Mizunara, *J. Jpn. Wood Res. Soc.,* 3, 125, 1957.

87. Schwenker, R. F., Jr. and Pascu, E., Pyrolytic degradation products of cellulose, *Ind. Eng. Chem., Chem. Eng. Data Series 2,* 83, 1957.

88. Klason, P., von Heidenstam, G., and Norlin, E., Theoretical investigations on the coking of wood, *Z. Angew. Chem.,* 22, 1205, 1909.

89. Widell, T., Thermal investigation into the carbonization of wood, *Acta Polytech. Chem. Incl. Metall. Ser.,* 1, No. 6, 35, 1949.

90. McNaughton, G. C., Ignition and charring temperatures of wood, *Wood Products,* 50, 21, 1945.

91. Klason, P., Theory of the dry distillation of wood, *J. Prakt. Chem.,* 90, 413, 1914.

92. Martin, S., The mechanisms of ignition of cellulosic materials by intense radiation, R & D Report, USNR DL-TR-102-NS081-001, 1956.

93. Sergeeva, V. N. and Vaivads, A., Thermographic study of pyrolysis of wood and its constituents, *Latv. PSR Zinat. Akad. Vestis,* 86, 103, 1954.

94. Kollman, F., Occurrence of exothermic reactions in wood, *Holz Roh Werkst.,* 18, 193, 1960.

95. Amy, L., The physio-chemical basis of the combustion of cellulose and ligneous materials, *Cah. Cent. Tech. du Bois No. 45,* 1961.

96. Soltes, E. J., Chemicals and energy from wood — a practical approach, paper presented at University of Maine, Pulp and Paper Summer Institute, Orono, 1977.

97. Gluckman, M. J., Processes and equipment for converting wood waste to chemicals or energy, internal report, St. Regis Paper Company, W. Nyack, N.Y., 1974.

98. Choi, T. C., Pyrolysis of cellulosic material using heated rollers, M. S. thesis, University of Minnesota, Minneapolis, 1976.

99. Kittelson, D. B., Murphy, T. E., Choi, T. C., and Noring, J., Progress report on pyrolysis of crop and forestry residue, in *Recovery of Energy from Farm Solid Wastes and Timber Production Residues,* University of Minnesota, Minneapolis, 1975.

100. Soltes, E. J. and Elder, T. J., Thermal degradation routes to chemicals from wood, special paper, 8th World Forestry Congress, Jakarta, Indonesia, 1978.

101. Azarniouch, M. K., Davy, M. F., Dubeauclard, P. L., Elgee, H., and Thompson, K. M., Feasibility study of production of chemical feedstock from wood waste, Pulp and Paper Research Institute of Canada, Pointe Claire, Quebec, 1975.

102. Levelton, B. H. and O'Connor, D. V., An evaluation of wood waste energy conversion systems, a study commissioned by the British Columbia Wood Waste Energy Coordinating Committee, Environment Canada, Western Forest Products Lab, Vancouver, 1978.

103. Sanner, W. S., Ortuglio, C., Walters, J. G., and Wolfson, D. E., Report of Investigations No. 7428, U.S. Bureau of Mines, Washington, D.C., 1970.

104. Hammond, V. L., Mudge, L. K., Allen, C. H., and Schiefelbein, G. F., Energy from forest residuals by gasification of wood wastes, *Pulp Pap.,* 48, 54, 1974.

105. Knight, J. A., Tatom, J. W., Bowen, M. D., Colcord, A. R., and Elston, L. W., Pyrolytic conversion of agricultural wastes to fuels, paper presented at Am. Soc. Ag. Eng. Annu. Meeting, St. Joseph, Mich., 1974.

106. Knight, J. A., Pyrolysis of pine sawdust, paper presented at American Chemical Society National Meeting, San Francisco, 1976.

107. Knight, J. A., Bowen, M. D., and Purdy, K. R., Pyrolysis—a method for conversion of forestry wastes to useful fuels, paper presented at Forest Products Research Society Meeting, Atlanta, November, 1976.

108. **Bailie, R. C.**, Technical and economic assessment of methods for direct conversion of agricultural residue to usable energy, final report no. 12-14-1001-598, West Virginia University, Morgantown, 1976.

109. **Finney, C. S. and Garrett, D. E.**, The flash pyrolysis of solid wastes, *Energy Sources,* 1, 192, 1974.

110. **Sass, A.**, Garrett's coal pyrolysis system, *Chem. Eng. Prog.,* 70, 72, 1974.

111. **Weiss, A. H.**, Conversion of solid waste to liquid fuel, *Text. Res. J.,* 42(9), 526, 1972.

112. **Willstatter, R. and Kalf, L.**, Reduction of lignin and of carbohydrate with hydriodic acid and phosphorus, *Ber.,* 55B, 2637, 1922.

113. **Freed, V. H., Barbour, J. F., and Groner, R. R.**, Hydrogenation of solid wastes, Oregon State University, Corvallis, 1969.

114. **Lindblad, A.**, Preparation of oils from wood by hydrogenation, *Ing. Vetenshaps Akad. Handl.,* 107, 7, 1931.

115. **Routala, O.**, Wood chips. I. Motor fuels from wood, *Acta Chem. Fennica,* 3, 115, 1930.

116. **Fischer, F., Schrader, H., and Treibs, W.**, Chemical breakdown of cellulose by pressure oxidation, *Gesammelte Abh. Kennt. Kchle,* 5, 211, 1921.

117. **Fischer, F. and Schrader, H.**, Effect of heating celluose lignin under pressure in presence of water and aqueous alkalies, *Gesammelte Abh. Kennt. Kohle,* 5, 332, 1921.

118. **Fischer, F., Schrader, H., and Treibs, W.**, Effect of heating under pressure the alkaline solutions obtained in the pressure oxidation of cellulose and lignin, *Gesammelte Abh. Kennt. Kohle,* 5, 311, 1921.

119. **Berl, E. and Biebesheimer, H.**, Origin of petroleum, *Ann.,* 504, 38, 1933.

120. **Berl, E. and Schmidt, A.**, Genesis of coals, *Ann.,* 496, 283, 1932.

121. **Wallin, J. H.**, Decomposition of waste cellulose liquor and other organic materials, U.S. Patent 1,608,075, 1927.

122. **Gillet, A. and Colson, P.**, From cellulose to coal, *Ind. Chim. Belge,* (Suppl. 1), 402, 1959.

123. **Gillet, A. and Colson, P.**, From cellulose to petroleum and anthracite, *Proc. Symp. Nature of Coal,* Jealogra, India, 35, 1959.

124. **Heinemann, H.**, Petroleum-type hydrocarbons from cane sugar, *Pet. Refiner,* 29, 111, 1950.

125. **Fierz-David, H. E.**, The liquefaction of wood and cellulose and some general remarks on liquefaction of coal, *Chem. Ind.,* 44, 942, 1925.

126. **Fierz-David, H. E. and Hannig, M.**, The distillation of cellulose, wood and similar substances under hydrogen pressure, *Helv. Chim. Acta,* 8, 900, 1925.

127. **Frolich, D., Spaulding, H. B., and Bacon, T. S.**, Destructive distillation of wood and cellulose under pressure, *Ind. Eng. Chem.,* 20, 36, 1928.

128. **Berl, E.**, Role of carbohydrates in the formation of oil and bituminous coals, *Bull. Assoc. Pet. Geol.,* 24, 1865, 1940.

129. **Berl, E.**, Production of oil from plant material, *Science,* 99, 309, 1944.

130. **Berl, E.**, The artificial formation of substances similar to bituminous coal and petroleum, *Proc. 3rd Int. Conf. Bituminous Coal,* 2, 820, 1932.

131. **Boomer, E. H., Argue, G. H., and Edwards, J.**, Destructive hydrogenation of cellulose and wood, *Can. J. Res.,* 13B, 337, 1935.

132. **Appell, H. R., Wender, I., and Miller, R. D.**, Conversion of urban refuse to oil, U.S. Bureau of Mines, Washington, D.C., Technical Rep. 25, 1969.

133. **Appell, H. R., Wender, I., and Miller, R. D.**, Hydrogenation of municipal solid wastes with carbon monoxide and water, paper presented at National Industrial Solid Waste Management Congress, Houston, March, 1970.

134. **Lindemuth, T.**, Biomass liquefaction program, paper presented at 2nd Annu. Fuels Biomass Symp., Rensselaer Polytechnic Institute, Troy, N.Y., 1978.

135. **White, E. W.**, Charcoal: fuel for thought, *Am. For.,* 84, 21, 1978.

136. **Knight, J. A.**, Utilization and/or stabilization of pyrolytic oil from pyrolysis of agricultural, municipal and other wastes, EPA Grant R-804-416-010, Quarterly Report No. 3, Georgia Institute of Technology, Atlanta, 1977.

137. **Soltes, E. J. and Wiley, A. T.**, Energy self-sufficiency for the Texas forest products industry: a problem analysis, paper presented at Society of American Foresters Meeting, Washington, D.C., 1978.

138. **Goldstein, I. S.**, Conversion of wood into useful chemicals, NSF Grant PFR 77-12243, North Carolina State University, Raleigh, 1978.

Chapter 6

CHEMICALS FROM CELLULOSE

Irving S. Goldstein

TABLE OF CONTENTS

I. INTRODUCTION

The gasification and pyrolysis processes discussed in the two preceding chapters are nonselective methods of converting total biomass to chemicals. They are necessarily drastic, involving high temperatures and in some cases high pressures as well. These techniques are basically no different from similar ones used in the processing of coal and yield analogous products.

Instead of subjecting the biomass to reaction conditions which cause simultaneous conversion of all its components, an alternative approach is to choose reactions which more or less selectively convert one of the components at a time to useful products or intermediates. In some cases such reactions may be selective enough that the components other than the target one are virtually unaffected, while in other cases it might be necessary to bring about at least partial separation of the biomass components before reaction. Preliminary separation may also be necessary where a biomass component interferes with the desired reaction, for example lignin with enzymatic hydrolysis of cellulose as elaborated in Section IV below.

In this chapter the selective conversion of cellulose to chemicals is considered. Since the reactions such as hydrolysis are also capable of converting the hemicellulose components of the biomass to sugars which would afford an undesirable mixture of products, it is expedient to first remove the hemicelluloses by prehydrolysis before conversion of the cellulose. The conversion of hemicelluloses and lignin to chemicals is treated in Chapters 7 and 8.

The structure of cellulose as a linear polymer of glucose units provides a conceptually simple mechanism for its chemical utilization, namely depolymerization to glucose. Hydrolysis to glucose can be effected by either acid or enzymes, but in neither case is it as facile as the hydrolysis of starch. These hydrolysis reactions as well as the barriers to hydrolysis resulting from the highly ordered crystalline structure of cellulose or the presence of lignin are discussed below. Thermal depolymerization of cellulose is also considered.

The concluding section of this chapter will outline the various routes by which glucose can be converted into a great number of the organic chemicals of present and potential commercial importance.

II. ACID HYDROLYSIS

It has been known for over 150 years that cellulose can be converted to glucose by acid hydrolysis; since Braconnet[1] first treated wood cellulose with strong sulfuric acid in the cold followed by heating with dilute acid. The reactions have been studied extensively on pure cellulose for structure determinations as well as on cellulosic composites such as wood for commercial saccharification. It took almost 100 years before the first commercial plants began production during World War I.[2] Further evolution of process technology, most notably in Germany and Japan, has been along two paths. These are dilute acid hydrolysis at elevated temperatures and strong acid hydrolysis at lower temperatures.

A. Dilute Acid Hydrolysis

A discussion of the mechanism and kinetics of the reaction will be followed by descriptions of various process configurations based on dilute acid hydrolysis.

1. Mechanism and Kinetics

While it might be inferred from the ready acid hydrolysis of simple glycosides that the acetal linkages in cellulose should also be easily cleaved, this is not the case, for the heterogeneous hydrolysis rate of cellulose is several orders of magnitude less than that of glycosides. This extreme resistance to hydrolysis is not attributable to the high polymeric nature of cellulose since starch does not present the same difficulties. It narrows down to the β-glycosidic linkages and the resultant crystalline structure of cellulose. Harris[3] points out that it is not simply lack of accessibility to the crystallites, since it might be expected that hydrated protons could move freely throughout the hydrophilic cellulose crystals. Rather, the rigidity of the anhydroglucose rings held tightly in the crystal structure might prevent the normal rapid glucoside hydrolysis. The slow hydrolysis permits the secondary degradation of the glucose produced to become significant.

Although the kinetic mechanism of the heterogeneous hydrolysis of cellulose with dilute acids has not been established, empirical equations have been developed which allow accurate representation of the actual experimental results. Saeman[4] has shown that the formation of glucose from wood cellulose may be represented by consecutive first-order reactions:

$$\text{Cellulose} \xrightarrow{k_1} \text{Glucose} \xrightarrow{k_2} \text{Degradation Products} \qquad (1)$$

where k_1 and k_2 are first-order rate constants. The maximum concentration of glucose is a function only of the ratio (k_r) of the rate constants, with the calculated maximum potential yield increasing from about 10% at $k_r = 0.2$ to about 50% at $k_r = 2.5$ and about 80% at $k_r = 14$.

Saeman[4] also found that the energy of activation for the hydrolysis step was 42.9 kcal/mol while that for glucose degradation was 32.8 kcal/mol. It follows that the ratio of the rate constants can be increased at higher operating temperatures thereby increasing the yield of glucose. This is shown graphically in Figure 1. Increases in catalyst concentration also improve the maximum yield of sugar as shown in Figure 2. The solid lines are based on experimental results while the dotted lines represent extrapolated values.

Degradation reactions of the glucose in acidic solution result in many products, most of which are highly reactive. Because they are present in only very low concentrations they are difficult to isolate. Thus the kinetics are not simple first order and the mechanism for glucose degradation is complex. Low yields of hydroxymethylfurfural (10 to 20%) and higher yields of levulinic acid (30 to 50%) have been obtained from glucose in dilute mineral acid solutions.[3]

Grethlein[5] has confirmed the kinetic projections of Saeman[4] concerning the selectivity of wood cellulose hydrolysis to glucose. The selectivity (k_r) increases with temperature since the activation energy for hydrolysis is greater than the activation energy for glucose degradation. He too found that selectivity also increases with acid concentration. Grethlein has shown the utility of the kinetic model by developing the relationships between yield, residence time, temperature, and acid concentration. One of these curves, for 2% acid, is shown in Figure 3. Although predicted values of 65% yield at 260 to 270°C for 1 sec are indicated, the highest actual yield reported was 52% glucose from newsprint at 232°C for 16 sec using 1.2% sulfuric acid catalyst. Results with corn stover indicate easier hydrolysis reflecting its lower cellulose crystallinity.

2. Batch Reactor

A single-stage batch process for hydrolysis of wood was used for the first commercial plants in the U.S. operated at Georgetown, S. C. from 1913 to 1923 and at Fullerton, La. after 1916. The wood was processed on a 1-hr cycle in a rotary digester. In

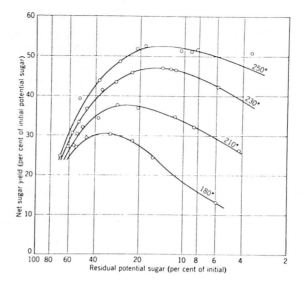

FIGURE 1.   Effect of reaction temperature on net sugar yield. (From Harris, J. F., Saeman, J. F., and Locke, E. G., *The Chemistry of Wood,* Browning, B. L., Ed., John Wiley & Sons, New York, 1963, 555. With permission.)

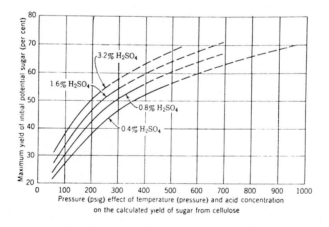

FIGURE 2.   Sugar yields from cellulose at various reaction conditions. (From Harris, J. F., Saeman, J. F., and Locke, E. G., *The Chemistry of Wood,* Browning, B. L., Ed., John Wiley & Sons, New York, 1963, 555. With permission.)

the pilot plant development work for these plants carried out at the U.S. Department of Agriculture Forest Products Laboratory, the yield of sugar produced in 15 min increased from 14% at 155°C to 23% at 175°C at sulfuric acid concentrations of 0.5 to 2%. Higher temperatures caused decreases in sugar yield.[6] The hydrolysis solution contained about 12% total solids, nearly 9% reducing sugar, and 6% fermentable sugar.[2] Sugars from softwoods were 70% fermentable and those from hardwoods about 30% fermentable. It is obvious that the easily hydrolyzed hemicelluloses comprised the major part of the wood components removed.

The process used in these plants was known as the American process. Both plants

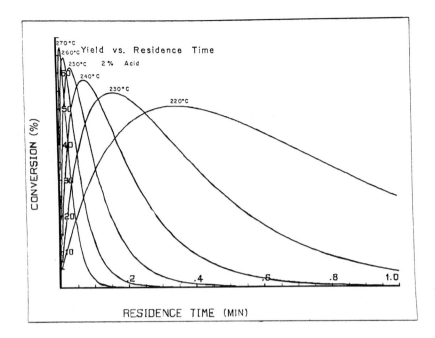

FIGURE 3.   Relationship between yield, residence time, and temperature for hydrolysis in a 2% sulfuric acid solution. (From Grethlein, H. E., *Proc. 2nd Annu. Fuels From Biomass Symp.*, Rensselaer Polytechnic Institute, Troy, N.Y., 1978.)

were closed after World War I because they were unable to compete with cheap molasses and were unable to obtain cheap wood residues nearby.

3. Percolation Processes

The next advance in wood hydrolysis was the development of the Scholler process in Germany during the 1920s. This process is based on the rapid removal from the reaction zone of the sugars formed during hydrolysis to minimize their further decomposition. This is accomplished by use of a stationary digester with the dilute acid being injected at the top, percolating through the wood, and being withdrawn through a filter at the bottom. Sugar production and removal go on simultaneously, and the sugar is separated and cooled as rapidly as possible to retard decomposition. The percolation is repeated with fresh charges of acid until all the cellulose is hydrolyzed.[2,6-8]

Operating conditions using 0.5 to 0.8% sulfuric acid range from a beginning temperature of 140°C at 5 atm to a maximum of about 190°C at a pressure of 12 atm. As many as 20 cycles of fresh acid percolation are used to reduce the lignin-rich residue to about one-third of the original wood, providing a 3.5 to 4% sugar solution with up to an 80% yield based on original total carbohydrate content. The first six cycles, up to about 160°C, contain for the most part hemicellulose hydrolysis products. The actual yield of glucose is about 50% of the original cellulose content of the wood.[8]

Before and during World War II five plants were in operation using the Scholler process, three in Germany, one in Switzerland and one in Korea. Similar technology with operating temperatures of 170 to 190°C yielding solutions containing 2.5 to 3.5% sugars for a yield of 40 to 50% of the original raw material is being applied in the Soviet Union with the first plant having been started in Leningrad in 1935 and "tens" of plants still in operation today on wood and agricultural residues.[9] Experimental continuous equipment with "efficiency five times higher than the efficiency of noncontinuous equipment applied at present" has been developed.

Ant-Wuorinen[10] used sulfur dioxide as a catalyst in a percolation process claimed to be more rapid than the Scholler process and causing less decomposition of the sugar. A modification of the Scholler process called the Madison process was developed at the U.S. Department of Agriculture Forest Products Laboratory.[7] After an initial low-temperature hydrolysis period, the acid solution is pumped in continuously at the top of the stationary digester and hydrolyzate is removed at the bottom with no interruption until the hydrolysis is completed. After 2½ hr, the yield of sugar with a concentration of 5.5% is 65% of the potential sugar in the wood.

The most recent modification of the Scholler process[8] separates each cycle into two phases, a reaction step and a washing step. During the reaction step the cellulose is hydrolyzed to sugar under stringent conditions, while in the wash step the sugars formed are dissolved away from the wood under milder conditions. With this scheme, it is possible to obtain sugar solutions of 9 to 10% concentration which can be quite sharply separated into a xylose fraction and a glucose fraction based on the different composition of the wash liquors as the hydrolysis progresses. A mixed fraction is almost completely avoided.

A two stage process developed in Sweden during World War II has been described by Cedarquist.[11] The first hydrolysis stage uses 0.5% sulfuric acid at 190°C for 3 min. After washing the hemicellulose sugars out, the residue is heated with 0.75% sulfuric acid at 215°C for 3 min. The second stage solution consists entirely of glucose. Overall yield of sugar is 48 to 50% of the wood at a high concentration (10 to 12%).

4. Plug Flow Reactor

In the above processes operating temperatures have been lower than the most favorable level indicated by the kinetic model because of equipment design limitations. At higher temperatures the rate of hydrolysis is increased as well as the selectivity favoring hydrolysis over degradation. With a sufficiently high rate of reaction, a continuous process becomes possible with its attendant advantages of higher production rate per unit volume, steady state operation and high sugar concentrations.

Grethlein[5,12] has designed such a continuous process approaching the ideal in which an acidified slurry of cellulose is heated "instantaneously" to the desired reaction temperature, maintained at temperature for the desired reaction time, and quenched "instantaneously" to provide the optimum glucose yield. This has been accomplished in an experimental laboratory scale plug flow reactor consisting of a glass lined stainless steel tube about ¼ in. in inside diameter operating at 500 to 600 psig. With reaction times in the 15-sec range, glucose yields of approximately 50% from wood cellulose have been obtained at about 230°C and acid concentrations over 1%, in good agreement with the kinetic model shown in Figure 3. Corn stover, presumably because of the lower crystallinity of its cellulose, gives comparable glucose yields at lower temperatures (215°C) and acid concentrations.

Continuing process development on the plug flow reactor is addressing the critical areas for ultimate commercial application of maximum slurry concentration and maximum slurry particle size.

5. Appraisal

Dilute acid hydrolysis of cellulose to glucose at elevated temperatures is feasible on a commercial scale and capable of providing sugar solutions of 10 to 12% concentration in 50% yield. Furthermore, process design improvements such as the plug flow reactor show promise of reducing capital and operating costs. Lignin does not interfere with the hydrolysis.

The maximum yield of glucose attainable by this method, however, is limited as a practical matter to about 50% from cellulose with a crystallinity as high as that in wood. Although higher yields are possible for very short reaction times at higher temperatures (Figure 3), heat transfer requirements for both heating to the reaction temperature and quenching will limit the extent to which the reaction time can be shortened.

Cellulose with lower crystallinity, either in its native state or as a result of prior processing or pretreatment, is more amenable to hydrolysis and could provide higher yields. An unexplored factor that could affect the extent of yield improvement is the propensity for the amorphous regions of the cellulose to recrystallize in the first stages of hydrolysis with dilute acids at elevated temperatures.[13]

B. Strong Acid Hydrolysis

The yield limitations inherent in the high temperature dilute acid hydrolysis of crystalline cellulose are not present in strong acid hydrolysis at lower temperatures. Even though a patent using fuming hydrochloric acid was issued as long as 100 years ago in 1880, subsequent technical activity and commercial application were greater for the dilute acid processes. More recently, improvements in corrosion resistant materials and acid recovery techniques have given renewed impetus to strong acid hydrolysis. Discussions of mechanisms and descriptions of various processes follow.[14,15] With two exceptions involving formic and hydrofluoric acids,[2] the emphasis has been on hydrochloric and sulfuric acid hydrolysis.

1. Mechanism

The basic principle on which strong acid hydrolysis processes depend is the disruption of the crystalline structure of the cellulose by solution or swelling in the acid. Subsequent hydrolysis of the cellulose can then be accomplished at temperatures low enough to avoid degradation, thereby affording almost quantitative yields of glucose.

After dissolution, the first step is the degradation of the cellulose to oligosaccharides which are then further hydrolyzed to glucose at rates which vary with acid concentration and temperature. If the hydrolysis in highly concentrated hydrochloric acid is followed polarimetrically, the solutions are optically inactive at first but show some activity after an hour and only reach the rotation value of glucose after 24 to 48 hr.[16]

Minimum acid concentrations necessary for dissolving the cellulose are critical, and because of the formation of various cellulose-acid complexes the kinetics may vary significantly with only small changes in acid concentration. In 16 hr at room temperature, 41% hydrochloric acid caused 100% hydrolysis, 40% HCl 73%, 39% HCl 45%, and 38% HCl caused only 22% hydrolysis. At the 41% concentration, the reaction constant increased with time; at 38 and 39% concentrations, the reaction constants decreased with time, while at 40% acid the reaction followed first order kinetics.[17]

Only minor degradation of cellulose occurs in 4 hr with 35% hydrochloric acid. Increasing the acid concentration to 40% gives quantitative hydrolysis in 4 hr at 40°C and in only 10 min at 60°C. Up to about 30°C the cellulose is only degraded to oligosaccharides. It is only at higher temperatures that the oligosaccharides are hydrolyzed to glucose by a first-order reaction whose velocity constant increases rapidly with temperature.[18]

For sulfuric acid solutions, the swelling and solubility of cellulose at 20°C increase until a maximum at 62% sulfuric acid is reached. Over the temperature range 20 to 40°C the swelling and solubility maxima occur at about 62 to 63% acid. At higher or lower temperatures, the swelling maximum decreases.[19]

Solubility and hydrolyzability of cellulose in sulfuric acid do not always follow the same trends, with the presence of water and ethanol often increasing solubility but not hydrolysis. The addition of glucose to the acid lowers the swelling. This effect can influence the hydrolysis when reaction products complex with the acid unless the original acid-cellulose ratio is high enough to compensate. It appears that the effective concentration of sulfuric acid in the hydrolysis of cellulose must remain above 60%.[20]

It has been found that the acid concentration decreases steadily during hydrolysis with both hydrochloric and sulfuric acids, and that the reaction comes to an end with the establishment of an equilibrium caused by the complexing of the acid with the sugars formed during the hydrolysis.[21] Here again the acid to cellulose ratio must be large enough to allow the reaction to proceed to completion.

A further complication which makes a posthydrolysis necessary in many processes is the formation of gentiobiose, a 6-β-glucosido-glucose dimer, by condensation of the glucose hydrolysis product on continued exposure to strong acids.[22] A trisaccharide has also been obtained. The degree of these reversion reactions depends on the water content and glucose concentration of the hydrolyzate.

The kinetics of the hydrolysis of cellulose in phosphoric acid solution has been measured by following the decrease in the degree of polymerization by viscometry.[23] The hydrolysis is first order with the degradation constant independent of the degree of polymerization between 130 and 1500. Condensation and reversion products of lower polymers, which are water soluble and show the properties of oligosaccharides, obscure the reaction mechanism at lower degrees of polymerization.[24]

2. Hydrochloric Acid Processes
a. Rheinau Processes

The original Rheinau process based on the use of supersaturated hydrochloric acid was an outgrowth of the recognition in 1880[25] that cellulose can be converted into glucose by the action in the cold of gaseous hydrogen chloride on wet wood, and that the spent hydrochloric acid could be recovered by vacuum distillation of the resultant glucose solution. By 1917 a mature technical process had been developed,[26] which was further refined and scaled up by Bergius.[27]

As operated in commercial plants in Germany at Mannheim-Rheinau (from 1932) and Regensburg (from 1938) until the end of World War II, the process used a countercurrent stream of 41% hydrochloric acid at room temperature or below. Fresh acid was contacted with previously treated wood chips to complete hydrolysis and wash out previously formed sugars. The acid solution was then passed successively over chips exposed to less treatment and finally over fresh chips. By the use of several reactors, sugar concentrations of about 30% could be obtained while the acid concentration fell to 30 to 32%. The residence time for a batch of chips was over 24 hr because of the slow reaction at the low temperatures.

The acid hydrolyzate was then subjected to vacuum distillation at 35 to 40°C to form a thick sugar syrup and recover the hydrochloric acid. Spray drying removed more acid and yielded a solid containing about 90% carbohydrate.

These sugars were for the most part oligomers because of both incomplete hydrolysis and condensation of glucose under the acid conditions. According to Bergius, tetramers predominated. The oligosaccharides were then hydrolyzed (posthydrolysis) at about 120°C for 30 min using 1 to 2% hydrochloric acid. Virtually quantitative yields of sugar of 70% of the weight of the wood were obtained. With softwoods 42% glucose based on wood and with hardwoods 48% glucose based on wood were obtained, for yields of about 85 to 95% based on cellulose.

In one modification of the Rheinau process[14] a prehydrolysis with 1% hydrochloric acid at 130°C followed by careful main hydrolysis and optimum posthydrolysis yielded a sugar solution of 90% pure glucose after ion exchange to remove salts. A one-step crystallization gave 99% pure glucose in the form of the hydrate in 85% yield.

Process improvements including the use of three evaporators for acid recovery reduced steam consumption to one third of the former value. The first evaporator at 100 mm provided 95% hydrogen chloride. The second evaporator at 50 mm yielded 30% hydrochloric acid. The third evaporator allowed 10% hydrochloric acid to be recovered by the injection of steam.

In another modification the prehydrolysis which had previously been carried out in a special reactor with dilute acid under pressure at high temperatures was incorporated into the main hydrolysis tower and carried out at 20°C with hydrochloric acid graduated in concentration.

In a final modification known as the Udic-Rheinau process, which did not progress beyond the semi-works plant stage,[14,15] prehydrolysis was carried out with 32% hydrochloric acid and the main hydrolysis with 41% hydrochloric acid at the same temperature (20°C) in the same digester. After acid recovery as above, the sugar solutions were diluted to 12% and posthydrolyzed at a temperature somewhat above 100°C. Both the prehydrolysis and hydrolysis sugar solutions were then filtered, decolorized, and freed of salts by ion exchange. After concentration to 70% in a multiple-effect evaporator, crystalline xylose and glucose could be recovered. The mother liquors could be hydrogenated to polyalcohols, and the highly reactive lignin residue was used for plastics.

b. Hydrogen Chloride Gas Processes

The greater ease of recovery of hydrogen chloride gas than hydrochloric acid solutions has stimulated the development of a number of processes based on the gas. None has progressed beyond the pilot plant stage.

In the Prodor process,[28] sawdust mixed with concentrated hydrochloric acid was moved by paddles through descending shelves while a stream of hydrogen chloride gas ascended countercurrently. After 8 hr of hydrolysis, the hydrogen chloride was recovered in a stream of hot air.

In the Hereng process,[29] wood chips impregnated with 30% hydrochloric acid were prehydrolyzed in a reactor fitted with inclined trays for the chips to move from top to bottom. After prehydrolysis the chips were reimpregnated with 30% acid and were reintroduced into the reactor to face an ascending countercurrent stream of hydrogen chloride gas. The hydrolyzed material was dried with anhydrous hydrogen chloride gas, extracted with water, and the weakly acid sugar solution heated to hydrolyze the oligomers to glucose.

Extensive pilot plant development was carried out in Japan from 1953 to 1959 on the Noguchi-Chisso process.[15,30,31] Pulverized wood was first prehydrolyzed by dilute acid to remove the hemicelluloses. Prehydrolysis conditions could range from 5% hydrochloric acid at 100°C for 3 hr to 2% hydrochloric acid at 130°C for 30 min. After washing and drying, the residual cellulose and lignin were treated with a small amount of hydrochloric acid and cooled to a low temperature (below 10°C) during the absorption of sufficient hydrogen chloride gas to provide 60 to 70% concentrated hydrochloric acid. The material was then heated to 40°C in a fluidized bed using diatomaceous earth or lignin to prevent caking. After this main hydrolysis, the recovery of the hydrogen chloride was accomplished by blowing hot air or hydrogen chloride at 125 to 139°C through the material for a few seconds.

The oligomeric sugar residue was posthydrolyzed with 3% hydrochloric acid at 100°C for 4 to 5 hr or with 6 to 7% acid for 1 hr. At 120 to 125°C and 0.5 to 1% acid, posthydrolysis was complete in 30 to 45 min. The yield of sugars was 95% of the theoretical, and crystalline glucose was obtained in 85% yield.

An important feature of the process was the short reaction time for the main hydrolysis. Impregnation with acid could require 30 to 60 min, but the hydrogen chloride absorption took less than 1 min, the main hydrolysis about ten min and the hydrogen chloride recovery only a few seconds, for a total elapsed time of about 1 hr.

The Russian Chalov process[32,33] appears to be very similar in principle. The amount of hydrogen chloride absorbed is directly proportional to the moisture content of the wood and inversely proportional to the temperature. Hydrogen chloride concentrations of up to 60 to 65% can be obtained. The heat generated in absorption of the hydrogen chloride and during swelling of the wood is removed by cycling cooled hydrogen chloride to keep the temperature below 35°C. Wood saturated with hydrogen chloride at 20°C and heated at 40°C was completely hydrolyzed before it had attained the reaction temperature. The hydrolysis rate and sugar yield increased with the moisture content of the wood. No decomposition of sugars occurred at 40°C. When the wood was heated at 60°C, the hydrolysis proceeded more rapidly, but sugar decomposition occurred after 10 min.

3. Sulfuric Acid Processes

Although concentrated sulfuric acid readily converts cellulose into glucose, its application for glucose production has not been extensively studied because of the greater difficulties in acid recovery. One strong sulfuric acid process did reach semiworks operation, the Japanese Hokkaido process.[34]

After prehydrolysis by usual means, such as 1.5% sulfuric acid at 140°C, the wood was pressed and dried. The main hydrolysis was carried out with 80% sulfuric acid at room temperature. To maintain a low ratio of acid to wood the pulverized wood was mixed with the acid by spraying them together. The product was immediately filtered and washed, yielding 96% sugar when a 1:1 ratio of acid and wood was used. A sulfuric acid concentration as high as possible (30 to 40%) was required in the filtrate to permit recovery of 80% of the acid by diffusion dialysis with an ion-exchange membrane. The sugar solution which still contained 5 to 10% sulfuric acid was heated at 100°C to posthydrolyze the glucose oligomers. After neutralization with lime and filtration of the calcium sulfate produced, the sugar solution was evaporated and the glucose crystallized in 85% yield.

In the Nihon-Mokuzai-Kagaku process,[15] all the sulfuric acid was neutralized with lime and the resultant gypsum used in gypsum board manufacture. This process was proved in a 5 ton/day pilot plant.

Karlivan[9] believes that mechanochemical treatment in the presence of small amounts of concentrated sulfuric acid, the Riga hydrolysis method, is the most promising method of sugar production from biomass. Catalyst consumption can be reduced to 10 to 20% of the raw material weight, eliminating the necessity for its recovery. The catalyzed biomass is subjected to intense mechanical action by a special device in the reaction zone of a screw conveyor hydrolyzer. In 30 sec the polysaccharide hydrolysis was 75% at an energy consumption of 0.3 kwhr/kg sugar compared to 900 sec for 90% hydrolysis in a vibrating mill at an energy consumption of 2.5 kwhr/kg. A potential advantage of the process is its suitability for comparatively small scale operations.

4. Appraisal

Strong acid hydrolysis of cellulose is capable of providing yields of crystalline glucose of 85% and even higher glucose yields in solutions suitable for fermentation. These improved yields over dilute acid hydrolysis are offset by higher capital costs for corrosion resistant equipment and higher operating costs associated with acid recovery and loss.

III. ENZYMATIC HYDROLYSIS

Enzymatic hydrolysis of cellulose to glucose is both older and newer than acid hydrolysis. Older, because the microbial decomposition of biomass has been an integral part of the carbon cycle since its inception. Newer, because it has only been very recently that research on the mechanism of the microbial degradation of cellulose has provided the conceptual framework on which to base an enzymatic conversion plant.

The pioneering work at the U.S. Army Natick Laboratories of Reese, Siu, and Mandels was originally undertaken to provide the fundamental knowledge which would allow better protection of Army cellulosic material in tropical environments.[35] With the discovery of fungal strains of high cellulolytic activity, it became conceptually possible to consider the positive aspects of enzymatic hydrolysis of cellulose in waste disposal and the production of chemicals, energy, and food, instead of merely the negative aspects in the decomposition of textiles and wood. This change in emphasis from preventing decomposition by the enzymes to promoting it is an excellent example of the unpredictability of the specific benefits which may result from a given research project.

Several volumes of symposium proceedings have been devoted to enzymatic hydrolysis of cellulose.[36-39] A plant for enzymatic hydrolysis requires two discrete stages, production of enzymes, and saccharification of cellulose. These are considered separately following a discussion of the mechanism of enzymatic hydrolysis.

A. Mechanism

The structure of cellulose as a linear polymer of glucose is such a simple concept that it might be supposed that a single, simple enzyme would suffice to depolymerize cellulose to glucose. However, just as the acid hydrolysis of crystalline cellulose was shown to be more complex than simple glycosidic hydrolysis, the enzymatic hydrolysis of cellulose is also a complex phenomenon.

Our present conception of the nature of cellulase, the enzyme system capable of hydrolyzing cellulose, stems originally from the suggestion by Reese et al.[40] that the cellulase system contains a component, C_1, which allows digestion of highly ordered forms of cellulose as well as C_x enzyme components capable of acting on easily accessible forms of cellulose. According to this scheme the enzymatic hydrolysis of cellulose proceeds as follows:

$$\text{Crystalline Cellulose} \xrightarrow{C_1} \text{Amorphous Cellulose} \downarrow C_x's$$

$$\text{Glucose} \xleftarrow{\beta-\text{Glucosidase}} \text{Cellobiose} \qquad (2)$$

C_1 was designated in that way because it was believed to act first on the crystalline cellulose to modify it in such a way that other enzymes could act on it further. The C_x enzymes represent a complex of endoglucanases which randomly hydrolyze noncrystalline cellulose, soluble cellulose derivatives, and β-1,4-oligomers of glucose. β-Glucosidase converts cellobiose and cellotriose to glucose.

Since this early speculation, the separation of the enzymes of the cellulase complex has been accomplished by many workers using fractionations based on such techniques as adsorption chromatography on a variety of substrates, affinity chromatography, gel filtration, isoelectric focussing, and electrophoresis. When a fungal cellulase was separated into C_1, C_x, and β-glucosidase components, none of the individual enzymes

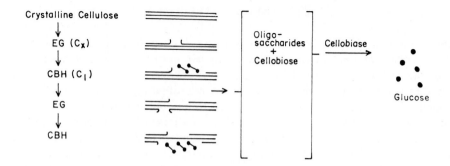

FIGURE 4.   Current concept of enzymatic hydrolysis of cellulose to glucose. CBH = cellobiohydrolase and EG = endoglucanase. (From Montenecourt, B. S. and Eveleigh, D. E., *Proc. 2nd Annu. Fuels From Biomass Symp.*, Rensselaer Polytechnic Institute, Troy, N.Y., 1978.)

could solubilize cotton fiber to any significant extent, yet when recombined in their original proportions, all the cellulase (cotton solubilizing) activity of the original culture filtrate was recovered.[41]

Although the original concept of C_1 was purely as a nonhydrolytic chain separating enzyme, the present consensus is that C_1 is an exoglucanase and in some cases a cellobiohydrolase as well. In this revised role for C_1 the process of solubilization is initiated by randomly acting C_x enzymes which generate new chain ends by splitting the cellulose chains. The new chain ends are then hydrolyzed by the endwise acting exoglucanase C_1. Solubilization occurs by the sequential action of endo- and exo-enzymes, reversing the roles of C_1 and C_x enzymes first suggested by Reese.[40]

No matter which enzyme initiates the attack on cellulose, C_1 or C_x, both must be present for extensive hydrolysis of highly ordered cellulose to occur, and the mechanism is clearly a synergistic one. Indeed Leatherwood[42] has hypothesized that an affinity factor C_1 and a hydrolytic factor C_x combine to form an active cellulase complex. Wood and McCrae[41] suggest that highly ordered cellulose is hydrolyzed only by the rapid sequential action of two enzymes, C_1 and C_x, which have formed a loose complex on the surface of the cellulose.

Whatever the specific order and mode of their activity, the cellulase complex contains the components enumerated below. Figure 4 shows the favored current concept of enzymatic hydrolysis.[43]

1. C_1 is an enzyme whose action is still open to question, with some still considering it a special enzyme with the peculiar ability to act on highly crystalline substrates, while the majority now consider C_1 an exoglucanase and cellobiohydrolase which acts only after endo-enzymes have acted.
2. C_x is a group of endo-β-1,4-glucanases which randomly cleave the cellulose chain preferentially at internal linkages rather than at the terminal linkages which are favored by the exoglucanases. As many as six of these endoglucanases have been isolated from a single organism.[41]
3. β-Glucosidase, more specifically cellobiase, which acts on the β-dimer of glucose, cellobiose. β-Glucosidases and exo-β-1,4-glucanases act in common on glucose oligomers from cellobiose to cellohexaose, with the small oligomers being most rapidly hydrolyzed by the β-glucosidase and the larger ones by the exo-glucanases.

The kinetics of enzymatic hydrolysis involving as it does the complex system of cellulase components is understandably complex also. An important factor is the degree of order or crystallinity of the substrate. Another is its lignin content. The more accessible regions are preferentially hydrolyzed at a faster rate than the highly ordered regions. Whether or not the cellulase enzymes have any access to the cellulose at all, and the extent of that access, can determine not only the rate of hydrolysis but in the case of highly lignified cellulose can completely prevent hydrolysis. The rate of hydrolysis decreases with increases in degree of crystallinity and lignification.

Those microorganisms such as wood-destroying fungi capable of hydrolyzing lignified cellulose have even more complex enzyme systems than that ascribed to cellulase above. Eriksson[44] has reported the presence of five endo-β-1,4-glucanases, one exo-β,1-4-glucanase, two β-1,4-glucosidases, one oxidoreductase and one oxidase in a white-rot fungus. Presumably the latter two enzymes are involved in removal of the lignin which prevents access of the cellulase to the cellulose, but as yet the exact sequence of reactions that result in the biological degradation of lignin through solubilization and/or depolymerization is not known. Nor have the enzymes involved been completely identified.[45] Recent studies indicate that oxidation of lignin side chains and oxidative cleavage of the aromatic nuclei in the polymer (see Chapter 2, Figure 4) are involved.[46]

Mechanisms of enzyme production and action in cellulose hydrolysis which operate effectively in nature may not be acceptable for economical large scale hydrolysis in a commercial reactor. The objectives are fundamentally different. The organism evolved producing low concentrations of glucose sufficient to sustain its growth over prolonged periods. In a commercial plant it is necessary to produce high concentrations of glucose as rapidly as possible.

Natural feedback mechanisms which must be modified in commercial practice include induction of cellulase production, catabolite repression of cellulase production, and catabolite inhibition of cellulose hydrolysis. Mutant strains of organisms, selected for rapid production of high concentrations of cellulase and insensitive to catabolite repression, will be of major importance in future progress. Furthermore, the ratio of enzymes produced by an organism may not be optimum for glucose production. For example, the cellobiose produced during cellulose hydrolysis inhibits further enzymatic hydrolysis, especially on crystalline cellulose. The addition of supplemental β-glucosidase can reduce the cellobiose concentration and increase hydrolysis.[47]

B. Production of Cellulases

The future application of enzymatic hydrolysis of cellulose to glucose on a commercial scale will depend to a great extent on the availability of high activity cellulase at low cost. Considerable effort has been and is being expended on the improvement of the technology of cellulase production.[48-52]

Although cellulases are produced by insects, molluscs, protozoa, bacteria, and thousands of fungi, only the organisms in the last category seem to be suitable for large scale production of cellulase. Insects and molluscs grow too slowly and isolation of their cellulase is scarcely feasible. Protozoa and bacteria are difficult to grow, and the cellulase must be extracted from association with the cells. However, the cellulolytic fungi grow rapidly on simple media and secrete their cellulases into the medium so the enzymes can be separated easily by filtration.

Of the thousands of fungi which have been studied for their ability to degrade cellulose and the hundreds which have been found to produce cell-free enzymes that can be used to hydrolyze cellulose, one species, *Trichoderma viride*, stands out for the ability of its extracellular enzymes to hydrolyze crystalline cellulose and for the storage stability of these enzyme preparations.

While laboratory quantities of cellulase can be produced in shake flasks, for production of enzymes in quantity other techniques are required. In the Koji process used in Japan for production of *T. viride* cellulase, wheat bran is piled in trays and inoculated after steaming. A water mist spray and forced air circulation during growth are accomplished by frequent turning to ensure good aeration. The enzyme is precipitated from the extract after filtration. Submerged fermentation in a 400-l production fermenter at the U.S. Army Natick Development Center[51,52] has proven to be simpler than the semisolid fermentation used in Japan. A pilot plant scale enzyme production unit was also operated by Gulf Oil Chemical Company near Pittsburg, Kan.[53]

In the Natick development work a two-vessel system is used for enzyme production. A 30-l fermenter is used to produce an 8% inoculum to a 400-l production fermenter after maximum respiration activity is observed between 20 and 25 hr. Fermentation time for optimum enzyme production using cellulose as substrate is about 100 hr at a temperature of 29 to 34°C, pH 3.25 minimum, and dissolved oxygen content at least 15% of saturation. After fermentation, the cultures are harvested by filtration of the mycelia from the enzyme rich broth.

Higher levels of cellulose (5%) give enhanced yields of enzymes provided the pH is maintained above 3.25. The use of NH_4OH for pH control supplies nitrogen as cellulose is consumed. Temperature profiling by increasing from 29°C to 33 to 34°C after inoculation until the desired respiration activity is reached, and then stepping down to 29°C for the remainder of the fermentation shortens the fermentation time without affecting enzyme yields.

A major contribution to enhanced enzyme productivity has been through the use of hypercellulolytic mutants of *T. viride*.[47] The original Natick strain QM6a, now along with its mutants designated *T. reesei,* as grown in shake flasks showed a cellulase productivity of 3 Filter Paper Units[54]/l/hr. Strain QM9414, a mutant produced by irradiation with a linear accelerator, showed a fourfold increase to 12 units/l/hr in shake flasks and a 10- to 15-fold increase to 30 to 46 units/l/hr in the fermentor. Enzyme concentration has been increased from 0.5 Filter Paper Units/ml for QM6a in shake flasks to 10 units/ml for QM9414 in the fermentor. A *T. reesii* mutant developed at Rutgers by ultraviolet radiation of QM6a has been designated Rut-NG-14.[43] This mutant not only produces nearly five times as much enzyme activity in shake flasks as QM9414, but also allows the attainment of a concentration of 15 units/ml in the fermentor. Another mutant, MCG77, produces 72 units/l/hr at a concentration of 5 units/ml in a shorter fermentation period.[47] This marked progress in enzyme production with a 30-fold increase in enzyme concentration and a 24-fold increase in productivity has been accomplished in less than 10 years.

The best culture broths presently attainable contain more than 2% soluble protein which is mostly cellulase. While further increases in cellulase concentration will be difficult, there are still many opportunities for further increases in cellulase productivity. For example, a second Rutgers mutant, Rut-C30, derived from Rut-NG-14 by ultraviolet treatment, is resistant to catabolite repression of cellulase production by glucose while still retaining hypercellulolytic activity.[43]

Cellulase productivity can be further increased by selection of mutants for still greater cellulase yield, mutants which produce cellulase without the requirement of the solid cellulose inducer which makes fermentor operation more complex and expensive, and mutants which produce high concentrations of cellulase at higher temperatures. A high temperature cellulase complex from *Thielavia terrestris*, which is active between 60 and 70°C, has been recently reported,[55] but its activity relative to that of *T. reesii* cellulase is still under study. Other thermophilic cellulase producing organisms have been studied by Pye.[56]

C. Saccharification of Cellulose

Although the complete hydrolysis of cellulose to glucose by enzymes is both conceptually and experimentally feasible, many obstacles still remain to its routine practice on a commercial scale.[47] The factors that influence the rate and extent of cellulose hydrolysis include the enzyme concentration, the susceptibility of the substrate to hydrolysis as determined by its degree of crystallinity and lignin content, inhibition of the cellulase components by the hydrolysis products, and inactivation of the enzymes during the reaction. Unfortunately several of these effects act in concert causing the rate of hydrolysis to fall off rapidly as the reaction proceeds.

As would be expected from the catalytic role of the enzymes in cellulose hydrolysis, the rate of sugar formation is a function of the enzyme concentration. A plot of sugar produced against the log of enzyme concentration is a straight line, with the rate of saccharification increasing as the enzyme concentration is increased.

The most important factor governing the rate and extent of hydrolysis is the susceptibility of the substrate to the enzyme systems used. Although amorphous cellulose is hydrolyzed rapidly, the crystalline portions of the cellulose are hydrolyzed at a much slower rate. The presence of lignin can completely inhibit hydrolysis by preventing access of the enzymes to the cellulose. Since most biomass is more or less lignified, this factor has important implications for the ultimate utility of enzymatic hydrolysis. The enhancement of substrate susceptibility is covered in some detail in Section IV.

Inhibition of cellulase activity by the hydrolysis products formed, cellobiose and glucose, provides a useful feedback mechanism in nature, but is undesirable when the objective is the rapid production of concentrated sugar solutions. Of the two hydrolysis products, cellobiose is a stronger inhibitor than glucose, and is especially effective in inhibiting the hydrolysis of crystalline cellulose. Glucose inhibits cellobiase, so cellobiose accumulates as the reaction proceeds. The normal abundance of cellobiase in *Trichoderma reesii* cellulase preparations is adequate in nature where sugars are metabolized by the organism as they are produced. However, for glucose production this cellobiase concentration is inadequate to withstand glucose inhibition.

It has been found that the addition of supplemental cellobiase provides a marked increase in the hydrolysis rate, even though the cellobiase itself has no effect on cellulose.[57] The selection of mutants with cellulase enzyme components resistant to end product inhibition would also enhance the hydrolysis of cellulose.[43] Another approach to avoid the inhibition effect is to ferment the glucose to ethanol in the same reactor as it is formed. In such a process[58] accumulation of reducing sugars was observed at the beginning stages of the combined hydrolysis-fermentation, but after this initial production the concentration of glucose and cellobiose fell to almost zero.

In addition to the expected increase in hydrolysis rate with enzyme concentration described above, experiments have shown that the ultimate yield of sugar is dependent on the initial enzyme concentration. This loss in enzyme activity is associated with thermal inactivation. At lower temperatures the reaction rate is lower, but the eventual conversion is greater than at higher temperatures with an initial higher reaction rate.[59] The optimum reaction temperature is estimated to be about 40°C. The enzymes which act on crystalline cellulose are most rapidly inactivated.[47]

In combination the above factors lead to a rapid decrease in hydrolysis rate with time. After initial rapid hydrolysis of amorphous cellulose, the residue becomes increasingly crystalline. Inhibition by reaction products and enzyme inactivation are both more severe for the cellulase components which act on crystalline cellulose so hydrolysis of the crystalline residue becomes even slower.

Enzyme recovery after hydrolysis would be very important for any commercial process.[59] Enzyme in solution could be recovered by adsorption on fresh cellulose, but if the enzyme was partially inactivated, fresh enzyme would have to be added as well. Enzyme adsorbed on solid hydrolysis residues can also be recycled, but its activity remains uncertain. Indications are that a large fraction of the enzyme activity is retained at lower temperatures, so some degree of recycle should be possible.

Several pilot plants have been operated for the enzymatic hydrolysis of cellulose to glucose.[47,53,59] Although sugar solutions of 30% concentration had been obtained in 15 days and 20% concentration in 2 days in early experiments in test tubes, the best pilot scale production of sugar reported to date has been a 6.5% solution in 4 hr for 26% yield or a 10% solution in 25 hr for 40% yield using enzymes from Natick strain QM9414 at 45°C in a 100-ℓ reactor.[60] Higher conversions, but usually at lower concentrations, have been obtained under longer hydrolysis times using a variety of substrates and recycling the solids.[47,59] The Gulf Oil Chemical Company pilot plant is reported to combine saccharification of cellulose and fermentation to glucose simultaneously in the same reactor.[53]

D. Appraisal

Although enzymatic hydrolysis of cellulose can provide 100% yield of glucose, the reaction is much slower than acid hydrolysis, requiring days rather than hours or minutes for completion. The reactor capacity must be much greater than in acid hydrolysis for a given production rate, but on the other hand the severe corrosion problems and expensive corrosion-resistant materials associated with acid hydrolysis are avoided. However, recycle of the more expensive enzyme catalyst is more critical than in acid hydrolysis.

Enzymatic hydrolysis is severely restricted by substrate susceptibility and can be completely prevented by the presence of lignin. Inasmuch as most of the biomass is highly lignified, it is mandatory that an inexpensive pretreatment to improve enzyme accessibility be developed in order for enzymatic hydrolysis to reach its full potential importance.

IV. REMOVAL OF PHYSICAL BARRIERS TO CELLULOSE HYDROLYSIS

The foregoing discussions have amply identified two physical barriers to facile cellulose hydrolysis. Cellulose crystallinity impedes hydrolysis by either dilute acid or enzymatic hydrolysis. And the presence of lignin can completely prevent enzymatic hydrolysis. The goal of removing one or both of these barriers has led to considerable research effort which has been reviewed from both the theoretical and practical perspectives.[61-65] The various approaches taken may be conveniently grouped into two categories, chemical and physical.

A. Chemical Methods of Barrier Removal

The treatment of biomass with chemicals to enhance its digestibility (enzymatic hydrolysis) by ruminant animals dates back to the beginning of this century. This early work used alkaline swelling agents such as sodium hydroxide and ammonia to increase the nutritive value of straw. Such processes were used extensively in Europe during wartime shortages of other fodder.

Swelling of cellulose causes significant increases in the hydrolysis rate. Mercerized cellulose is hydrolyzed by acid at a 40% faster rate than the original cellulose, and the complete dissolution and regeneration of cellulose in the viscose process gives a sixfold increase in the rate of acid hydrolysis.[66] Liquid ammonia is also known to change the crystal structure of cellulose.[61]

Application of these swelling treatments to wood at lower alkali concentrations appears to have limited success in enhancing ruminant digestibility, probably in relation to the lignin content. Although 80% of the carbohydrate in 80 mesh spruce wood was converted to sugars by alternate swelling in 2N sodium hydroxide followed by digestion with *T. viride* cellulase,[67] softwoods as a group were unresponsive to treatment with 1% sodium hydroxide to enhance in vitro rumen digestibility. In contrast certain hardwoods such as basswood and aspen showed 55% digestibility after 1% NaOH treatment while others such as elm and oak showed little improvement.[68] A plot of lignin content against digestibility showed a linear relationship, with digestibility after treatment increasing as original lignin content decreased. Similarly aspen appears to be unique in its response to ammonia treatment with a digestibility of 50% compared to 2% for Sitka spruce and 10% for red oak.[69] The main consequence of the alkaline swelling treatments has been postulated to be saponification of intermolecular ester bonds, thus allowing the wood to swell to a greater extent and permitting increased enzymatic access.[61]

Since lignin is a major barrier not only to direct access of enzymes to the cellulose, but also to the extensive decrystallization of cellulose by swelling agents, it would appear that delignification should provide a ready solution to the problem of barrier removal. This is a technically successful solution as shown by the high yields obtained from chemical wood pulps in enzymatic hydrolysis,[47] and by the large-scale use of chemical wood pulps for animal feed in Scandinavia during World War II. However, the value of conventional pulps prepared for papermaking is too great for their use as fodder or raw material for commodity chemicals except under emergency situations.

Since the strength and color of delignified wood would not be of importance for a chemical raw material as they are in papermaking, much greater latitude exists for the development of a less expensive process of delignification than the conventional ones. Furthermore, even partial delignification affords enhanced accessibility of enzymes to the cellulose. Honey mesquite, which had been partially delignified from an original lignin content of 29 to 16%, underwent almost complete carbohydrate utilization by rumen fluid organisms although cellulase from *T. reesii* (Natick strain QM9123) caused only 12% weight loss.[70] Baker[71] has measured in vitro rumen digestibility as a function of delignification and shown that for hardwoods digestibility increases rapidly with delignification reaching a high plateau at 50% lignin removal. Softwoods which have an originally higher lignin content show a lag phase before a linear increase in digestibility, confirming the dependency on lignin content.

The treatment of wood with gaseous sulfur dioxide can in principle disrupt the lignin-carbohydrate association and enhance the carbohydrate digestibility without selective removal of either component. In actual treatments of wood with SO_2, over half the lignin (measured as Klason lignin) was lost from hardwoods and as much as one third of the lignin was lost from softwoods. Enzymatic hydrolysis of hardwood carbohydrates was quantitative while hydrolysis of softwood carbohydrates was about 75% of that present. The lignin was apparently depolymerized by the SO_2 treatment since no weight loss occurred.[72]

Steaming can also improve the digestibility of wood presumably by the comparable hydrolytic effect of acetic acid liberated by the cleavage of acetyl groups in the wood. This treatment is also more effective on hardwoods, with aspen showing the greatest response as might be expected from its lowest lignin content.[73] Steaming of various lignocellulosic materials such as straw and wood at high pressure followed by sudden decompression to atmospheric pressure has been recently reported to yield a high degree of cellulose digestibility.[74]

A final approach to chemical pretreatment to enhance cellulose hydrolysis is the selective dissolution of the cellulose from the lignocellulosic raw material, followed by its precipitation in amorphous, highly reactive form. Solvents which have been found to be effective in simultaneously removing the cellulose from the lignin and decrystallizing it include cadoxen (tris [ethylenediamine] cadmium hydroxide), CMCS (aqueous solution of sodium tartrate, ferric chloride, sodium hydroxide, and sodium sulfite), and concentrated sulfuric acid.[75,76]

B. Physical Methods of Barrier Removal

High energy cathode rays have been shown to depolymerize cellulose at irradiation levels in excess of 10^6 equivalent roentgens (R), and at a dosage of 5×10^8R cellulose is converted to water-soluble materials.[77] The higher dosages caused considerable carbohydrate decomposition as well. The rate of dilute acid hydrolysis of the irradiated cellulose showed an increase in the amount of easily hydrolyzed material as well as an increase in the hydrolysis rate of the crystalline portion. Maximum glucose yield from wood pulp was over 70%. Rates of hydrolysis at 10^8R were 10 to 17 times greater and overall sugar yields two to three times greater than for untreated controls.

These dosage rates of approximately 100 Mrad are too expensive for commercial application. Treatment of waste newspaper with high energy ionizing radiation from an electron beam accelerator at lower dosage rates (5 to 50 Mrad) was found to enhance the rate of subsequent hydrolysis with dilute sulfuric acid.[78] At the most cost effective dosage of 10 Mrad, a maximum yield of glucose of about 50% was obtained in 15 min at 230°C and 0.87% acid compared to a maximum yield for unirradiated pulp of about 35% in 25 min at 2.25% acid.

Reduction of the particle size of wood has a beneficial effect on enzymatic hydrolysis presumably by breaking down the lignin sheath around the carbohydrates and exposing carbohydrate substrate to the enzymes. Thus 75% of the carbohydrates of birch and 53% of the carbohydrates of pine can be utilized by thermophyllic bacteria if the wood is reduced below 110 mesh in size.[79] Vibratory ball milling is even more effective, since it not only exposes carbohydrate surfaces to the enzymes but decreases crystallinity as well. The carbohydrates of milled spruce and aspen were essentially completely hydrolyzed by cellulase compared to less than 10% digestion for the initial 60 mesh sawdust.[67] Vibratory ball milled red oak (240 min) showed total accessibility of its carbohydrates (93% convertible to sugar) compared to an initial value of 6% for the oak sawdust. The rate of dilute acid hydrolysis was increased fivefold and the maximum yield of sugar obtainable under batch hydrolysis was increased 60 to 140% by ball milling.[80]

Another milling technique effective in increasing the susceptibility of cellulosic material to enzymatic hydrolysis is differential speed two roll milling.[81] Processing times of 10 min or less gave the following percent increases in susceptibility over untreated controls: cotton 1100, maple chips 1600, white pine chips 600, newspaper 125. It is believed that both decrystallization and depolymerization of the cellulose occur during the milling by analogy with experience with rubber.

Other physical treatments of cellulosic materials to increase their reactivity include exposure to both high and low temperatures and to high pressure. Heating at 200°C for prolonged periods (32 hr) provided a 35% increase in dilute acid hydrolysis rate and 27% increase in sugar yield.[82] Freezing of cellulosic materials in aqueous suspension at −75°C reduced the degree of polymerization and increased reactivity.[83] Compression of cotton hydrocellulose at 8000 kg/cm² doubled the amount dissolved during ethanolysis.[84]

C. Appraisal

Access of enzymes to the carbohydrate of lignocellulose can be provided by chemical means including partial or complete delignification, lignin depolymerization, and selective dissolution of the cellulose, or by physical means such as disruption of the lignin matrix in which the cellulose is embedded. Reduction of the cellulose crystallinity, which increases the hydrolysis rate with both acids and enzymes, often accompanies the lignin disruption or removal, but can be effected independently as by irradiation. Some pretreatments are effective only on selected susceptible substrate materials and all can benefit from further development to increase cost effectiveness.

V. OTHER CELLULOSE CONVERSION PROCESSES

While hydrolysis to glucose is the most obvious and most investigated cellulose conversion scheme, other processes are also capable of providing useful low molecular weight products from cellulose. A major product of cellulose pyrolysis is 1,6-anhydro-β-D-glucopyranose (levoglucosan) whose structure is shown in Chapter 5, Figure 3. Pyrolysis under vacuum gives yields of levoglucosan of up to 30% at 300°C under nitrogen.[85] Dry distillation of cellulose at 400 to 500°C gives about 80% of a tar that contains mainly levoglucosan. This may be converted to glucose by mild acid hydrolysis in 50% yield based on the original cellulose.[86]

Direct thermal decomposition of prehydrolyzed hardwoods under reduced pressure gives a yield of levoglucosan of 10% of the original wood.[9] The reactive anhydride ring and three hydroxyl groups of levoglucosan have led to its use as a monomer for the synthesis of a variety of synthetic polysaccharides, polyurethanes, polyesters, and polyepoxy resins.[87]

A study has been initiated to assess the role of catalysis in the thermochemical conversion of biomass to fuels and chemicals.[88] Many of the advances in catalysis which have been made in recent years are likely to have a significant effect on the chemical structures of cellulose, and could provide other useful chemicals by direct catalytic conversion of cellulose. Oxidative hydrolysis of lignocellulose with air yields 30% glucose as well as a mixture of organic acids.[89]

VI. CHEMICALS FROM GLUCOSE

The opportunities for further conversion of glucose to useful organic chemicals are multiple and varied. The examples cited below are the most obvious, but chemical ingenuity will undoubtedly develop many new schemes for the chemical utilization of glucose if it should become an inexpensive intermediate in biomass conversion.

A. Ethanol and Derivatives

The fermentation of sugars to ethanol is one of man's most ancient arts. As elaborated in detail in Chapter 3, commercially proven techniques can provide yields of 85 to 95% of the theoretical along with by-product yeast and carbon dioxide.

$$C_6H_{12}O_6 \xrightarrow{\text{Fermentation}} 2C_2H_5OH + 2CO_2 + \text{Yeast} \qquad (3)$$
$$\text{Glucose} \qquad\qquad \text{Ethanol} \quad \text{Carbon Dioxide}$$

Ethanol is in its own right an important industrial chemical with about 1.5 billion lb being produced synthetically each year in the U.S. by hydration of ethylene. The reverse reaction, the dehydration of ethanol to ethylene, while uneconomic in competition with cheap ethylene from petroleum, proceeds in 96% yield and was practical commercially many years ago. Similarly, butadiene can be readily obtained from ethanol in 70% yield by processes which have been proven commercially.[90]

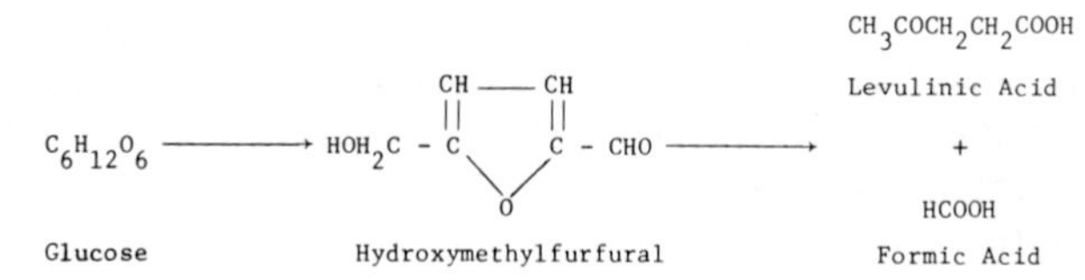

$$CH_3COCH_2CH_2COOH$$
$$\text{Levulinic Acid}$$

$$C_6H_{12}O_6 \longrightarrow HOH_2C - C \overset{CH \longrightarrow CH}{\underset{O}{\big|\big| \quad \big|\big|}} C - CHO \longrightarrow +$$

$$HCOOH$$

Glucose   Hydroxymethylfurfural   Formic Acid

FIGURE 5.   The transformation of glucose in hot mineral acids.

$$\begin{array}{lll} C_2H_5OH \longrightarrow & CH_2 = CH_2 & + \; H_2O \\ \text{Ethanol} & \text{Ethylene} & \text{Water} \end{array} \quad (4)$$

$$\begin{array}{lll} 2C_2H_5OH \longrightarrow & CH_2 = CH-CH = CH_2 + H_2O \\ \text{Ethanol} & \text{Butadiene} \qquad\qquad \text{Water} \end{array} \quad (5)$$

The conversion of glucose via ethanol to ethylene and butadiene represents the largest volume potential utilization of cellulose for chemicals because of the importance of ethylene as an intermediate for petrochemicals and polymers and of butadiene in the production of synthetic rubber. In 1977 the U.S. produced 25 billion lb of ethylene (the most important organic chemical of commerce) and 3 billion lb of butadiene.[91] Other potentially major derivatives of ethanol include acetaldehyde, acetic acid (2.6 billion lb), and acrylonitrile (1.6 billion lb).

Ethylene is chiefly used for polyethylene (10 billion lb in 1977) and with benzene in the production of styrene (almost 2 billion lb of ethylene). Other important uses are for the production of ethylene oxide (4.4 billion lb), ethylene glycol (3.5 billion lb), and vinyl chloride (5.8 billion lb). Ethylene is such a versatile intermediate that many chemicals and polymers now made from propylene could alternatively be made from ethylene if ethylene from biomass were to become cheaper than propylene from fossil hydrocarbons.

B. Lactic Acid Derivatives

Another route for glucose conversion to chemicals would be by direct fermentation to lactic acid in 85 to 95% yield instead of to ethanol.[6] By dehydration the lactic acid can be converted in one step to acrylic acid.[92]

$$\begin{array}{llll} C_6H_{12}O_6 \longrightarrow & 2CH_3CHOHCOOH \longrightarrow & CH_2 = CHCOOH & + \; CH_3CHO \\ \text{Glucose} & \text{Lactic Acid} & \text{Acrylic Acid} & \text{Acetaldehyde} \end{array} \quad (6)$$

C. Other Fermentation Products

Other organic chemicals readily obtainable from glucose through fermentation[6,14,93] include butanol, isopropanol, 2,3-butanediol, glycerol, acetone, acetic acid, and butyric acid.

D. Levulinic Acid

One of the principal side reactions in the high temperature dilute acid hydrolysis of cellulose is the further conversion of glucose under the hydrolysis conditions to first hydroxymethylfurfural and then levulinic acid as shown in Figure 5. Hydroxymethylfurfural is a very reactive compound which could find application as an intermediate in the production of a large variety of chemical products.[94] Prominant among these is levulinic acid, which is stable under the conditions of its formation.[95] This material, because of its polyfunctionality, could well serve as an intermediate for a new family of polymers. Levulinic acid was produced commercially for a while for processing into diphenolic acid, but the loss of the diphenolic acid outlet caused the venture to be abandoned.

E. Hydrogenation Products

The hydrogenation of glucose to sorbitol for use as a humectant and as an intermediate in the syntheses of ascorbic acid and surfactants has been practiced commercially for decades. When higher temperatures are used in glucose hydrogenation, the principal product is 1,2 propanediol with smaller amounts of glycerol and ethylene glycol. However, under these same conditions sorbitol is hydrogenated to glycerol in 40% yield with lesser amounts of propylene glycol, ethylene glycol, and erythritol being formed as well.[96]

REFERENCES

1. Braconnet, H., *Ann. Chem. Phys.*, 12(2), 172, 1819.
2. Hall, J. A., Saeman, J. F., and Harris, J. F., Wood saccharification: a summary statement, *Unasylva*, 10(1), 7, 1956.
3. Harris, J. F., Acid hydrolysis and dehydration reactions for utilizing plant carbohydrates, *Appl. Polym. Symp.*, 28, 131, 1975.
4. Saeman, J. F., Kinetics of wood saccharification, *Ind. Eng. Chem.*, 37, 43, 1945.
5. Grethlein, H. E., Acid hydrolysis of cellulosic biomass, *Proc. 2nd Annu. Fuels From Biomass Symp.*, Rensselaer Polytechnic Institute, Troy, N.Y., 1978.
6. Stamm, A. J. and Harris, E. E., *Chemical Processing of Wood*, Chemical Publishing, New York, 1953, chap. 16.
7. Lloyd, R. A. and Harris, J. F., Wood hydrolysis for sugar production, Report No. 2029, U.S. Department of Agriculture, Forest Products Laboratory, Madison, Wis., reprinted 1963.
8. Eickemeyer, R. and Hennecke, H., New technical and economic possibilities for wood saccharification, *Holz-Zentralbl.*, 86, 1374, 1967.
9. Karlivan, V. P., New aspects of the production of chemicals from biomass, presented at World Conference on Future Sources of Organic Raw Materials, Toronto, July 10—13, 1978.
10. Ant-Wuorinen, O., New method of wood saccharification with sulfurous acid, *Sven. Papperstidn.*, 45, 149, 1942.
11. Cedarquist, K. N., Some remarks on wood hydrolyzation, presented at Seminar on The Production and Use of Power Alcohol in Asia and the Far East, Lucknow, India, October 23—November 6, 1952.
12. Grethlein, H. E., A comparison of the economics of acid and enzymatic hydrolysis of newsprint, *Biotechnol. Bioeng.*, 20, 503, 1978.
13. Brenner, F. C., Frilette, V., and Mark, H., Crystallinity of hydrocellulose, *J. Am. Chem. Soc.*, 70, 877, 1948.
14. Wenzl, H. F. J., *The Chemical Technology of Wood*, Academic Press, New York, 1970, chap. 4.
15. Locke, E. G. and Garnum, E., Working party on wood hydrolysis, *For. Prod. J.*, 11, 380, 1961.
16. Bechamp, A., *Ann. Chim.*, 48, 458, 1856.
17. Petkevich, A. A., Ochneva, N. V., Korotkov, N. V., Revzina, E. D., Chalov, N. V., Leshchuk, A. E., and Goryachikh, E. F., Hydrolysis of wood with concentrated HCl in a pilot battery of diffusers, *Zh. Tr. Gos. Nauchn. — Issled Inst. Gidrolizn i Sulfitno Spirt Prom.*, 8, 47, 1960.
18. Lebedev, N. V. and Bannikova, A. A., Hydrolysis of cellulose with concentrated HCl at different temperatures, *Zh. Tr. Gos. Nauchn. — Issled Inst. Gidrolizn i Sulfitno Spirt Prom.*, 9, 7, 1961.
19. Vyrodova, L. P. and Sharkov, V. I., Effect of the concentrated sulfuric acid ratio and the presence of sugars on the solubility of cellulose, *Zh. Tr. Gos. Nauchn. — Issled Inst. Gidrolizn i Sulfitno Spirt Prom.*, 12, 40, 1964.
20. Sakai, Y., Combination of sulfuric acid with cellulose during hydrolysis with a small amount of concentrated sulfuric acid, *Bull. Chem. Soc. Jpn.*, 38(6), 863, 1965.

21. Chalov, N. V., Leshchuk, A. E., Kozlova, L. V., and Volkova, T. M., Characteristics of the hydrolysis of polysaccharides with 65—90% sulfuric acid at the equilibrium stage, *Gidroliz. Lesokhim. Prom.,* 19(2), 2, 1966.
22. Berlin, H., Identify of isomaltose with gentiobiose, *J. Am. Chem. Soc.,* 48, 1107, 1926.
23. Ekenstam, A., The behavior of cellulose in mineral acid solutions, *Chem. Ber.,* 69, 549, 1936.
24. Hiller, L. A. and Pacsu, E., Cellulose studies. V. Reducing end-group estimation, *Textile Res. J.,* 16, 318, 1946.
25. Dauziville, E. S., German Patent 11, 836, 1880.
26. Goldschmidt, T. A. G. and Hagglund, E., German Patent 391, 969, 1917.
27. Bergius, F., Conversion of wood to carbohydrates and problems in the industrial use of concentrated hydrochloric acid, *Ind. Eng. Chem.,* 29, 247, 1937.
28. Vernet, G., "Prodor" process for the manufacture of cellulose alcohol, *Chimie Industrie Spec.,* No. 654, May 1923.
29. Desforges, J., Hydrolysis of wood, *Chim. Ind. (Paris),* 67, 753, 1952.
30. Kusama, J. and Ishii, T., Wood saccharification by gaseous HCl, *Kogyo Kagaku Zasshi,* 69(3), 469, 1966.
31. Kusama, J., Saccharification of cellulose-containing material, U.S. Patent 3,067,065, 1962.
32. Chalov, N. V., Hydrolysis of vegetable raw material, U.S.S.R. Patent 119,491, 1959.
33. Chalov, N. V. and Leshchuk, A. E., Continuous hydrolysis of wood with 46-48% HCl, *Izv. Vyssh. Uchebn. Zavad. Lesn. Zh.,* 5(1), 155, 1962.
34. Oshima, M., *Wood Chemistry Process Engineering Aspects,* Noyes Development, New York, 1965.
35. Reese, E. T., History of the cellulase program at the U.S. Army Natick Development Center, *Biotechnol. Bioeng.,* Symp. No. 6, 9, 1976.
36. Hajny, G. J. and Reese, E. T., Eds., *Cellulases and Their Applications,* Advances in Chemistry Series 95, American Chemical Society, Washington, D.C., 1969.
37. Wilke, C. R., Ed., *Cellulose as a Chemical and Energy Resource,* John Wiley & Sons, New York, 1975.
38. Gaden, E. L., Jr., Mandels, M. H., Reese, E. T., and Spano, L. A., Eds., *Enzymatic Conversion of Cellulosic Materials: Technology and Applications,* John Wiley & Sons, New York, 1976.
39. Ghose, T. K., Ed., *Bioconversion of Cellulosic Substances into Energy, Chemicals and Microbial Protein,* Indian Institute of Technology, New Delhi, 1978.
40. Reese, E. T., Siu, R. G. H., and Levinson, H. S., Biological degradation of cellulose derivatives and its relation to the mechanism of cellulose hydrolysis, *J. Bacteriol.,* 59, 485, 1950.
41. Wood, T. M. and McCrae, S. I., The mechanism of cellulase action with particular reference to the C_1 component, in *Bioconversion of Cellulosic Substances Into Energy, Chemicals and Microbial Protein,* Ghose, T. K., Ed., Indian Institute of Technology, New Delhi, 1978, chap. 3.
42. Leatherwood, J. M., Cellulase complex of *Ruminoccus* and a new mechanism for cellulose degradation, in *Cellulases and Their Applications,* Hajny, G. J. and Reese, E. T., Eds., Advances in Chemistry Series 95, American Chemical Society, Washington, D. C., 1969, 53.
43. Montenecourt, B. S. and Eveleigh, D. E., Hypercellulolytic mutants and their role in saccharification, *Proc. 2nd Annu. Fuels From Biomass Symp.,* Rensselaer Polytechnic Institute, Troy, N.Y., 1978.
44. Eriksson, K. E., Enzyme mechanisms involved in cellulose degradation by the white-rot fungus, *Sporotrichum pulverulentum,* presented before Cellulose, Paper and Textile Division, American Chemical Society, Appleton, Wisc., 1978.
45. Kirk, T. K., Lignin-degrading enzyme system, *Biotechnol. Bioeng.,* Symp. No. 5, 139, 1975.
46. Kirk, T. K. and Chang, H-m., Characterization of heavily degraded lignins from decayed spruce, *Holzforschung,* 29, 56, 1975.
47. Mandels, M., Dorval, S., and Medeiros, J., Saccharification of cellulose with *Trichoderma* cellulase, *Proc. 2nd Annu. Fuels From Biomass Symp.,* Rensselaer Polytechnic Institute, Troy, N.Y., 1978.
48. Mandels, M. and Weber, J., The production of cellulases, in *Cellulases and Their Applications,* Hajny, G. J. and Reese, E. T., Eds., Advances in Chemistry Series 95, American Chemical Society, Washington, D.C., 1969, 391.
49. Mandels, M., Microbial sources of cellulases, *Biotechnol. Bioeng.,* Symp. No. 5, 81, 1975.
50. Sternberg, D., Production of cellulase by *Trichoderma, Biotechnol. Bioeng.,* Symp. No. 6, 35, 1976.
51. Nystrom, J. M. and Allen, A. L., Pilot scale investigations and economics of cellulase production, *Biotechnol. Bioeng.,* Symp. No. 6, 55, 1976.
52. Nystrom, J. M. and Diluca, P. H., Enhanced production of *Trichoderma* cellulase on high levels of cellulose in submerged culture, in *Bioconversion of Cellulosic Substances into Energy, Chemicals and Microbial Protein,* Ghose, T. K., Ed., Indian Institute of Technology, New Delhi, 1978, 293.
53. Mooney, J. R., The cellulose project, *The Orange Disc,* Vol. 12, 12th ed., Gulf Building, Pittsburgh, Pa., 1977, 2.

54. Mandels, M., Andreotti, R., and Roche, C., Measurement of saccharifying cellulase, *Biotechnol. Bioeng.*, Symp. No. 6, 21, 1976.
55. Skinner, W. A. and Tokuyama, F., Production of cellulase by a thermophilic *Thielavia terrestris*, U.S. Patent 4,081,328, 1978.
56. Pye, E. K., Thermophilic degradation of cellulose for production of liquid fuels, *Proc. 2nd Annu. Fuels From Biomass Symp.*, Rensselaer Polytechnic Institute, Troy, N.Y., 1978.
57. Sternberg, D., Vijayakumar, P., and Reese, E. T., β-Glucosidase: microbial production and effect on enzymatic hydrolysis of cellulose, *Can. J. Microbiol.*, 23, 139, 1977.
58. Takagi, M., Abe, S., Suzuki, S., Emert, G. H., and Yata, N., A method for production of alcohol directly from cellulose using cellulase and yeast, in *Bioconversion of Cellulosic Substances Into Energy, Chemicals and Microbial Protein*, Ghose, T. K., Ed., Indian Institute of Technology, New Delhi, 1978, 551.
59. Andren, R. K., Aspects of enzymatic saccharification of cellulose at the pilot plant level, in *Bioconversion of Cellulosic Substances Into Energy, Chemicals and Microbial Protein*, Ghose, T. K., Ed., Indian Institute of Technology, New Delhi, 1978, 397.
60. Blodgett, C., personal communication to M. Mandels cited in Reference 47.
61. Tarkow, H. and Feist, W. C., A mechanism for improving the digestibility of lignocellulosic materials with dilute alkali and liquid ammonia, in *Cellulases and Their Applications*, Hajny, G. J. and Reese, E. T., Eds., Advances in Chemistry Series 95, American Chemical Society, Washington, D.C., 1969, 197.
62. Cowling, E. B., Physical and chemical constraints in the hydrolysis of cellulose and lignocellulosic materials, *Biotechnol. Bioeng.*, Symp. No. 5, 163, 1975.
63. Millett, M. A., Baker, A. J., and Satter, L. D., Pretreatments to enhance chemical, enzymatic and microbiological attack of cellulosic materials, *Biotechnol. Bioeng.*, Symp. No. 5, 193, 1975.
64. Millett, M. A., Baker, A. J., and Satter, L. D., Physical and chemical pretreatments for enhancing cellulose saccharification, *Biotechnol. Bioeng.*, Symp. No. 6, 125, 1976.
65. Halliwell, G., Cellulosic materials and their pre-treatment, in *Bioconversion of Cellulosic Substances Into Energy, Chemicals and Microbial Protein*, Ghose, T. K., Ed., Indian Institute of Technology, New Delhi, 1978, 81.
66. Millett, M. A., Moore, W. E., and Saeman, J. F., Preparation and properties of hydrocelluloses, *Ind. Eng. Chem.*, 46, 1493, 1954.
67. Pew, J. C. and Weyna, P., Fine grinding, enzyme digestion and the lignin-cellulose bonds in wood, *Tappi*, 45(3), 247, 1962.
68. Feist, W. C., Baker, A. J., and Tarkow, H., Alkali requirements for improving digestibility of hardwoods by rumen microorganisms, *J. Anim. Sci.*, 30(5), 832, 1970.
69. Millett, M. A., Baker, A. J., Feist, W. C., Mellenberger, W., and Satter, L. D., Modifying wood to increase its *in vitro* digestibility, *J. Anim. Sci.*, 31(4), 781, 1970.
70. Goldstein, I. S. and Villarreal, A., Chemical composition and accessibility to cellulase of mesquite wood, *Wood Sci.*, 5(1), 15, 1972.
71. Baker, A. J., Effect of lignin on the *in vitro* digestibility of wood pulp, *J. Anim. Sci.*, 36(4), 768, 1973.
72. Moore, W. E., Effland, M. J., and Millett, M. A., Hydrolysis of wood and cellulose with cellulytic enzymes, *J. Agric. Food Chem.*, 20(6), 1173, 1972.
73. Bender, F., Heaney, D. P., and Bowden, A., Potential of steamed wood as a feed for ruminants, *For. Prod. J.*, 20(4), 36, 1970.
74. Iotech Corporation, Ontario, poster session at World Conference on Future Sources of Organic Raw Materials, Toronto, 1978.
75. Ladisch, M. R., Ladisch, C. M., and Tsao, G. T., Cellulose to sugars: new path gives quantitative yield, *Science*, 201, 743, 1978.
76. Tsao, G. T., Ladisch, M., Ladisch, C., Hsu, T. A., Dale, B., and Chou, T., Fermentation substrates from cellulosic materials, part 1, Production of fermentable sugars from cellulosic materials, in *Annual Reports in Fermentation Processes*, Vol. 2, Perlman, D., Ed., Academic Press, New York, 1978.
77. Saeman, J. F., Millett, M. A., and Lawton, E. J., Effect of high-energy cathode rays on cellulose, *Ind. Eng. Chem.*, 44, 2848, 1952.
78. Brenner, W., Rugg, B., Arnon, J., Kumar, R., and Rogers, C., New approaches for the acid hydrolysis of cellulose, Report NYU/DAS-77-30, New York University, New York, 1977.
79. Virtanen, A. I. and Nikkila, O. E., Cellulose fermentation in wood dust, *Suom. Kemistil. B*, 19, 3, 1946.
80. Millett, M. A., Effland, M. J., and Caulfield, D. F., Influence of fine grinding on the hydrolysis of cellulosic material — acid vs. enzymatic, presented before Cellulose, Paper and Textile Division, American Chemical Society, Appleton, Wisc., 1978.
81. Tassinari, T. and Macy, C., Differential speed two roll mill pretreatment of cellulosic materials for enzymatic hydrolysis, *Biotech. Bioeng.*, 19, 1321, 1977.

82. **Millett, M. A. and Goedken, Y. L.,** Modification of cellulose fine structure — effect of thermal and electron irradiation pretreatments, *Tappi,* 48(6), 367, 1965.
83. **Bykov, A. N. and Frolov, S. S.,** *Khim. Volokna,* 2(3), 33, 1961.
84. **Sharkov, V. I. and Levanova, V. P.,** Study of "amorphous" cellulose, *Zh. Prikl. Khim. (Lenin.)* 33(11), 2563, 1960.
85. **Shafizadeh, F. and Chin, P. P. S.,** Thermal deterioration of wood, in *Wood Technology: Chemical Aspects,* Goldstein, I. S., Ed., ACS Symposium Series No. 43, American Chemical Society, Washington, D.C., 1977, 57.
86. **Shafizadeh, F.,** Development of pyrolysis as a new method to meet the increasing demands for food, chemicals and fuel, presented at Eighth World Forestry Congress, Jakarta, Indonesia, 1978.
87. **Pernikis, R. Y.,** *Oligomers and Polymers From Sugar Anhydrides,* Institute of Wood Chemistry, Riga, U.S.S.R., 1976.
88. **Garten, R. L.,** Catalytic conversion of biomass to fuels, *Proc. 2nd Annu. Fuels From Biomass Symp.,* Rensselaer Polytechnic Institute, Troy, N.Y., 1978.
89. **Schaleger, L. L. and Brink, D. L.,** Chemical production by oxidative hydrolysis of lignocellulose, *Tappi,* 61(4), 65, 1978.
90. **Faith, W. L., Keyes, D. B., and Clark, R. L.,** *Industrial Chemicals,* 3rd ed., John Wiley & Sons, New York, 1965.
91. *Chem. Eng. News,* 56(24), 48, 1978.
92. **Phillips, B.,** Renewable resources in chemical perspective, presented at 175th national meeting, American Chemical Society, Washington, D.C., 1978.
93. **Hajny, G. J.,** Microbiological utilization of wood sugars, *For. Prod. J.,* 9(5), 153, 1959.
94. **Harris, J. F., Saeman, J. F., and Zoch, L. L.,** Preparation and properties of hydroxymethylfurfural, *For. Prod. J.,* 10(2), 125, 1960.
95. **McKibbins, S. W., Harris, J. F., and Saeman, J. F.,** Kinetics of the acid catalyzed conversion of glucose to 5-hydroxymethyl-2-furaldehyde and levulinic acid, *For. Prod. J.,* 12(1), 17, 1962.
96. **Clark, I. T.,** Hydrogenolysis of sorbitol, *Ind. Eng. Chem.,* 50, 1125, 1958.

Chapter 7

CHEMICALS FROM HEMICELLULOSES

Norman Storm Thompson

TABLE OF CONTENTS

I. INTRODUCTION

The organic components of vegetable materials constitute a large reservoir of raw materials available for limited exploitation. According to Burwell,[1] the annual biomass production for food, lumber, and paper in the U.S. would furnish 25% of the present energy requirements. For the most part, the collection of all this material is difficult and expensive and is most often considered to be a scavanging operation which uses raw materials having no value for conventional end uses.[2] Those that do arrive at a mill, are frequently burned. The lignin and extractives components of this waste have the greatest heat content. It is estimated that the calorific content of lignin is 6.0 kcal/g, whereas that of the polysaccharides is only 4.2 kcal/g.[3] Because of their greater abundance, however, polysaccharides contribute 60 to 70% of the heat value of the vegetable material.

Sjöström[4] claims the heating value of the carbohydrate residues in the spent kraft liquor amounts to only 25% of that contributed by the dissolved lignin. This energy loss reflects not only the loss in heat value due to the smaller quantities of carbohydrate in the liquor, but also a loss in the heat of combustion of the alkali-degraded carbohydrate residues.

The carbohydrate residues from the sulfite process and from many high yield pulping processes are of greater potential for the recovery of raw materials for other end uses than those from the kraft process. Most of these residues are not competitive for use as a raw material for basic chemical products. They can be used for the production of specialized chemical products by taking advantage of the unique physical molecular structures of the various organic components.[5] The production of furfural from pentosans is an exception to this generalization.

The polysaccharides comprising the hemicelluloses vary according to the definition employed. In this review, only industrially important noncellulosic polysaccharide components of normal fibrous tissues of plants will be discussed. This includes arabinogalactans, but does not include pectins derived from fruit. The study does not include the carbohydrates found in abnormal wood, storage cells (such as starches and amyloids), commercial gums and mucilages (such as galactomannans, gum arabic, etc.), nor the polysaccharides of bacterial or synthetic origin. The polysaccharides that will be considered here will be those ordinarily encountered in common agricultural and pulping wastes.

II. TYPES OF HEMICELLULOSES

The high molecular weight components of typical gymnosperms and angiosperms are given in Table 1. The major component of these plants is cellulose, which serves as the principal building substance of the cell wall. All contain lignin, which acts as a binder and stiffening agent. The angiosperms contain xylan as the principal hemicellulose component as well as lesser quantities of glucomannan and other miscellaneous polysaccharides.[7,9,10]

Glucomannan is the major hemicellulose component of most conifers. Notable exceptions are incense cedar where xylan predominates[11] and some species of larches where an arabinogalactan predominates in certain sections of the tree.[12] Frequently, unusual hemicellulose-like substances are found in tension and compression wood, in plants that are infected with tumors, as well as in growing tissues. These components will not be considered since they do not yet have any industrial significance.

Numerous unique chemical substances have been isolated from hemicelluloses and their derivatives, but almost all are laboratory curiosities and have been prepared as a

Table 1
ESTIMATED QUANTITIES OF HIGH MOLECULAR
WEIGHT COMPONENTS OF TYPICAL ANGIOSPERMS
AND GYMNOSPERMS[a]

	Angiosperm			Gymnosperm
Component	Bamboo	Wheat straw	Hardwood	Softwood
Lignin	23	21	24	29
Cellulose	48	50	48	43
Glucomannan	Trace	Trace	3	17
Xylan	28	27	22	8
Miscellaneous polymers, pectin, arabinan, galactan, etc.	3	2	3	3

[a]   Polysaccharide contents estimated from data in References 6 through 9.

result of investigating the chemical and physical properties of the polymers. Some constitute the source of unique sugars of limited esoteric industrial value and are described by Whistler and Wolfrom.[13] In addition to being sources of unique sugars, hemicelluloses can give rise to homologous series of oligomers as a result of partial acid hydrolysis. Most common are a xylose series,[14] xylose-uronic acid series,[7,15-18] galactose-containing fragments,[19] others containing glucose and mannose,[7] and oligomers derived from pectin.[20] Derivatives of pure hemicelluloses have been prepared and their physical properties evaluated (such as common esters and ethers,[7] carboxymethyl ethers,[21,22] and special oxidation products[23-25]).

The chemicals of industrial importance derived from hemicelluloses are much fewer in number. All are derived from agricultural and wood industry wastes as well as from the waste liquors of different pulping processes. Since the availability of agricultural waste is seasonal, storage facilities must be provided, and the cost of production suffers. The use of waste from the wood industries for chemical production must compete with the increasing use of these residues as fuel as well as with the alternative chemical sources provided by agricultural residues or the petroleum technology. It cannot be predicted which end use will predominate in the future. The use of hemicellulose residues of the spent effluents from pulping operations is also a source of chemicals. This source requires complex isolation and purification procedures which raise the ultimate cost of the product. Nevertheless, continuous careful reevaluation of economic desirability is necessary.

III. XYLOSE-DERIVED CHEMICALS

A. Introduction

Xylose-containing polysaccharides, next to cellulose and lignin, may well be the most plentiful organic substances on earth. In most instances, xylose is the major constituent in these polymers. As mentioned previously, xylans are very plentiful in dicotyledonous and monocotyledonous plants where the polymer occurs principally as a variation of a 4-O-methyl-D-glucuronopyranosyl-β-D-xylan.[7] In the case of hardwoods, the polymer contains about 8 to 17% acetyl groups although the heartwood or long stored wood may have lost much through saponification. The xylan of hardwood is a family of polymers in which a slight variation in the acid to xylose ratio may occur. In the case of birch xylan, galacturonic acid units are thought to be a nonterminal compo-

X = β−D−XYLOPYRANOSYL UNIT

A = URONIC ACID

Ar = L−ARABINOFURANOSYL UNIT

OAc = ACETYL

R = IN GRASSES, R IS A SIDE CHAIN CONTAINING XYLOSE AND ARABINOSE, OR GALACTOSE XYLOSE AND ARABINOSE UNITS.

FIGURE 1.   A schematic representation of xylan.

nent.[26] The xylans from grasses and forage crops are much more complex and many act as mixtures of related groups of polymers.[27] Varying amounts of glucose, arabinose, and uronic acid units are incorporated into these polymers. Attachment of the sugar branches to the xylan chain is often by (1-3) glycosidic bonds, while uronic acids are generally linked by α-(1-2)-glycosidic bonds. Galacturonic acid units are also components of xylans of some nonarborescent plants.

The xylans from conifers are almost as complex and in pine exist as two closely related substances.[28] Unlike hardwood xylans, they do not contain acetyl groups but they do contain L-arabinofuranosyl units linked by (1-3) glycosidic bonds to the main chain of β-1-4 linked anhydroxylopyranosyl units. A schematic representation of xylans from different sources is given in Figure 1. In only a few instances (such as from esparto) can a pure xylan containing only xylose be isolated from a natural source.

The principal sources of xylose-rich raw materials capable of economic utilization are found in the wastes and residues of the angiosperms. A significant proportion of the xylan of monocotyledonous and dicotyledonous plants is easily extracted with a variety of solvents[7] and is easily accessible to attack by acids and other reagents. The less acceptable sources of xylan are coniferous woods and many types of seed hulls. In these instances, the xylose content is frequently low. Because of penetration difficulties, the raw stock cannot be processed satisfactorily unless it has been thoroughly crushed. Even after milling, very little xylan can be extracted from most coniferous woods until the lignin has been removed. The chemical composition of some of these potential raw materials is illustrated in Tables 1 and 2.

B. Xylan

Industrial procedures based on alkaline extraction have been developed to isolate xylan from agricultural wastes.[7,33-35] Efforts have been made to develop industrial uses for the xylan present in corn fiber. The hemicellulose is extracted with lime water to give a water soluble homogenous hydrocolloid with rheological properties similar to other gums used as thickeners, stabilizers, and emulsifiers.[36] This utilization has not been successful. Other chemicals can be derived from xylans which have greater industrial potential at the present time. A summary of the pure chemicals which can be derived from xylan and which have some industrial potential is given in Table 3. Details concerning the manufacture of these products will be given in succeeding sections.

C. Xylose

Koch[33] developed the earliest published procedure for preparing crystalline xylose from hardwood. His procedure involved alkaline extraction of the wood and isolation

Table 2
SOURCES OF XYLOSE-RICH RESIDUES FROM
SPENT PULPING LIQUORS

	Spent pulping liquors			Prehydrolysis liquors	
	Hardwood sulfite	Softwood sulfite	Masonex ®[a]	Gum	Pine
Xylose			24	33.4	12.6
Arabinose	16	6	4.8	2.4	5.2
Galactose			2.8	2.4	9.9
Glucose			11	5.6	8.9
Mannose	4	14	9.5	2.8	14.2
Uronic acid	12	12	N.A.[b]	N.A.[b]	N.A.[b]
Lignin	55	55	+	22.8	7.7
Reference	(29)	(29)	(30)	(31)	(32)

[a] Trade Name, a product of Masonex Corporation, Chicago, Ill.
[b] Not analyzed.

Table 3
POTENTIALLY SIGNIFICANT PURE CHEMICALS DERIVED FROM XYLAN

Product	Source	Process	Possible uses
D-xylose	Xylan component of wood and agricultural wastes; spent pulping liquors	Acid hydrolysis	Secondary products, food additives, detergents after esterification, polyurethane from methyl ether
Xylitol	Xylose	Reduction	Sweetener, humectant, plastic plasticizer
Xylonic, tartaric, trioxyglutaric acids	Xylose	Oxidation of spent sulfite liquors	Binders, sequestering agents
Furfural	Xylose, xylan of vegetable materials	Acidic dehydration	Plastics, solvents, numerous chemical products

of the intermediate xylan by alcohol precipitation. Xylose was isolated after hydrolysis of this purified intermediate. This approach has not been pursued by many inventors in the patent literature because it is not now an economically feasible route.[34] Extraction processes have proven invaluable in understanding hemicellulose properties, however,[7,35] and the approach may yet prove economically advantageous for the preparation of xylose.

Direct hydrolysis of easily accessible xylan is the more plausible route to xylose manufacture, and variations of it abound in the patent literature. Bertrand[37] first hydrolyzed oat straw with mineral acid for this purpose. Subsequently several investigations by Hudson et al.[38,39] at the National Bureau of Standards led to the development of a pilot process for the isolation of xylose from cotton seed bran.[40] The complex purification procedures needed to prepare crystalline xylose by this process set limitations on its economic feasibility and defined many of the problems to be solved. Except for the war years (1914 to 1918, 1939 to 1945), little effort was directed toward improving the technology in the West although several industrial installations were set up in the U.S.S.R. as part of their economic development programs.

The principal difficulties associated with the production of pure xylose from acceptable sources arise from the many different types of impurities in the hydrolyzates. These can include other sugars derived from miscellaneous accessible polysaccharides, a wide range of extractives, lignin, lignin-carbohydrate complexes, unsaponified sugar acetates, and various organic acids. These impurities are very difficult to remove and will impair crystallization, contaminate ion exchange resins, unbalance neutralization schemes, and poison hydrogenation catalysts. Many of them were ignored in the past for certain end uses. The recent interest in preparing pure xylitol and glucitol from these hydrolyzates has led to many schemes to overcome the difficulties of purification. The schemes vary according to the source of xylose — whether they are derived from wood or agricultural wastes or from spent pulping liquor. The following types of schemes have been proposed in the literature. One or several of these schemes can be present in a given patent description.

1.   Isolation and purification of xylan before hydrolysis
2.   Extraction of raw material before hydrolysis to remove impurities
3.   Separation of pentose-rich products from hexose components by means of (a) mechanical devices, (b) differential hydrolysis, and (c) optimization of acid penetration
4.   Ion exchange resins for purification of hydrolyzates
5.   Dialysis and reverse osmosis
6.   Use of solvent partition to separate xylose from contaminants
7.   Special derivatives for the purification of xylose.

Little effort has been devoted to the isolation of purified xylan for hydrolysis.[33] The economic disadvantages compared to direct hydrolysis might be overcome under special circumstances. A preextraction of aspen with kraft white liquor before cooking might accomplish this by facilitating liquor penetration, minimizing alkali consumption by xylan during cooking, and permitting the isolation of purified xylan for other useful purposes.[41] Xylans have also been isolated from spent hardwood neutral sulfite cooking liquors by Nelson[42] and Inada et al.[43]

If less concentrated alkali is used for extraction, very little xylan is dissolved, acetyl groups and certain lignin-carbohydrate bonds are cleaved, and miscellaneous polysaccharides and extractives are removed from the raw material.[44] Recently, Friese[45] patented a process to take advantage of this whereby an alkali metal hydroxide solution (less than 4%) is used to saponify acetyl groups. Acid hydrolysis then produces a less contaminated xylose hydrolyzate. In a subsequent patent, Friese et al.[46] recommend the use of 1.1 to 1.2 mol of hydroxide per mole of chemically bound acetic acid. Fahn and co-workers have patented a two-stage process for xylose manufacture in which alkaline and acidic stages are carried out in a single reaction vessel.[47] A variation of this process involves the pretreatment with sodium chlorite at 50°C for 2 hr. The mildly alkaline salt saponifies some esters, generates chlorine dioxide as the pH decreases, and probably causes some delignification.[48]

Many of the contaminants present in the hydrolyzates can be avoided by other obvious preventative techniques. The glycosidic bonds between the anhydroxylose units of xylan are less resistant to acid hydrolysis than are the glycosidic bonds between the components of hexosans. Presumably, it should be possible to preferentially hydrolyze the former under proper conditions. The technique of sequential hydrolysis is especially useful in wood saccharification schemes where the separation of glucose from xylose can be important. Nee and Yee,[49] for example, have used this difference to develop a modification of the Theinau saccharification process.[50] Eickemeyer[51] has described an apparatus for decomposing cellulosic materials which involves a preliminary hydrolysis of hemicellulose. The equipment described by Knauth[52] and Kiminki et al.[53] does the same on a continuous basis. Funk[54] has proposed a saccharification scheme for the continuous production of a variety of degradation products including xylose from vegetable material. These schemes have been discussed in several publications.[55,56]

The patent of Funk[54] also describes a method of impregnating the raw material with steam to ensure good penetration of reagents and to effect preferential hydrolysis of the xylan with a dilute mixture of organic and inorganic acids. Nobile's modification[57] softened the raw material with steam, passed it through a disk defibrator and concentrated the filtrate. Sugar mixtures rich in xylose were obtained by hydrolysis of the concentrate and the woody residue could be used for other purposes.

Maspoli[58] ensured penetration of catalyst by steaming the raw material while venting, then adding an acidic solution to cause the steam to condense within the pores of the wood. Hydrolysis was then conducted at elevated temperatures. Harris et al.[59] studied the effect of acid concentration on the recovery of xylose. Steiner and Lindlar[60] reacted hardwood shavings with oxalic acid to take advantage of the ease of hydrolysis of xylan.

The difficulty of purifying xylose-rich hydrolyzates is solved by many means. The potentially high profit anticipated from new xylitol end uses encouraged the use of relatively expensive ion-exchange resins. This research was pioneered by Leikin and co-workers in the U.S.S.R.[61,62] Jaffe et al.,[30,63,64] Steiner and Lindlar,[60] and Nobile[57] have patented variations of the use of ion-exchange resins to purify xylose-containing hydrolyzates and to recover acids. It is claimed by Friese[65] that the use of stone fruit shells as a source of xylose eliminated the need for these complex purification schemes since the hydrolyzate is almost pure xylose.

Spent cooking liquors (SSL) from the pulping of hardwoods by sulfite, neutral sulfite, or thermal type pulping processes provides an inexpensive source of crude xylose. Many purification schemes have been invented to exploit this source. Originally, the removal of reducing sugars was accomplished to minimize the contamination of useful lignosulfonic acids by reducing substances.[66] Hexoses were removed by fermentation to yield alcohol and yeast as useful by-products.[67] Early research by Boggs[68] attempted separations of reducing substances by preparing and separating diisopropylidene derivatives by molecular distribution. A more economic approach involved the attempts to separate sugars from other components of dried SSL using solvents such as methanol-acetone mixtures by DeHaas et al.,[69] and ethanol-methanol mixtures by Paabo and Uessen.[70] More complex mixtures of alcohol were used by Stranger-Johannessen[71] to extract xylose (after hexose removal by fermentation) while other similar schemes have been patented by other inventors.[72,73] A very complex purification procedure invented by Melaja and Hämäläinen[74] and leading ultimately to the production of xylitol will be discussed later.

D. Xylitol and Other Polyols

Interest has increased in the use of xylitol as a sweetener because it is not cariogenic. Another desirable feature is the cooling sensation felt on initial chewing because of its high endothermic heat of solution. The unconditional use of xylitol as a food additive has not yet been granted by the U.S. Department of Agriculture (USDA), although it has been tested in a variety of food products. Noncrystalline xylitol was first prepared by Fisher and Stobel,[75] and Bertrans[76] by sodium amalgam reduction. Wolfrom and Kohn first isolated metastable crystals after carefully purifying the product reduced with a nickel catalyst supported by kieselguhr.[77] The stable form was isolated a year later.[78] Xylitol next became an undesirable contaminant. The separation of xylitol was explored by hydrolysis plants in the U.S.S.R. in order to provide pure glucitol for their vitamin industry. Leiken and co-workers[79,80] demonstrated that mixtures of these alcohols could be separated by fractional crystallization from ethanol solutions.

At that early stage, satisfactory catalysts, optimum reduction conditions, and catalyst deactivation were continuing problems of the industry.[81] Nemanov[82] found the reactivation of the catalyst at the Chinkent hydrolysis plant in the U.S.S.R. could be achieved using a vibratory mill to shake off an inactivating film. The information in the patent literature suggests this was not a satisfactory solution. Although catalyst compositions are frequently trade secrets, some data are available. Vasyunina et al.[83] recommended the use of nickel catalysts at elevated temperatures (170°C) and pressures (50 atm); others recommended active nickel catalysts precipitated onto active charcoal or kieselguhr,[84-86] and even more complex systems composed of silica and nickel with traces of copper, manganese, and chromium components[87] have been suggested. Steiner and Lindlar use conventional Raney nickel for their reductions.[60]

More recent research is concerned with preparing pure xylitol (rather than glucitol) from the reduction mixtures for use as food or for medicinal purposes. Steiner and Lindlar[60] passed the reduced hydrolyzates over a bed of cation-exchange resin before crystallizing xylitol. Kohno et al.[88] did so before reduction of the hydrolyzate. Many of the patents discussed for the preparation of pure xylose did so to facilitate the preparation of pure xylitol. Most recently, Malaja and Hämäläinen[74] patented a very complex method of purifying hardwood hydrolyzates by ion exclusion and color removal followed by a chromatographic fractionation to purify xylose. Another fractionation by ion-exchange resins was used to purify xylitol after reduction and before crystallization.

The reduction of wood hydrolyzates can provide other pure chemicals, although the economic attractiveness of the process is doubtful. For example, a process has been patented in which the products from conventional catalytic hydrogenation containing hexitols and penitols are "cracked" in the presence of an activated nickel-charcoal catalyst at elevated temperatures such as 100 to 350°C and pressures such as 100 to 350 atm hydrogen.[89] The two-stage reduction is necessary to stabilize sensitive carbonyl groups before "cracking" under strenuous conditions. The glycols separated from this mixture in low yields by chromatography on an ion-exchange resin column are sorbitol, mannitol, xylitol, erythritol, glycerol, 1,2-propylene glycol, ethylene glycol, and 1,3-propylene glycol.

E. Furfural

Furfural is the most important chemical derived from xylan or any hemicellulose. The production, properties and uses of furfural are reviewed in an American Chemical Society monograph.[90] The importance and potential of furfural is reflected in the large amount of literature reviewed in *Chemical Abstracts.* It is not possible to include all this material in this review.

The production of furfural was developed almost simultaneously by LaForge at the National Bureau of Chemistry[91] and by the Quaker Oats Company[90,92,93] at the end of World War I. The latter process initially employed oat hulls and became a commercial success. As developed by Dunlop it involves the use of large rotating digesters which are charged with sulfuric acid and rotated slowly while steam is introduced to the desired temperature. A vapor outlet valve is opened to permit the removal of furfural as a steam distillate. The distillate is passed through a fractionating column, and the enriched furfural distillate is further purified to give the commercial product.

133

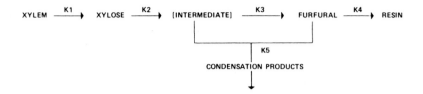

FIGURE 2. The formation of furfural from xylan.

Economically attractive sources of furfural must combine high pentosan content with high tonnage sources, ease of collection, availability throughout the year, suitable physical properties (density and accessibility), and freedom from unwanted side products. Common raw materials considered for furfural production are cornstalks and cobs, and the hulls of seeds and cereals such as oats, cottonseed, soybean, and rice, to name but a few.

The kinetics of furfural production are complex, and many of the kinetic details have been summarized recently.[94] The initial reaction involves the hydrolysis of xylan followed by the loss of water by the pentose component. The reaction is not a simple dehydration but proceeds to furfural through an intermediate unknown species. As shown in Figure 2, the reaction is complicated by secondary reactions leading to resin and colored condensation products. These harmful products are minimized by removing the furfural as soon as possible after formation and by proper storage. Oshima,[94] after reviewing the kinetic studies of Schonemann,[95] Yagi and Nemoto,[96] and Root et al.,[97] represents these reactions as five successive and parallel first-order reactions shown in Figure 2. The acidic catalysts used in most industrial processes are sulfuric acid by Quaker Oats,[90] calcium phosphate by Petrole-Chemie Engineering,[98] hydrochloric acid,[99,100] and no external acidic catalyst by the Rosenlew or Lemo Process. The use of salts as catalytic agents has been occasionally reported in the literature.[101,102]

Elaborate saccharification schemes often include furfural as one of their primary products besides organic acids,[103-105] activated charcoal,[54,55] and other products.[56] The wood and wood processing industries have been considered as sources of furfural[59] but have not been able to compete with available agricultural residues because of their higher collection costs and lower pentosan contents even though wood offers advantages of higher density and year round availability. According to Harris,[106] it would require only 2000 tons/day of hardwoods to supply the entire U.S. production of 175 million lb/year (in 1975). If this optimistic yield is correct, the integration of the process into pulp mill operation would give hardwoods some advantage over agricultural residues. A single-stage process with the residue used as fuel is presently the logical choice only because little is known of the potential for two-stage processes.

An alternative to using waste wood as a source of furfural is the use of spent sulfite pulping liquors.[106] This process is unlikely to be used in the U.S. where kraft pulping predominates. In this process, the dehydrating reactions are carried out at 240°C for 10 sec and at sulfuric acid concentrations of 0.5 to 16 g/100 g of liquor. Since the yield approaches that expected from ideal conditions, and because of the moderate acid requirements, the economics of producing furfural from SSL is favorable. Other investigations dealing with the production of furfural from SSL are to be found in the literature. The hemicellulose extracts from viscose steeping liquors have also been suggested as potential sources of furfural.[107]

The production of furfural throughout the world was reviewed recently by Panicker;[108] the principal producers in order of capacity were U.S., Dominican Republic, Italy, France and China, U.S.S.R., Spain, Japan, Argentina, and finally, Finland.

Panicker recommended the industry to tropical countries where agricultural waste and inexpensive labor are available in abundant quantities. These factors have encouraged the development of processes requiring less technological skill than the original process developed by Quaker Oats. Table 4 shows a comparison of processes of industrial significance taken from reviews by Panicker,[108] Zin,[109] and Eisner.[110]

IV. MANNOSE-DERIVED CHEMICALS

Most coniferous plants, unlike the plants in the monocotyledonous and dicotyledonous classes, contain glucomannans as the major hemicellulosic component instead of xylans.[10] The polymers from conifers contain between 2½ to 3 anhydromannose units per anhydroglucose unit.[111,112] The glucomannans from hardwoods are present in small amounts and contain a higher proportion of glucose than softwood glucomannans. Partial acid hydrolysis and methylation experiments show that the backbone of the polymer consists of linear chains of (1-4)-β-D-linked anhydroglucose and anhydromannose units. A random arrangement of mannose and glucose units has been demonstrated in the polymer from many conifers.[111] The glucomannan polymers within a single coniferous tree range in galactose content from nil to 1 part/3 mannose units. Some researchers classify them as a single polymer with varying galactose contents, whereas others speculate that two distinct galactose-rich and galactose-poor polysaccharides exist within a single tree. For the most part, the galactose units are linked by α-(1-6)-glycosidic bonds to both glucose and mannose units of the main chain, although evidence exists suggesting other linkages as well.[111,113] It is not known whether the galactose units are randomly distributed on the chain or whether they are distributed in patterns as has been observed for certain galactomannans, although they differ from one coniferous species to another.

Acetyl groups are present in the molecule, but are frequently absent from the polymers in the heartwood.[111] The location of the acetyl groups on isolated glucomannans has been determined; the information is of little consequence since the groups can probably migrate on the polymer during aging of the tree and certainly do migrate during polymer isolation. The loss of the acetyl groups induces crystallization, and for this reason can make the polymer less accessible to reactants.[114] A simplified representation of a galactoglucomannan from a conifer is as follows (Figure 3).

Only traces of polysaccharides can be isolated from most lignified conifers[111] except larch,[115] by direct extraction. Polysaccharides are extracted after delignification and the glucomannan component can be separated from the mixture only with great difficulty using proper fractionation techniques.[111,116]

There is little demand for mannose and mannose-derived chemicals unless their price becomes competitive with the corresponding glucose-derived chemicals. Mannose itself can be isolated from the hydrolyzate of ivory nut mannan.[117] Mannitol has been in ancient and somewhat greater demand as a sweetening agent[118] and for other uses shown in Table 5. Commercial production is based on the mild alkaline epimerization and reduction of invert sugar and glucose where mannitol is easily recovered from the reaction mixture by crystallization.[119-121]

Many of the mannose-containing polymers of wood are degraded to soluble components during the acidic treatments necessary for the production of certain types of pulp. They are found, for example, in the spent sulfite liquors from softwood pulping operations and contribute to the high hexosan content necessary for alcohol and yeast production. According to Herrick et al.,[122] the production of sulfite pulp in North America makes available approximately 400,000 ton/year of soluble mannose in the SSL. Other potential sources of the sugar are the prehydrolysis liquors from prehydrolyzed kraft pulping processes and the hydrolyzates of waste softwood (Table 5).

Page 135

Table 4
COMPARISON of DIFFERENT PROCESSES FOR THE PRODUCTION OF FURFURAL[108-110]

	Dnepropetrovsk	Savo	Agrifurane	Rossi	Sebev	Quaker Oats
Raw material	?	Particle size important	Particle size not critical	?	?	Particle size not critical
Yields furfural, % theoretical	77	47	65	60	50	49
Agitation	?	None	None	?	?	Much required
Catalyst	?	Acetic acid generated by process	$1,2 \rightarrow 1.5$ kg superphosphate	H_2SO_4 about ⅓ Quaker Oats requirement	H_2SO_4 about 1.5 Quaker Oats requirement	$0.2 \rightarrow 0.4$ kg H_2SO_4 kg furfural
Steam pressure	?	15 kg/cm²	9 kg/cm²	?	?	4.2 kg/cm²
Operation principal	Continuous	Continuous	Semibatch	Continuous	Continuous	Batch
Potential levulinic acid production	?	0	+	?	?	+
Utilization of ligno-cellulose	Phenols, carbon	Fuel	Fertilizer	Fuel	Fuel	Fuel

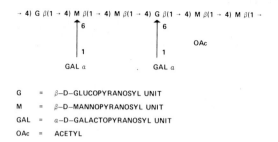

FIGURE 3.   A schematic representation of glucomannan.

Table 5
POTENTIALLY SIGNIFICANT PURE CHEMICALS DERIVED FROM GLUCOMANNAN

Product	Source	Process	Possible uses
Mannose	Softwood, spent liquors from pulping softwoods	Acid hydrolysis	Food products, other chemicals
Mannitol	Mannose	Reduction	Sweetening agent, nonhygroscopic carrier for medicines, resins, drying oils, plasticizers, emulsifiers
Methyl mannoside	Mannose, residues from softwoods or softwood pulping liquors	Methanolysis	Polurethane, polyester, polyethers
Sodium mannose bisulfite	Mannose, residues from softwoods or softwood pulping liquors	Bisulfite addition	Synthesis of medicinals, germicides, dyes, carrier for sulfite reagents
Mannoheptonic acid	Mannose, residues from softwoods or softwood pulping liquors	Kiliani synthesis	Additives in alkaline washing compounds, detergents scale inhibitors, additive to cement

The isolation of sugars and other components from coniferous spent sulfite liquors was pioneered by Wiley et al.[123] and Boggs.[124] Their procedure involves reacting dried SSL with acetone and acid. The acetone-soluble isopropylidene derivatives were separated into pentose-rich and hexose-rich fractions by molecular distillation. Diacetone mannose crystallized on cooling when the latter fraction was taken up in hot water. An alternative approach[125] was to extract dried SSL with alcohol, hydrogenate, and crystallize mannitol from a mixture of alditols. Concentrated SSL can be reacted with NaCN (Kiliani synthesis).[126] The modified liquor may be worked up to provide a crystalline aldonate derived from D-mannose by procedures involving ion exchange, or electrodialysis after selective precipitation of the modified lignosulfonates.

Herrick and co-workers[122,127] continued research in this area looking for ways to isolate mannose derived chemicals economically. They concentrated softwood SSL to 50% solids, reacted it with sodium metabisulfite, and recovered sodium mannose bisulfite as crystals in 93% yield of the mannose content of the liquor. Similar results were obtained when sodium metabisulfite was added to pinewood hydrolyzates prepared with sulfuric or sulfurous acids. The glycosidation of the carbohydrate residues in dried SSL and prehydrolysis liquors was accomplished with methanolic HCl by Hamilton et al.[128,129] Methyl α-D-mannopyranoside was isolated in excellent yield. Her-

Table 6
FACTORS AFFECTING THE ECONOMIC PRODUCTION OF MANNOSE

	Formation of			
	Diacetone derivatives	Aldonic acid	Bisulfite adduct	Methyl glycoside
Process equipment	Complex	Simple	Simple	Simple, but must be resistant to acid
Process complexity	Safety problems	Safety problems	Not complex	Pressurized equipment necessary
Character of raw material	Dried	Concentrated	Concentrated	Dried
Ease of crystallization	Moderate	Moderate	Easy	Difficult
Yield	Nearly quantitative	Not quantitative	Nearly quantitative	Nearly quantitative
Difficulty of mannose regeneration	Difficult	Impossible	Easy	Difficult
Reuse of SSL	Yes	Yes	Yes	Uncertain

rick et al.[122,127] critically analyzed the advantages and disadvantages of the isolation procedures they investigated. The data suggest mannose derived chemicals would not be competitive with the analogous glucose compounds. A portion of this evaluation, combined with similar factors affecting other processes for mannose isolation, is reviewed in Table 6.

V. ARABINOGALACTAN

Arabinogalactan is a unique water-soluble polysaccharide found in many (but not all[8]) coniferous species. Highest concentrations of the polymer (ranging from 1 to 25%) are found in the heartwood and branches of most species of larch.[130] Arabinogalactan was discovered in 1898 by Trimble[131] and has been the object of considerable research ever since.[132] Its unique properties, such as complete miscibility with water and low viscosity at high solids content, has made its commercial utilization a reality.[133] In this respect, the utilization of arabinogalactan as a polymer differs from the utilization of xylan in the form of derivatives.

Investigations of the structure of arabinogalactan have been numerous and conflicting opinions of its structure are held by different investigators. The review by Timell[134] suggests the following simplified structure for larch arabinogalactan (Figure 4). The structure shown assumes a ratio of 6:1 galactose to arabinose for most species. The ratio seems to be dependent upon molecular weight since a ratio of about 7:1 exists for higher molecular weights and about 3:1 for lower molecular weight fractions.[135] The backbone of the polymer is composed of (1-3) linked β-D-galactopyranosyl residues to which are attached various fragments to its six positions. The majority of these side chains are composed of (1-6) linked β-D-galactopyranosyl residues of varying length. Other branches are composed of 3-O-α-L-arabinofuranosyl-L-arabinofuranose units also linked to the six position of the main chain. Some D-galactose units are thought to bear two side chains, the second being at the C-4 position. The low and high molecular weight fractions differ in their extent of branching as well as in arabinose content.

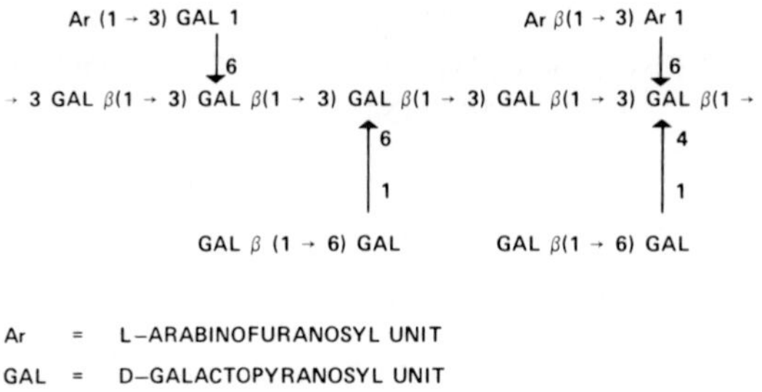

Ar (1 → 3) GAL 1   Ar β(1 → 3) Ar 1

↓6   ↓6

→ 3 GAL β(1 → 3) GAL β(1 → 3) GAL β(1 → 3) GAL β(1 → 3) GAL β(1 →

↑6   ↑4

|1 |1

GAL β (1 → 6) GAL   GAL β(1 → 6) GAL

Ar   =   L–ARABINOFURANOSYL UNIT

GAL   =   D–GALACTOPYRANOSYL UNIT

FIGURE 4.   A schematic representation of larch arabinogalactan.

Many procedures for the isolation of arabinogalactan have been published. The commercially acceptable procedure employed in the U.S. by St. Regis Paper Company involves a countercurrent extraction of chipped larchwood waste.[136] The final extract contains about 8 to 10% polysaccharide which is dried on a drum drier. This product is marketed as technical grade "Stractan" by Stein Hall Co.[137] Further purification is achieved by removing iron salts and phenolic compounds with activated magnesium oxide.[138] Final purification, if necessary, is achieved by treating this solution with activated charcoal.

According to the Federal Food, Drug and Cosmetic Act, arabinogalactan may be used as an emulsifier, stabilizer, and binding agent for a wide variety of food and cosmetic products. It is also used in the lithographic industry, in photographic processing, and as an emulsifier. It has been proposed as a source of mucic acid.[139-141] Austin[142] reinvestigated the production of mucic acid from this source and concluded that his optimum of conditions of reaction would provide a marginally economic product.

VI. MISCELLANEOUS CHEMICALS

The necessity of eliminating various emissions from pulping operations has resulted in the recovery of many potentially useful chemicals from hemicelluloses as well as from lignin. Volatile gases are collected in many pulping operations and, in some, gases are fed back to the furnaces for recovery as heat. One of the components is methanol, which can be derived from pectic substances as well as from lignin. At present, these effluents are more valuable as fuel than as chemicals.

The Sonoco Products Company extracted substances contributing to much of the basic oxygen demand (BOD)[143] of black liquor of semichemical or neutral sulfite pulping process. A solvent combination such as methanol and methyl acetate, for example, avoids the formation of emulsions and gives good phase separations. The BOD is reduced about 80% and recoveries of 95% of the acetic and formic acid contents and 10 to 20% of the sugar contents are achieved. Fractionation gives good quality acetic acid derived chiefly from the hemicellulose esters of the incoming wood.

ACKNOWLEDGMENTS

The author would like to thank Miss K. Ewing for arranging the computer searches of *Chemical Abstracts* and the *Abstract Bulletin* of the Institute of Paper Chemistry. He is also indebted to Mrs. F. Arnell for suggestions of style and typing of the manuscript.

REFERENCES

1. Burwell, C. C., *Science,* 199, 1041, 1978.
2. Locke, E. G. and Johnson, K. B., *Ind. Eng. Chem.,* 46(3), 478, 1954.
3. Ledig, F. T. and Linzer, D. I. H., *Chem. Technol.,* 10, 18, 1978.
4. Sjöström, E., *Tappi,* 61(2), 151, 1978.
5. Maisel, D. S., *Tappi,* 62(1), 51, 1978.
6. Aspinall, G. O., *Adv. Carbohydr. Chem.,* 14, 429, 1959.
7. Timell, T. E., *Adv. Carbohydr. Chem.,* 19, 247, 1964.
8. Dickey, E. E. and Thompson, N. S., unpublished research, 1964.
9. Karnik, M. G., Morak, A. J., and Ward, K., *Tappi,* 46(2), 130, 1963.
10. Hamilton, J. K. and Thompson, N. S., *Pulp Pap. Mag. Can.,* 59(10), 233, 1958.
11. Thompson, N. S., Heller, H. H., Hankey, J. D., and Smith, O., *Pulp Pap. Mag. Can.,* 67(12), T541, 1966.
12. Mitchell, R. L. and Ritter, G. J., *J. For.,* 49(2), 112, 1951; U.S. For. Prod. Lab., Rep. 1771, 1950.
13. Whistler, R. L. and Wolfrom, M. L., *Methods in Carbohydrate Chemistry,* Vol. 1, Academic Press, New York, 1962.
14. Tu, C. V. and Whistler, R. L., *J. Am. Chem. Soc.,* 74(17), 4334, 1952.
15. Hamilton, J. K. and Thompson, N. S., *J. Am. Chem. Soc.,* 79, 6464, 1957.
16. Bearce, W. H., *J. Org. Chem.,* 30, 1613, 1965.
17. Timell, T. E., *Sven. Papperstiden.,* 65(11), 435, 1962; *Can. J. Chem.,* 40(1), 22, 1962.
18. Gyaw, M. O. and Timell, T. E., *Can. J. Chem.,* 38(10), 1957, 1960.
19. Haq, S. and Adams, G. A., *Can. J. Chem.,* 39(8), 1563, 1961.
20. Banerji, N. and Thompson, N. S., *Cellul. Chem. Technol.,* 2, 655, 1968.
21. Schmorak, J. and Adams, G. A., *Tappi,* 40(5), 378, 1957.
22. Dudkin, M. S., Kaga, E. A., and Grenshoun, S. I., *Latv. PSR Zinat. Akad. Vestis Kim. Ser.,* 5, 63, 1964.
23. Aspinall, G. O. and Cairncross, I. M., *J. Chem. Soc.,* 3998, 1960.
24. Rogers, J. K. and Thompson, N. S., *Carbohydr. Res.,* 7, 66, 1968.
25. Germino, F. J., U.S. Patent 3,297,604, 1967.
26. Samuelson, O. and Wictorin, L., *Sven. Papperstidn.,* 69, 777, 1966.
27. Gramera, R. E. and Whistler, R. L., *Arch. Biochem. Biophys.,* 101(1), 75, 1963.
28. Richards, G. N. and Whistler, R. L., *Carbohydr. Res.,* 31(1), 47, 1973.
29. Zoch, L. L., Harris, J. F., and Springer, E. L., *Tappi,* 52(3), 486, 1969.
30. Jaffe, G. M., Szkrybalo, W., and Weinert, P. H., U.S. Patent 3,784,408, 1974.
31. Casebier, R. L., Hamilton, J. K., and Hergert, H. L., *Tappi,* 56(3), 135, 1973.
32. Casebier, R. L., Hamilton, J. K., and Hergert, H. L., *Tappi,* 52(12), 2369, 1969.
33. Koch, *Pharm. Zig. Russ.,* 25, 657, 1886.
34. Risso, W. and Forner, G., D. D. R. Patent 20,624, 1961.
35. Nelson, R. and Schuerch, C., *J. Polym. Sci.,* 32, 435, 1956.
36. Whistler, R. L. and Shah, R. N., Recent advances in the industrial use of hemicelluloses, in *Modified Cellulosics,* Ravell, R. M. and Young, R. A., Eds., Academic Press, New York, 1978, 341.
37. Bertrand, G., *Compt. Rend.,* 129, 1025, 1899.
38. Hudson, C. S. and Hardy, T. S., *J. Am. Chem. Soc.,* 39, 1038, 1917.
39. LaForge, T. and Hudson, C. S., *Ind. Eng. Chem.,* 10, 925, 1918.
40. Schreiber, W. T., Geib, N. V., Wingfield, B., and Acree, S. F., *Ind. Eng. Chem.,* 22(5), 497, 1930.
41. Springer, A. and Thompson, N. S., *Pap. Puu,* 53(9), 499, 1971.
42. Nelson, P. F., *Appita,* 22(4), 97, 1969.
43. Inada, O., Samashima, K., and Kondo, K., *Japan Tappi,* 29(11), 589, 1975.
44. Thomas, B. B., Ph.D. thesis, The Institute of Paper Chemistry, Appleton, Wis., 1944.
45. Friese, H., U.S. Patent 3,954,497, 1976.
46. Friese, H., Buckl, H., and Bernd, B., U.S. Patent 3,990,904, 1976.
47. Fahn, R., Bernd, B., and Buckl, H., U.S. Patent 4,072,538, 1978.
48. Friese, H., U.S. Patent 3,970,712, 1976.
49. Nee, C. I. and Yee, W. F., *J. Appl. Chem. Biotechnol.,* 27, 662, 1977.
50. Schoenemann, K., The New Rheinau Wood Saccharification Process, paper presented to the Congress of The Food and Agricultural Organization of the United Nations, Stockholm, 1953.
51. Eickemeyer, R., U.S. Patent 3,787,241, 1974.
52. Knauth, H. H., U.S. Patent 4,023,982, 1977.
53. Kiminki, K., Kulmala, R., and Sipilä, R., U.S. Patent 4,029,515, 1977.
54. Funk, H. F., U.S. Patent 3,523,911, 1970.

55. Kalninsh, A. Y. and Vedernikov, N. A., *Appl. Polym. Symp.*, 28, 125, 1975.
56. Funk, H. F., *Appl. Polym. Symp.*, 28, 145, 1975.
57. Nobile, L., U.S. Patent 3,479,248, 1969.
58. Maspoli, R., Canadian Patent 1,010,859, 1977.
59. Harris, D. F., Saeman, J. F., and Locke, E. G., *For. Prod. J.*, 9, 248, 1959.
60. Steiner, K. and Lindlar, H., U.S. Patent 3,586,537, 1971.
61. Leikin, E. R., Gutiva, S. L., Meshkova, V. Ya., *Sb. Tr. Gos. Nauch.-Issled. Inst. Gidroliz. i. Sulfitno-Spirt. Prom.*, 11, 77, 1963.
62. Leikin, E. R., *Gidroliz. Lesokhim. Promst.*, 19(5), 16, 1966.
63. Jaffe, G. M., Szkrybalo, V., and Weinert, P. H., Canadian Patent 919,664, 1973.
64. Jaffe, G. M., Szkrybalo, V., and Weinert, P. H., Canadian Patent 945,090, 1973.
65. Friese, H., U.S. Patent 3,579,380, 1971.
66. Luner, P., Dubey, G. A., Boggs, L. A., and Wiley, A. J., *For. Prod. J.*, 8(2), 82, 1958.
67. Whitmore, L. M. and Wiley, A. J., *Chem. Eng. Prog.*, 54(12), 80A, 1958.
68. Boggs, L. A., *Tappi*, 40(9), 752, 1957.
69. DeHaas, G. G., Clark, L. H., and Lang, C. J., U.S. Patent 3,337,366, 1967.
70. Paabo, G. J. and Uessen, A. M., U.S. Patent 3,542,590, 1970.
71. Stranger-Johannessen, P., U.S. Patent 3,687,804, 1969; French Patent 2,010,689, 1970.
72. Paabo, G. J. and Uessen, A. M., U.S. Patent 3,639,171, 1972.
73. Gasche, U., Lindlar, H., Rittishauser, M., and Steiner, K., U.S. Patent 3,700,501, 1972.
74. Melaja, A. J. and Hämäläinen, L., U.S. Patent 4,008,285, 1977.
75. Fisher, E. and Stobel, R., *Ber.*, 24, 536, 1891.
76. Bertrand, G., *Bull. Soc., Chim.*, 30(5), 551, 1891.
77. Wolfrom, M. L. and Kohn, E. S., *J. Am. Chem. Soc.*, 64, 1739, 1942.
78. Carson, J. F., Waisbrot, S. W., and Jones, F. T., *J. Am. Chem. Soc.*, 68, 1777, 1943.
79. Leikin, E. R., *Sb. Tr. Gos. Nauch.-Issled. Inst. Gidroliz. i. Sulfitno-Spirt. Prom.*, 11, 86, 1963.
80. Leikin, E. R., Soboleva, G. D., and Meshkova, V. Ya., *Sb. Tr. Gos. Nauch.-Issled. Inst. Gidroliz. i. Sulfitno-Spirt. Prom.*, 12, 189, 1964.
81. Apel, A. and Rossler, G., U.S. Patent 2,917,390, 1959.
82. Nemanov, E. A., *Gidroliz. Lesokhim. Promst.*, 19(4), 23, 1966.
83. Vasyunina, N. A., Balandin, A. A., Mamotoy, Yu., and Grigoryan, E. S., U.S.S.R. Patent 165,163, 1964.
84. Soboleva, G. D., Meshkova, V. Ya., Cheromukhim, I. K., and Grankina, L. G., U.S.S.R. Patent 167,845, 1965.
85. Anon., British Patent 838,766, 1960.
86. Specht, H. and Dewein, H., German Patent 1,066,567, 1959.
87. Riem, T., German Patent 1,064,489, 1959.
88. Kohno, S., Yamatsu, I., and Veyama, S., U.S. Patent 3,558,725, 1971.
89. Anon., British Patent 842,743, 1960.
90. Dunlop, A., and Peters, F. N., *The Furans*, Reinhold Publishing New York, 1953.
91. LaForge, T., *Ind. Eng. Chem.*, 13, 1024, 1921; *Ind. Eng. Chem.*, 16, 130, 1924.
92. Dunlop, A. P., Furfural, in *Kirk-Othmer Encyclopedia of Chemistry*, Vol. 10, John Wiley & Sons, New York, 1966, 237.
93. Dunlop, A. P., Industrial uses of cereals, Symp. Proc., Pomeranz, Y., Ed., American Society of Cereal Chemicals, St. Paul, Minn., 1973, 229.
94. Oshima, M., *Wood Chemistry*, Noyes Data, N.J., 1965.
95. Schonemann, K., *Chem. Eng. Technol.*, 29, 665, 1957; Second Meeting of the Working Panel on Wood Hydrolysis, F.A.O., Technical Panel on Wood Chemistry, Tokyo, 1960.
96. Yagi, M. and Nemoto, T., Annual Report of the Naguchi Institute, No. 10, 40, 1960.
97. Root, D. F., Saeman, J. F., Harris, J. F., and Neill, W. K., *For. Prod. J.*, 9, 158, 1959.
98. Anon., British Patent 842,689, 1960.
99. Chalov, N. V., Khol'kin, Yu. I., Khol'kina, A. S., and Kryukova, G. A., *Izv. Vyssh. Uchebn. Zaved. Lesn. Zh.*, 3(2), 153, 1960.
100. Koryakin, V. I. and Sokolova, A. I., *Gidroliz. Lesokhim. Promst.*, 13(5), 3, 1960.
101. Riklis, S. G. and Kolotilo, D. M., *Gidroliz. Lesokhim. Promst.*, 14(2), 3, 1961.
102. Suehiro, Y., Araki, K., Fujishima, K., and Sato, K., Japanese Patent 4,976, 1961.
103. Truba, T. I., Repka, V. P., Pysaetskaya, L. V., Panasyuk, V. G., and Maksineko, N. S., U.S.S.R. Patent 122,482, 1959.
104. Anon., German Patent 1,072,466, 1959.
105. Ramos-Rodriguez, E., U.S. Patent 3,701,789, 1972.
106. Harris, J. F., *Tappi*, 61(1), 41, 1978; TAPPI Conf. Papers/Forest Biology Wood Chemistry Conf., 1977.

107. Kurschner, K., Czechoslovakian Patent 93, 967, 1960.
108. Panicker, P. K. N., *Chem. Age India*, 25(11), 793, 1975; 26(2), 101, 1976; 26(6), 457, 1976.
109. Kin, Z., *Przegl. Papier.*, 21(5), 157, 1965.
110. Eisner, K., *Drevo*, 21(3), 79, 1966.
111. Timell, T. E., *Adv. Carbohydr. Chem.*, 20, 409, 1965.
112. Hamilton, J. K. and Thompson, N. S., *Pulp Pap. Mag. Can.*, 59(10), 233, 1958.
113. Rogers, J. K. and Thompson, N. S., *Sven. Papperstidn.*, 72, 61, 1969.
114. Katz, G., *Tappi*, 48(1), 34, 1965.
115. Peterson, F. C., Maughan, M., and Wise, L. E., *Cellul. Chem.*, 15, 109, 1934.
116. Lindberg, B., *Pure Appl. Chem.*, 5, 67, 1962.
117. Pigman, W., *The Carbohydrates*, Academic Press, New York, 1917, 249.
118. Wicker, R. J., *Chem. Ind. (London)*, 1966, 1708.
119. Creighton, H. J., *Can. Chem. Process Ind.*, 26, 690, 1942.
120. Hales, R. A., U.S. Patent 2,289,189, 1942.
121. Kasehagen, L. and Luskin, M. M., U.S. Patent 2,759,024, 1956.
122. Herrick, F. W., Casebier, R. L., Hamilton, J. K., and Wilson, J. D., *Appl. Polym. Symp.*, 28, 93, 19.
123. Wiley, A. J., Whitmore, L. M., and Boggs, L. A., *Tappi*, 42(5), 14A, 1959.
124. Boggs, L. A., *Tappi*, 40(9), 752, 1957.
125. Boggs, L. A., *Tappi*, 56(9), 127, 1973.
126. Boggs, L. A., Canadian Patent 811,048, 1969.
127. Casebier, R. L., Herrick, F. W., Gray, H. G., and Johnston, F. A., U.S. Patent 3,677,818, 1972.
128. Hamilton, J. K., Herrick, F. W., and Wilson, J. D., U.S. Patent 3,507,853, 1970.
129. Hamilton, J. K., Herrick, F. W., and Wilson, J. D., U.S. Patent 3,531,401, 1970.
130. Mitchell, R. L. and Ritter, G. J., *J. For. Prod. Res. Soc.*, 3, 66, 1953.
131. Trimble, H., *Am. J. Pharm.*, 70, 152, 1898.
132. Bouveng, H. O., *Sven. Kem. Tidskr.*, 73(3), 113, 1961.
133. Ettling, B. V. and Adams, M. F., *Tappi*, 51(3), 116, 1968.
134. Timell, T. E., *Wood Sci. Technol.*, 1, 45, 1967.
135. Swenson, H. A., Kaustinen, H. M., Bachhuber, J. J., and Carlson, J. A., *Macromolecules*, 2, 142, 1969.
136. Adams, M. F., U.S. Patent 3,337,526, 1963.
137. Anon., *Chem. Eng. News*, 41(19), 58, 1963.
138. Herrick, I., Adams, M. F., and Huffacker, E. M., U.S. Patent 3,325,473, 1967.
139. Acree, S. F., British Patent 160,777, 1921.
140. Acree, S. F., U.S. Patent 1,816,137, 1931.
141. Schorger, A. W., U.S. Patent 1,718,837 1929.
142. Austin, G. T., *Ind. Eng. Chem. Prod. Res. Dev.*, 8(4), 424, 1969.
143. Copenhaver, J. E., Biggs, W. A., and Baxley, W. H., U.S. Patent 2,744,927, 1956.

Chapter 8

CHEMICALS FROM LIGNIN

David W. Goheen

TABLE OF CONTENTS

I. INTRODUCTION

When one sets out to evaluate the Earth's biomass as a source of chemicals, it becomes apparent rather quickly that wood occupies a unique and special position. Woody plants not only synthesize the two most abundant organic materials (cellulose and lignin) on a grand scale, they do it in such a way that the two materials are concentrated in a form that is relatively easy to harvest in the boles of trees. Most, if not all, of the other products of photosynthesis such as microscopic algae, grass, straw, and vegetable crops, which are also produced on a grand scale, are much more dispersed and difficult to collect. Furthermore, they are extremely difficult to store in order to ensure continuous and year-round operation for any process based on their utilization. Wood, on the other hand, is relatively easy to collect and can be stored under certain conditions for very long periods. In fact, wood can be considered as the only feasible way that we have for storing incident solar energy for any length of time.

We, of course, must recognize that natural gas, petroleum, and coal are also stored solar energy. Their problem, although they are in a very useful form for mankind's needs, is that they accumulate at an extremely slow rate and represent an infinitesimal fraction of the solar energy that has been intercepted by the earth over millions of years. The entire energy content of all the known fossil fuels has been estimated at about 21 days worth of the total solar energy impacting the Earth's surface.

What this means in terms of the efficiency of solar radiation conversion to stored energy, is nearly impossible to estimate. The efficiency has to be vanishingly small, and therefore, the fossil fuels (so far as we are concerned) are irreplaceable. Trees, on the other hand, are capable of intercepting and storing solar energy in a useful form at efficiencies that may appear low (from 0.5 to 1.0%), but that are many times greater than the accumulation of gas, petroleum, and coal.

Nevertheless, and despite the drawbacks of irreplaceability, mankind has found the fossil bonanza of stored organic materials to be of great importance and value. So much so, that for the present generation, it is difficult to imagine a time when their use was not widespread.

Although most of the world's chemical industries are presently based on petrochemical feedstocks, there was a time only a few decades ago, when many of the chemical derivatives of petroleum were manufactured from other substances. Before 1930, for example, almost all of the industrial ethanol and many other solvents and organic chemicals used in the U.S. were prepared by fermentation processes based on agricultural and renewable crops.

Certain countries, and particularly the U.S. were blessed by nature with huge pools of petroleum that were easily and geographically accessible, easy to reach by relatively shallow drilling, and accompanied by immense quantities of natural gas. The ready availability of these inexpensive raw materials, with much of the energy required for conversion to finished products already taken care of by forces in the earths crust, caused what was at first a gradual shift in chemical feedstock sources. Just prior to World War II, the petrochemical industry began to come of age. Ways of producing a host of products from fossil gas and oil sources were investigated. These developments continued at an accelerating pace during the war years and following the war, when pent-up demand created an enormous market, and the shift to petrochemical feedstocks occurred very quickly. The demand for plastics, adhesives, and other chemical products was met by construction of hugh petrochemical plants that utilized plentiful, cheap, and easy-to-convert oil and natural gas. Thus, the chemical pricing trend for nearly all of the past 30 years was downward with production costs ever lower, owing to construction of ever larger plants. If there were truly no foreseeable limit to the amount of oil and gas available on a global scale, this trend would continue for a long time. That there is a finite limit was painfully learned in 1973 during and following the OPEC (Organization of Petroleum Exporting Countries) oil embargo on exports of petroleum from the Middle Eastern oil fields. Little attention was paid at the time to the fact that U.S. consumption of petroleum first exceeded U.S. production in 1966 to 1967. The difference was made up by importation, so that by 1971, one quarter of U.S. petroleum requirements was imported.[1] The figure increased each year by some 5%, so that by 1973 to 1974, nearly 40% of U.S. petroleum needs were imported, and today, this has reached 50%. In 1973, the oil embargo resulted in the economically painful combination of recession and inflationary price increases. What happened to the U.S. in 1967 to 1973 may very well happen to the world in 1990 to 2000. Sometime during that period, it was estimated that world production of petroleum will peak, and world demand will begin to exceed production capacity shortly thereafter.[2]

The downward trend in chemical prices that was so abruptly halted in 1973 is thus seen as a temporary and transient phenomenon that occurred only as the result of artificial pricing. It is true that the cost of finding and producing oil is really quite low, especially in accessible regions where the oil is not very deep in the earth. Thus, in terms of cost of production vs. selling price, the low unit petroleum costs in the quarter century following World War II made economic sense. What was not considered, and so no price was placed on it, was the nonreplenishability of petroleum. No value was assigned to the fact that oil is virtually irreplaceable. This is the paradox facing all of us. Oil costs are low, but its real value is very high, owing to its utility and irreplaceability. This has led to many of our industrial energy and feedstock problems of today. The artificially low costs of petroleum meant that there was little or no economic incentive to develop alternative sources, while increasing demand led to depletion of early exploitable petroleum reserves.

II. RENEWABLE RESOURCES

It is now quite clear that mankind will make greater demands on renewable resources in the future to satisfy needs for energy and particularly for carbon-based chemicals. Even though the production of chemicals from wood and other renewable sources dwindled almost to the vanishing point during the expansion of the petrochemical industry, some thoughtful persons contemplated the eventual depletion of fossil oil and natural gas. While many of these people have merely turned to consideration of yet another fossil source, coal, others have recognized the vast potential of replenishable vegetative matter, particularly that found in the forest biomass. Wood has long been one of mankind's most abundant and versatile raw materials for a great variety of construction and industrial uses. Even up to the time of World War II, wood and products obtained from wood provided a portion of the raw materials used in the chemical industry. With the advent of cheap petroleum and the many processes discovered to convert it, the use of wood as a source of feedstocks practically ceased.

More than 40 years ago, however, Germany, faced with a cut-off of imported petroleum, devised a national program designed to produce a variety of products based on wood as the raw material.[3] Along with fuels and construction materials, it was proposed that basic materials for chemical industries could be prepared. While only some of the goals were achieved during the difficult years of the war, enough was accomplished to demonstrate the enormous potential of the forest biomass. Sweden, on a somewhat more limited scale, was able to bolster its supplies for many critical chemical materials from wood.

The German and Swedish wartime experiences with wood were summarized in 1949 by Glesinger.[4] Most of the ideas and principles set forth by Glesinger were ignored during the 30-year petrochemical period, but today his proposals and suggestions need to be taken more seriously. As petroleum and natural gas become more scarce, and inevitably more expensive, many products from wood surely will become economically competitive with the same products produced from fossil feedstocks. Glesinger believed that with proper management, the world's forests could produce up to 14 billion tons of wood per year perpetually. This astounding figure is surely the result of unbounded optimism, but even one half of this amount could cause mankind to enter an "Age of Wood". It has been estimated that at present, the amount of wood processed per year on a world scale is about 2×10^9 m,[3] which corresponds to some 7 to 8 $\times 10^8$ tons.[5] This demand is expected to reach 5×10^9 m^3 or 1.75 to 2.0×10^9 tons by the year 2000. Glesinger believed that it would be possible to produce more than seven times this projected figure. Along with the marked increase in the potential amount of wood that he proposed could become available, he also felt that wood has three unique attributes which make it a very attractive raw material.

1. Wood is widespread. It can be found almost everywhere.
2. Wood is abundant. Forests cover nearly 25% of the earths land area. This is in contrast to the very limited areas where coal, iron ore, oil, and minerals are found.
3. Wood is inexhaustible. The forest is not a mine that is to be exploited and finally exhausted, but it is a cropland. This attribute has lately been questioned, and it has been suggested that repeated removal of wood and vegetative matter from the land will exhaust nutrients, and after a few cycles, tree growth will be stunted.[6] This is probably not a valid argument. It should be no more difficult (probably less so) to maintain nutrients in a forest plantation than it is to replace nutrients in agricultural practice. Fertilization has been a major part of agricultural efforts in all advanced countries for many years.

Wood has a fourth attribute that could also be added to Glesinger's three. It has a low ash content, little nitrogen, and practically no sulfur. With all of these attributes, wood can and should play a major role as a chemical feedstock source in the future.

III. CHEMICALS FROM LIGNOCELLULOSE

The technology for producing a large number of the basic commodities of modern industrial chemistry from wood is already known, but until the present period of rising petroleum prices, procedures for accomplishing the conversions have not been economically competitive with procedures for the same chemicals based on petroleum. Wood is composed of about 70 to 80% polymeric carbohydrates and 20 to 30% lignin. This matrix can be degraded in various ways (as shown in Figure 1) to produce practically all of the industrially important organic chemicals produced by the chemical industry.

Considering wood as a raw material source, the question naturally arises as to whether chemical feedstocks from these renewable sources should be recovered as by-products of the chemical pulping industry, or whether separate chemical plants should be set up to produce a number of basic chemicals directly from wood. Due to the fact that lignin, and to a lesser extent, hemicellulose, are available in large quantities from pulp mills, it would appear that procedures based on their conversion from pulping liquors would have a more immediate competitive edge. Thus, lignin (especially that from chemical pulp mills) appears to be an attractive alternate source of feedstocks. It should figure prominently during the interim period when the chemical industry

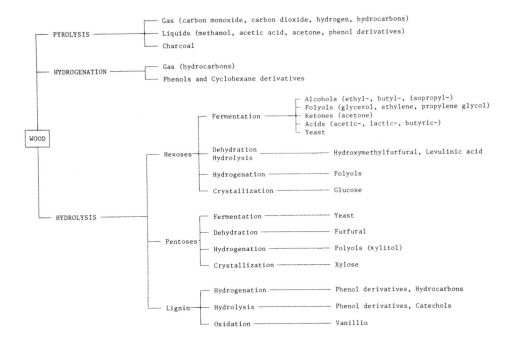

FIGURE 1. Chemicals obtainable from wood.[7]

begins to shift from a petroleum and natural gas base to coal and renewable materials. Wise and Jahn recognized the great potential value of lignin, especially that from the pulping industry, as one of the chief sources of organic material annually replenished by nature on a tremendous scale.[8]

However, up to the present, the primary use for lignin from pulp mills has been as a fuel during recovery of inorganic pulping chemicals. Many other uses have been suggested both in the high molecular weight polymer form and as low molecular weight fragments. The current and projected uses are summarized in Figure 2.

IV. FORMATION AND CHEMICAL STRUCTURE OF LIGNIN

Until early in the 19th Century, wood was considered to be a chemical entity. This belief was held until Payen showed that wood is composed of several components.[9,10] His studies were the first serious attempts to determine the chemical composition of lignified plants. He succeeded in isolating a pure cellulose and recognized that cellulose is the basic wood substance. By various chemical and solvent treatments, he was able to remove from the cellulose fibers a material that he never isolated, but that he called *matiére incrustante,* or encrusting material. Today we call this substance lignin. In the years following Payen's work, other investigators and inventors were able to react lignin with various chemical reagents to solubilize it and separate it from cellulose.[11-13] These procedures resulted in the establishment of the modern chemical pulping industry.

Ever since the discovery that lignin can be chemically modified and thereby separated from cellulose, many investigators have studied it both with a view toward establishing its chemical structure and using it as a raw material source for production of organic chemicals. The question of how lignin and cellulose are combined in wood occupied the attention of numerous investigators for a long time, with some agreeing

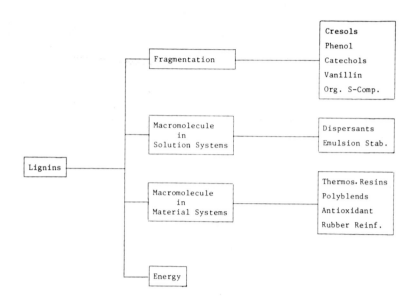

FIGURE 2.   Current and suggested uses for lignins.[7]

with Payen that lignin was merely an encrusting agent around cellulose fibers, while others believed that they were chemically combined. It has turned out that both schools of thought were at least partially correct. It was suggested at a relatively early stage in these studies that lignin has an aromatic nature. Erdmann[14] found catechol following fusion of wood with caustic, but was unable to detect it when he treated purified cellulose the same way. Czapek[15] and Grafe[16] studied a material (hydromal) that Czapek isolated in low yield from wood by a rather complicated procedure involving stannous chloride. Both of these investigators believed that hydromal was either an aromatic compound or a mixture of aromatic moieties. Grafe identified vanillin as one of the components and also found vanillin in the products of alkaline heating of sulfite spent lignin. It was not until the work of Lange[17] that the aromatic nature of lignin in wood was conclusively proven by ultraviolet spectroscopy. The results of purely chemical investigations were always ambiguous, and the extreme difficulty in isolating lignin in an unchanged form from its association with carbohydrates made a purely chemical solution to the problem very difficult.

For many years, little progress was made in establishing a chemical structure for the complex lignin polymer. It was only within the last 2 decades that the work on the biosynthesis and formation of lignin in wood by Freudenberg and his co-workers established a plausible chemical structure.[18] The work of Adler and co-workers on structural studies of isolated lignin preparations has extended the Freudenberg proposal and modified it to give a schematic structure that accounts for many, but not all, of the known structural details of the lignin molecule.[19]

It is now recognized by most lignin investigators that the material is produced in lignified plants by a rather unique process for a living system. Only the initial step in the process appears to be enzymatically controlled. As shown in Figure 3, the most important step in the synthesis consists of an enzymatically induced dehydrogenation of coniferyl alcohol (conifer trees, I) or sinapyl alcohol (deciduous trees, Ia) to the mesomeric structures (II, III, IV, and V). The radicals of the mesomorphs can couple in many possible ways in an apparently random manner, accompanied by the addition of water or phenolic or aliphatic hydroxyl groups to intermediate quinone methide

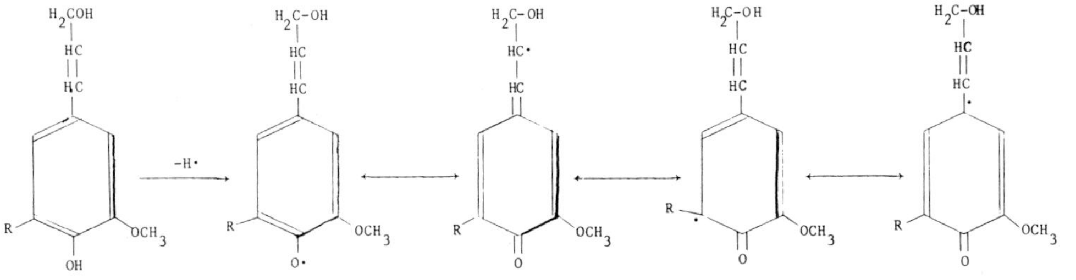

FIGURE 3. Dehydrogenation of coniferyl or sinapyl alcohol to a highly mesomeric free radical.

structures. The final lignin that results is a complex polymeric matrix with no simple repeating units as are found in other naturally occurring polymers such as proteins, hemicelluloses, and cellulose. Thus, simple hydrolytic procedures do not result in significant depolymerization of the complex lignin polymer, bound together not only with potentially hydrolyzable carbon-oxygen bonds, but also nonhydrolyzable carbon-carbon bonds. In fact, it has been shown that native lignin in wood is reactive under such conditions and may condense to form even more intractable, higher molecular weight moieties. A further complication to isolating lignin that should be soluble in a number of organic solvents by its aromatic nature, is the now well-established linkage to polysaccharides. Lignin, as it is deposited in the cell walls, is linked here and there to the hemicelluloses and even cellulose by ether linkages, possibly of a hemiacetal nature.[20] Thus, it can be seen that both the encrusting theory and the chemically bound theory of the early lignin investigators were partially correct.

After so many years of study, it is somewhat of a disappointment to realize that there is probably no absolute chemical structure for lignin, and that the schematic shown in Figure 4 is only an approximation that satisfies a number of the observations that have been made during the study of lignin.

As is shown, the lignin molecule in woody plants is a complex, highly branched polymer attached here and there to polysaccharides. It is composed of phenylpropane-based monomeric units linked together by several types of ether linkages and also various kinds of carbon-carbon bonds. The schematic of Figure 4 should be considered not as a structural formula, but rather as an outline. The molecular weight of lignin in wood is very high, essentially infinite. The Freudenberg and Adler schemes comprise only about 20 monomeric units and therefore cannot possibly describe all structural details.

Even when lignin is separated to yield 35 to 40% of the total lignin present in wood by the mildest known procedure, i.e., solvent extraction of milled wood,[21] the resulting material is amorphous and crumbly with little coherence and solidity. This is especially frustrating to potential users of lignin, as it has been demonstrated that the role of lignin in plants is as a filler and cementing agent comparable to the cement in reinforced concrete.[22] No way has been discovered yet to isolate lignin from wood and then (without extensive modification) use it as a resin or glue comparable to its function in wood.

Information as to how lignin is deposited in wood has been obtained using the technique of ultraviolet microscopy.[23] It has been shown that the concentration of lignin in the middle lamella between fibers is very high, but that 70% of the total lignin is

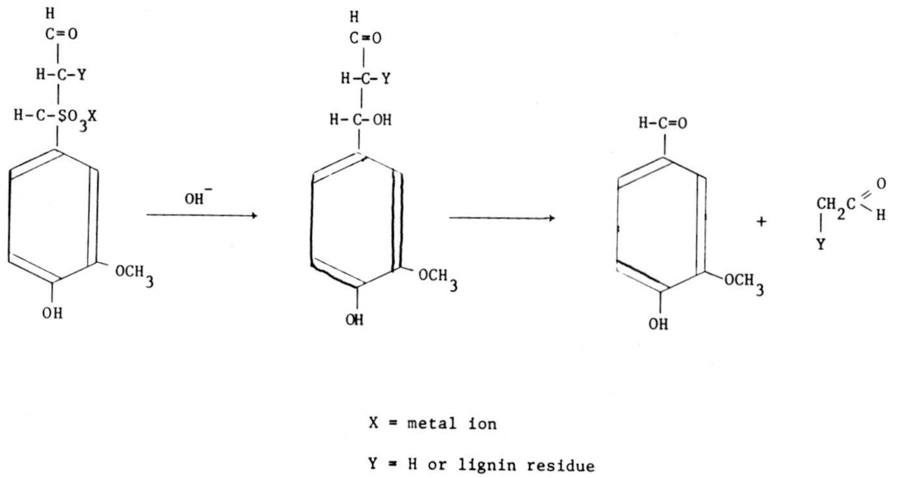

X = metal ion

Y = H or lignin residue

FIGURE 4.   Formation of vanillin by reverse aldol condensation.

scattered throughout the secondary cell wall of the tracheid. This is so because most of the volume of the woody substance is in the tracheid walls and only a low percentage of the volume is in the middle lamella. Thus, it can be seen that lignin is a network polymer deposited between the cells and throughout the cell walls. It is reactive, and therefore easily condensable, owing to its content of a number of reactive functional groups. It is insoluble because of the size of the polymer network and its bonding to carbohydrate.[24]

V. CHEMICALS FROM LIGNIN

Despite its general intractability, a number of chemical uses for lignin have been developed, both by reacting it as a high molecular weight polymer and by converting it to lower molecular weight materials by chemical transformations. With only a few exceptions, these conversions during the period of inexpensive oil and natural gas were economically noncompetitive with processes obtaining the same chemcicals from inexpensive fossil sources. As a result of the rapid price escalations of the last few years, nearly all of these formerly noncompetitive processes need to be reevaluated. Processes based on lignin isolated from chemical pulping liquors have had the most commercial success, and for the near future such lignins (available in extremely large quantities) appear to have the competitive edge over other lignin preparations resulting from chemical transformations of wood.

In chemical pulping of wood, approximately one half of the wood is usually recovered as fiber. The remainder of the original organic material is not recovered, but is used as a fuel mainly to evaporate water from the spent pulping liquors. In the most widely used pulping system (the kraft or sulfate process), it is essential to burn the liquors to recover the inorganic chemicals for recycling. The system could not operate economically without such recovery of pulping chemicals. Therefore, in considering the use of kraft lignins, it is imperative to keep in mind the heat value of the lignin removed from the liquor system, and that the energy equivalent be replaced by an alternate fuel whether it be low-grade petroleum residues, wood residues, or other biomass.

The amount of lignin potentially available for chemical conversion in spent pulping

Table 1
1977 U.S. AND WORLD PRODUCTION OF PULP AND LIGNIN

	Kraft pulp produced, 10^6 short tons	Kraft lignins in liquors, 10^6 short tons	Dissolving and sulfite pulp, 10^6 short tons	Lignin in liquors 10^6 short tons
U.S.	35	21	3.5	2.1
World	94	57	9.7	5.8

Adapted from Hass, L., Kalish, J., and Bayliss, M., 1977 Paper and pulp production for 142 countries, *Pulp Pap.*, 52, No. 7, 103, 1978.

liquors is very large. The production of chemical pulp and lignin for the U.S. and the world in 1977 is shown in Table 1. It must be remembered that this large amount of material is renewable and should be considered as a replacement of at least a part of the worlds finite store of petroleum. The tremendous amount of material has not escaped notice, and many schemes have been proposed to tap this vast source of organic chemicals. Of the many proposed schemes, it has been found that only a few have produced products that have successfully competed with petrochemical production. In the stepwise breakdown of the lignin molecule, it has been shown that the products that retain the unique and characteristic forms they had in the polymer, are those that are most nearly competitive. It has not yet been proven economically feasible to break the lignin completely into its smallest fragments and then recombine the fragments. This is, of course, in major contrast to petrochemical conversions where larger molecules are cheaply and routinely cracked into simple olefins that form the basis for a wide variety of chemical syntheses.

A. Hydrolysis and Similar Degradative Reactions

As has already been mentioned, simple acid-catalyzed hydrolysis reactions applied to lignin do not result in the production of significant amounts of depolymerization. Indeed, the opposite is true for acid-catalyzed heating of spent liquors. These give intractable mixtures of carbonaceous material with little or no commercial value. However, certain alkaline degradations that also involve hydrolytic reactions can split lignin and yield useful lower molecular weight fragments.

It has been known for a long time that vanillin can be obtained from the alkaline heating of lignin, especially lignin sulfonates in spent sulfite liquors.[16] The yield can be improved by the presence of oxygen during the alkaline hydrolysis. A number of successful commercial ventures have been based on this procedure,[26,27] and most of the vanillin now produced is based on lignin as the starting point.

The mechanism of the formation of vanillin by alkaline hydrolysis of spent liquors has been extensively studied over the years. The importance of an α-sulfonic acid function on the phenyl propane group was established by Tomlinson and Hibbert,[28] who proposed that the sulfonic group in the lignin sulfonate was replaced by a hydroxyl group. The resulting structure (Figure 4) can then form vanillin by means of a reverse aldol condensation reaction.

The effect of oxygen or other oxidizing agents on vanillin formation was studied by Freudenberg et al.,[29-31] who showed that yields of vanillin several times higher than those from alkaline hydrolysis alone could be obtained using air, oxygen, and especially nitrobenzene as the oxidizing agents. Yields of up to 20% of the lignin were claimed with the latter reagent, but this reaction has never been commercialized due to the difficulty in separating vanillin from the reduction products of the oxidant. Nitrobenzene has, however, been extremely useful and important in structural studies on lignin. Its oxidation products have been extensively used in establishing various linkages in the lignin polymer.

The vanillin processes of today use alkaline hydrolysis with lime and sodium carbonate in the best process with carefully controlled partial pressures of oxygen. Production costs are high, so that without substantial yield increases or simplification of the processing, it is unlikely that major industrial uses for vanillin will develop beyond its use in food flavoring and perfumery. Increasing amounts are, however, being used in the synthesis of pharmaceuticals such as L-dopa.

Many other products have been isolated in relatively low yields from alkaline and oxidative alkaline hydrolysis of lignin. A listing of such products has been compiled and published.[32]

The oxidations of lignin with strong oxidants such as ozone and alkaline potassium permanganate or with oxygen pressure and heat on lignin solutions and suspensions have been studied. In the past, such oxidations that result in extensive degradation and production of mixtures of simple acids like formic, acetic, lactic, oxalic, and other dibasic acids, have not been considered seriously as commercial sources of such products. With the increasing costs of converting hydrocarbons from petroleum and natural gas to the same products in large petrochemical complexes, it may very well be the time to reevaluate and see if production of these materials by conversion of renewable lignin residues is commercially feasible.

Such conversions were studied and patented by Grangaard in 1960.[33] He found that lignin residues in spent pulping liquors could be almost completely converted to various organic acids by controlling the temperature, pH, and oxygen partial pressure of solutions heated in pressurized reactors. Brink and Schaleger[34] also contemplated the formation of organic acids by wet oxidation of the lignaceous residues resulting from the acidic hydrolysis of wood. In this way, the lignin, which has always been a disposal problem in acid hydrolysis of wood for glucose production, can be completely converted to heat and organic acids. The organic acids can be recycled to aid the wood hydrolysis reaction, or can be recovered, purified, and sold in the chemical market place. In the future, production of these organic acids should become competitive with their preparation by petrochemical operations.

A different kind of alkaline degradation reaction is used to prepare dimethyl sulfide from kraft pulping liquors. It has been known for a long time that dimethyl sulfide is produced in relatively low amounts during kraft pulping. By means of a pressure reaction of kraft black liquor in the presence of sulfide ions, the yield of dimethyl sulfide (along with some methyl mercaptan) can be increased manyfold.[35] A commercial production based on improvements in the basic patent now produces dimethylsulfide from black liquor in commercial quantities.[36] It has been shown that the reactions shown in Figure 5 take place in the dimethyl sulfide process.[37]

As shown in the equations, hydrosulfide ions formed from sodium sulfide in the liquor, attack the methoxyl group of the lignin in a neuclophilic displacement reaction yielding a methyl mercaptide ion. This latter ion is even more nucleophilic than hydrosulfide and quickly attacks another methoxyl group of the lignin forming dimethyl sulfide, which is flashed from the lignin and collected. Only small amounts of methyl mercaptan can be obtained due to the fact that it is formed as an intermediate and is even more reactive than the initial hydrosulfide reactant.

In the same plant that produces dimethyl sulfide, the extremely useful and versatile solvent, dimethyl sulfoxide, is produced by oxidation of the sulfide. It is true that only a small portion of the lignin molecule is converted to a salable product by this procedure, but it does show that the potential for producing chemicals from lignin is very good if the proper reactions to cleave preformed functional groups from the polymer

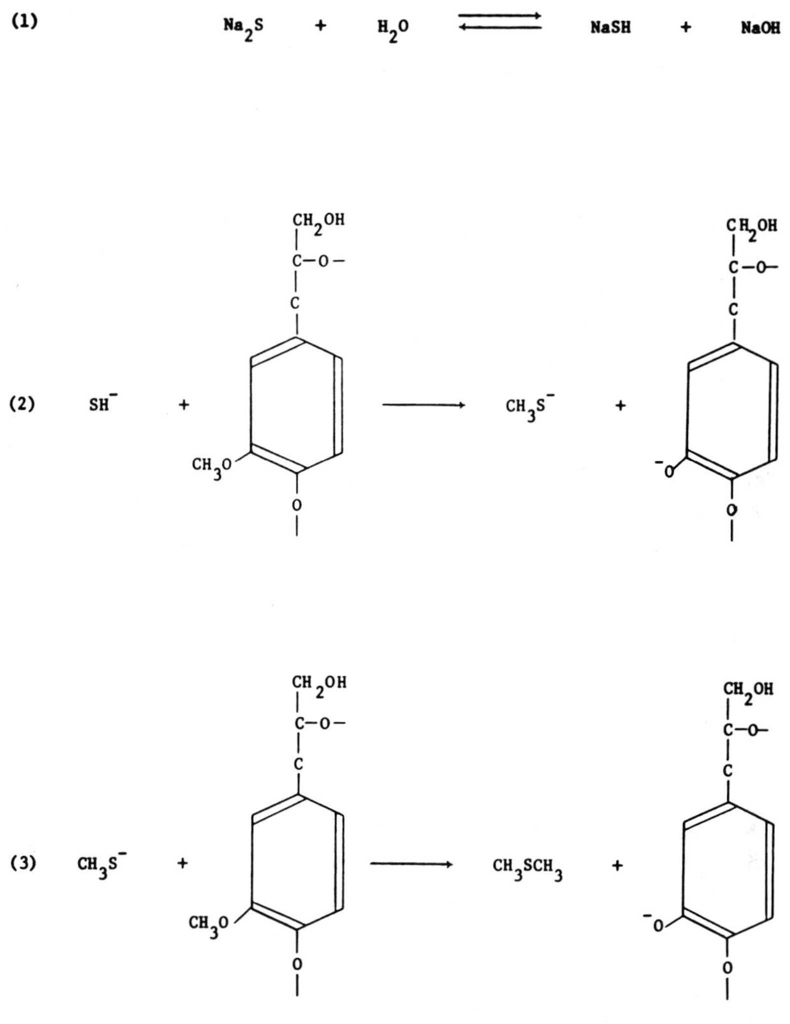

(1)   Na$_2$S + H$_2$O $\rightleftharpoons$ NaSH + NaOH

(2) SH$^-$ + [structure] $\longrightarrow$ CH$_3$S$^-$ + [structure]

(3) CH$_3$S$^-$ + [structure] $\longrightarrow$ CH$_3$SCH$_3$ + [structure]

FIGURE 5.   Reactions involved in dimethyl sulfide formation from lignin.

can be found. One of the elegant features of the dimethyl sulfide reaction is that the residue from the process is returned to the inorganic chemical recovery part of the pulp mill and is disposed of by burning.

Other compounds containing methyl groups can also be made by this same procedure. It has been shown that trimethylamine can be prepared if an alkaline solution of lignin is heated with ammonia. It is essential that sulfide ions be absent from the reaction mixture, because sulfide ions are many times more nucleophilic than ammonia, thus, if both are present, dimethyl sulfide is formed much more readily than the methylamine.[37]

A third type of hydrolytic reaction, generically related to the dimethyl sulfide process, has been studied by Enkvist and co-workers. These workers found that the dimethyl sulfide reaction could be extended to cleave not only the methoxyl groups of the lignin, but also other bonds in the polymer so that substantial quantities of ether-soluble materials could be obtained. By heating kraft black liquor with sodium sulfide and sodium hydroxide in controlled amounts to temperatures of 250 to 290°C, up to 33% of the lignin could be obtained as phenolic products.[38] A number of conditions were studied by these investigators, and their results show that substantial quantities of lower molecular weight products in addition to phenols can be prepared by the rather severe hydrolytic conditions they employ.

To get away from the necessity of pressure vessels, a modification of the process was also studied. It was found that when concentrated spent liquors containing 33% sodium hydroxide were heated to 300°C for 15 to 60 min in an open vessel, severe degradation of the lignin and other materials in the liquors took place.[39] From the initial heating, some dimethyl sulfide and phenolic material was volatilized. The solid obtained was completely soluble in water and was neutralized first to pH 6 liberating phenols, cresols, and guaiacols. Further neutralization yielded simple and complex acids. This procedure is the subject of a Finnish patent,[40] in which the production of monohydric and dihydric phenols, dimethyl sulfide, and phenolic carboxylic acids is claimed. In a further variation, 18% sodium hydroxide was added to 55% kraft black liquor and the solution was spray dried to greater than 90% dryness. The spray dried product was then quickly heated to 500°C, and this gave pyrocatechol and methyl and ethyl homologues at a rate corresponding to 48 kg/ton of air dried pulp from which the lignin was obtained.[41]

Although these procedures generally give complex mixtures of products, and separation of individual compounds may be dificult, this work demonstrates that there is the potential of producing industrially important amounts of phenols and acids from black liquor lignin. Further work in this area is certainly indicated, and Enkvist's work may yet be the basis for the establishment of plants for converting lignin to useful lower molecular weight organic compounds.

B. Hydrogenolysis

Although alkaline hydrolysis reactions yield appreciable amounts to chemical fragments when applied to lignin, the most promising approach to conversion of lignin to useful organic compounds involves the procedures of hydrocracking and hydroreforming that have been so successfully applied in petroleum technology. Ever since the first lignin investigators studied lignin and found phenolic materials such as breakdown products, there has been considerable effort devoted to the conversion of lignin into phenols. At first, the research was directed mainly toward lignin structural determination, but with the establishment of the certainty that lignin has an aromatic structure, serious efforts to produce aromatic compounds, principally phenols, were undertaken. Aside from the hydrolytic procedures already discussed and pyrolytic procedures to be discussed later, the main studies aimed at phenols production have involved reduction by hydrogenation. For almost 60 years, beginning with the studies of Willstatter and Kalb reported in 1922,[42] and Dorée and Hall reported in 1924,[43] and continuing right up to the present, various kinds of reductive reactions have been applied to all kinds of lignin preparations.

The first attempts were noncatalytic in nature. Willstatter and Kalb treated hydrolysis lignin with red phosphorous and hydriodic acid. This did not give phenols, but reduced the lignin extensively and a complex mixture of hydrocarbons was obtained. Dorée and Hall subjected lignin sulfonic acid to the action of nascent hydrogen generated from zinc and acid. Again, no phenols were obtained, and the main effect noted was desulfonation of the lignin sulfonic acid.

Some 10 to 15 years after the initial studies, a number of investigations using the newer techniques of catalytic hydrogenation were undertaken. These studies are typified by those of Harris et al.[44,45] and Adkins et al.[46] These investigators used active hydrogenation catalysts such as copper chromite with various lignins suspended in dioxane. Monomeric cyclohexyl alcohols and neutral products of unknown structure were obtained. Similar results were obtained by Hibbert et al.[47] in structural studies on lignin using similar techniques. These procedures showed that hydrogenolysis was possible, but that the use of high pressures in combination with active hydrogenation catalysts caused the hydrogenation to proceed too far and aromatic phenolic materials could not be isolated in any reasonable amounts.

The work of Lautsch[48-50] was somewhat similar in results to those reported by Harris and Adkins. In catalytic hydrogenations with various catalysts including metal oxides and sulfides, complex mixtures of phenols and hydrogenated phenols were obtained. Noncatalytic reduction of lignin sulfonates by heating them with caustic and alcohol also gave very complex reduction products.

In later studies, two groups of investigators carried out hydrogenolysis studies under conditions that were designed to minimize the reduction of intermediate phenols to alcohols and hydrocarbons. The first group, Inventa A.G. für Forschung and Patent Verwertung, obtained a number of patents, with Giesen[51,52] as the principal inventor, on the conversion of sulfite and hydrolysis lignins to phenolic materials using sulfur resistent catalysts and hydrogen. Another procedure used no catalyst, but heated lignin suspended in a phenolic carrying oil to 300°C in the presence of very high hydrogen pressures. From 30 to 50% of the lignin was obtained as distillable material, and the phenol content of the oils varied from 30 to 45% of the liquid product.

The second group of investigators to study hydrocracking rather than more extensive hydrogenation was the Noguchi Institute of Japan. Basing their work on experience gained from coal hydrogenation, they devised a sulfur resistant catalyst that would liquify a desulfonated lignin sulfonate suspended in a carrier medium of either phenol or a heavy oil obtained from the hydrocracking of lignin. The suspension of catalyst and desulfonated lignin was hydrogenated in stirred autoclave reactors at pressures of about 100 to 200 atm of hydrogen, and temperatures of from 370 to 430°C. The yields of monophenols were claimed to be as high as 44% of the charged lignin, with an additional 22% of heavy oil suitable for recycle. The catalyst was considered to be the critical part of the process and was covered in two Canadian patents.[53]

The Noguchi process was extensively studied by the Crown Zellerbach Corporation in the U.S., and a complete report of these studies has been published.[54] It was found that the claimed yields of monophenols were too high and that part of the phenol used as a suspending medium was being alkylated and was reported as product. The actual yield of monophenols from the Noguchi procedure was 21% of the lignin charged to the reactor. Later work in the U.S. (as yet unpublished in any detail) has increased this yield to 38%, with a 7 to 8% higher phenolic fraction. These are very respectable yields, and this process may very well be competitive in the near future with production of cresylic acids from petroleum and coal. By means of hydrodealkylation reactions that have been studied extensively in petroleum refining, it should be possible to remove alkyl groups from the mixture of alkylated phenols obtained from lignin hydrogenation and convert the mixture to phenol and benzene. These then could serve as raw materials for producing a whole series of aromatic chemicals, all based on the fundamental aromatic structure of lignin.

The preparation of aromatic chemicals from lignin hydrogenolysis appears to be the procedure most likely to be used for the first large-scale industrial conversion of lignin to useful chemicals. The application of this technique to lignin preparations obtained from wood hydrolysis and from new pulping processes should be very interesting, and the absence of sulfur in such lignins may make the processing requirements much more simple.

The argument has been raised that future expansion of coal converting plants will result in availability of enormous amounts of by-product chemicals, including phenol and cresylic acids. As a result, it has been proposed that aromatics from lignin will not be competitive. There are, however, a number of advantages to the use of lignin rather than coal.[55] Among these are

1. Lignin contains little or no nitrogen. Coal often has considerable nitrogen that must be removed during processing.
2. Lignin is already collected and processed in huge amounts so no additional environmental problems need to be addressed. Coal recovery has many adverse environmental features and future large-scale mining ventures particularly in water-deficient or limited areas are viewed with considerable alarm by those persons concerned with maintaining healthful environments.
3. Lignin is renewable and need never be depleted. Coal utilization, no matter how large the reserves actually are, will eventually deplete the resource.
4. Lignin is widespread and can be converted in widely scattered areas. Coal is confined to certain regions.
5. Lignin is composed of aromatic structures with oxygen functions already present. All that is required is the appropriate chemical bond cleavage and saturation of radicals so produced with hydrogen. A calculation, based on the structural formula proposed by Freudenberg, shows that lignin could theoretically be converted to *p*-cresol in nearly 60% yield, so it is really a condensed cresylic acid material. Coal is deficient in oxygen and hydrogen and is really a graphitic-like arrangement of carbon atoms.

Thus, lignin should still be considered as a reasonable potential feedstock for aromatics production, even in competition with increased coal usage in the future.

C. Pyrolysis

Although, as previously mentioned, modern industrial chemistry is largely based on petroleum, at one time destructive distillation of wood was the major industrial chemical source. In this way, large quantities of acetic acid, methanol, wood creosote, and charcoal were produced. The acetic acid was obtained from carbohydrate (cellulose) and methanol and creosote are from lignin. Charcoal preparation is still a fairly large industrial operation, but recovery of the accompanying volatile products is rarely undertaken at the present time. These products, especially the tars and oils generally called wood creosote, are exceedingly complex mixtures that are difficult to purify. Thus, it has become general practice to burn the oils obtained from wood pyrolysis or carbonization in plant boilers for production of steam. With the general rapid escalation of petroleum feedstock costs, it may be the time to reassess wood creosote as a source of chemicals. It should be noted that the issue of complexity has not prevented the separation of many industrially important chemicals from coal tar obtained by carbonization of coal. Certainly coal tar is just as complex as the tars and oils obtained from lignin through wood carbonization.

Efforts have been made to simplify the carbonization by pyrolyzing isolated lignins. Early studies on lignins obtained from acid hydrolysis of wood were carried out by Heuser,[56] Fischer,[57] and Tropsch.[58] In general, these authors obtained about 50% charcoal and 10 to 15% of tar with some acetone, acetic acid, and methanol. Noncondensible gases such as carbon dioxide, carbon monoxide, and methane made up the remainder of the products. In somewhat more elegant experiments, Fletcher and Harris[59] subjected hydrolysis lignin from Douglas-fir to dry distillation at 400°C. This gave a little more coke and a little less tar than the experiments reported by the earlier investigators. However, these authors separated the tar into neutral, acidic, and phenolic fractions, and identified phenol, o- and p-cresol, 2,4-xylenol, guaiacol, 4-methyl guaia-col, and 4-ethylguaiacol among the many products of the phenolic fraction.

Freudenberg and Adam[60] also carried out dry distillation experiments on lignin, but were able to improve on the yields of phenolic material by conducting the experiments in an atmosphere of hydrogen and in the presence of a variety of catalysts. This variation of the pyrolytic procedure resulted in a considerable increase in the amount of phenolic material produced. The best conversion was found to be 35% of the starting lignin as a complex phenolic mixture of ten identified phenols and a large fraction of higher unidentified phenolic products. Once again, the complexity of the mixture, with no single predominant product, prevented commercial application of the procedure. However, this process certainly deserves a reevaluation. It should be readily adaptable to continuous processing at relatively low pressures. With appropriate catalysts, the possibility of producing a less complex mixture of phenols should be enhanced. Integration of the procedure with existing pulping processes could greatly improve the economic aspects.

Thermal breakdown of lignin has been studied and reported in the U.S.S.R., but no commercialization of the procedures has been reported. The main reasons again being low yields (10 to 15%) of complex products.

An interesting pyrolysis of the lignin in kraft black liquors has been reported.[61] In this procedure kraft black liquor was dried and then subjected to rapid pyrolysis in a current of super heated steam at 400 to 500°C. Although the total yield of oil was relatively low (about 10% of the liquor solids), the main single ingredient was found to be guaiacol with significant, but lesser amounts of 4-methyl and 4-ethyl guaiacol. The yield of guaiacol was found to be 2.1% of the liquor solids or more than 6% of the lignin in the liquor solids. The residual material was obtained as a fine, soft char that could easily be suspended in additional black liquor for burning. Thus, the procedure need not upset conventional black liquor recovery furnace operation. As yet, no commercial application of this simple procedure has been made, but it is inconceivable that production of several thousand tons per day of valuable phenolics by the kraft pulping industry will continue to be ignored in the future.

Lignin has also been subjected to extremely high temperature and short time pyrolysis, and this has resulted in surprisingly high yields of acetylene.[62] Either lignin in wood or that isolated from kraft pulping was found to give up to 23% of its weight as acetylene when it was rapidly heated to at least 1200°C and the gaseous products rapidly quenched to room temperature to prevent acetylene decomposition. Although the largest yields of acetylene were obtained from isolated kraft lignin, substantial yields were obtained from ground wood and also from powdered cellulose. For regions that are deficient in natural gas, but rich in wood residues, this process could and should be considered for the preparation of the useful unsaturated hydrocarbon, acetylene, and possibly for the preparation of ethylene that also was produced in lesser yields than those for acetylene.

It would appear that pyrolysis and modified pyrolysis procedures have a great potential in processing of lignin to provide useful low molecular weight chemicals. The best procedures will undoubtedly combine pyrolysis with other degradative procedures. The most likely combination for commercial success appears to be pyrolysis combined with hydrogenolysis. Such processes should be considered for at least part of the chemical industrys future requirements of aromatic chemicals.

D. Condensation and Other Reactions

In addition to the breakdown of lignin into lower molecular weight chemicals, many uses of it in its polymeric form have been developed.[63] In these cases, the lignin is generally used as it is obtained from commercial pulping processes. Lignin is capable of reacting and condensing with various reagents to make its properties more useful.

It has been found possible to make flame retardant materials by halogenation.[64] Either the lignin in kraft liquors or that in sulfite spent liquors can be treated first with chlorine, and then with bromine and chlorine to produce a halogen content of 20 to 40% of the lignin by weight. The lignin precipitates, and can be used as a flame retardant material.

Lignins from both kraft and from sulfite liquors have been studied and used as ingredients and reactive extenders for resins in plywood and particle boards. Both ammonium and sodium lignin sulfonates have been used in combination with conventional phenolic resins as particle board binders. Up to 35% of the resin can be substituted with no significant reduction in board properties.[65] Concentrated sodium lignin sulfonate liquors have also been used in adhesives for gluing plywood veneers. The liquor is mixed with phenol formaldehyde resin, and the mixture is used to coat veneers. When done correctly under controlled pH and temperatures, up to 30% of the resin can be replaced to give plywood with no loss of properties.[66] Another procedure for making plywood adhesives from sulfite spent liquor lignin has been patented. Concentrated sulfite liquor is mixed with phenol in a weight equal to the liquor solids. The mixture is heated for 2 hr at 160 to 200°C, formaldehyde is added, and the pH adjusted to 10.5 with sodium hydroxide. The resin formed is claimed to be suitable for plywood manufacture.[67]

Lignin preparations have long been used as surface active agents. Improvements in these properties can be made by reaction of certain bifunctional compounds with kraft lignin. The bifunctional compounds contain a halogen or epoxide on one end of a hydrocarbon chain and a sulfonate or certain cationic groups on the other end. The product is a surface active agent with improved properties, as compared to regular lignin-based surfactants.[68]

Lignin can also be modified to act as an ingredient in polyols to prepare urethane polymers. Kraft lignin can be treated by copolymerization with maleic anhydride and oxyalkylation with propylene oxide. The lignin-polyester-ether polyols can be treated with a number of isocyanates. The urethanes so produced compare favorably with commercial polyurethanes in strength, swelling, chemical resistance and color.[69]

These chemical modifications of the lignin polymer give increasing evidence that additional utilization of lignin as a chemical in its polymeric form is possible. It may be that the most promising way of chemically utilizing lignin will be to find further ways to modify it as a polymer without resorting to more energetic processes necessary to break it down into low molecular weight fragments.

E. Chemicals From Lignin in Wood

Although lignin from pulping operations probably has a more immediate chance to compete with petroleum chemicals than does lignin in woody residues, such lignin should not be overlooked. Goldstein[70] has proposed that plants for producing chemicals could be built to produce ethanol, ethylene, phenol, and benzene entirely from wood. He has suggested that there are a number of places where sufficient wood not suitable for lumber or pulp production could be collected to provide a continuous supply of several thousand tons per day of raw material. Cellulose and hemicelluloses would be hydrolyzed to yield sugars that would be converted to ethanol by fermentation. The alcohol would be used as such or converted to the most widely used petrochemical building block, ethylene. The residual lignin could be treated by hydrogenolysis and hydroalkylation to yield cresylic acids, phenol, and benzene. With these chemicals a whole host of other products could be produced and a chemical industry based on wood and entirely separated from natural gas, petroleum, and coal could be developed.

In the long run, economics will govern the choice of feedstock selection. So long as there are supplies of oil and natural gas available, considering the tremendous investments in petrochemical plants already built, the fossil feedstocks will have a real economic advantage. Although, as has been shown, the basic chemicals for industry can be prepared from wood and lignin, considerable work must be carried out on the raw product before the final conversion. With the fossil materials, much of the work has already been accomplished by forces in the Earth's crust. So, if fossil supplies were not finite, there would be no economic incentive to seek alternate supplies. However, the fossil sources are being consumed at an alarming rate and certainly, at some date in the relatively near future, their exhaustion and scarcity will cause their price to rise to the point where alternate feedstocks will be able to compete.

Supplies of wood appear to be more than adequate to supplant the fossil sources. Even if the practices that Glesinger believed could increase wood production manyfold do not come to pass, there appears to be sufficient material available. It has been estimated that the annual unused forest biomass in the U.S. may be as much as 180 million t.[71] Such a mass of material is considerably larger than the total of all organic chemicals produced by the petrochemical industry. Thus, all of the chemical industry's feedstocks could be supplied by wood even without increasing the annual wood cut. It all depends on future supplies and, consequently, prices for oil and natural gas. No one can say for sure when the supplies will give out. The only certainty is that at some point scarcity will cause the price to rise to the point where chemicals from biomass, and particularly the forest biomass, will be economically competitive and will have the almost priceless advantage of being replaceable. There is no doubt that chemicals from lignin will be important for mankind's welfare in the future.

REFERENCES

1. Coheen, D. W., Silvichemicals, what future?, paper presented at the 73rd National Meeting, American Institute of Chemical Engineers, New York, 1972.
2. King, M., World resources of fossil organic raw materials, in *Resources of Organic Matter for the Future*, St. Pierre, L. E., Ed., World Conf. Future Sources of Organic Raw Materials, Multiscience, Montreal, 1978, 58.
3. Young, H. E., Forest Biomass Inventory — The basis for complete tree utilization, paper presented at TAPPI Conference, 1977.
4. Glesinger, E., *The Coming Age of Wood,* Simon and Schuster, New York, 1949.
5. Karlivan, V. P., New aspects on production of chemicals from biomass, in *Resources of Organic Matter for the Future,* St.-Pierre, L. E., Ed., World Conf. Future Sources of Organic Raw Materials, Multiscience, Montreal, 1978.
6. Anon., Biomass use for fuel may not be feasible, *Chem. Eng. News,* 55(36), 27, 1977.
7. Kringstad, K., The challenge of lignin, in *Resources of Organic Matter for the Future,* St.-Pierre, L. E., Ed., World Conf. Future Sources of Organic Raw Materials, Multiscience, Montreal, 1978.
8. Wise, L. E. and Jahn, E. C., *Wood Chemistry,* 2nd ed., Reinhold, New York, 1952, 412.
9. Payen, A., Report on the composition of the tissue proper of plants and wood (trans.), *Comp. Rend.,* 7, 1052, 1838.
10. Payen, A., Addendum to a note on the chemical composition of the tissue proper of plants and on the different aggregation states of this tissue (trans.), *Compt. Rend.,* 10, 941, 1840.
11. Watt, C. and Burgess, H., U.S. Patents 1,448 and 1,449, 1854.
12. Tilghman, B. C., U.S. Patent 70,487, 1866.
13. Dahl, C. F., U.S. Patent 296,935, 1884.
14. Erdmann, J., Concerning the concretions in pears, *Ann.,* 138, 1, 1866, and Ann. Supplement V, p. 223 (1867) (in Ger.).

15. Czapek, F. Z., *Physiol. Chem. Phys.,* 27, 141, 1899.
16. Grafe, V., Investigation of woody matter from a chemical-physiological standpoint (trans.), *Monatsh.,* 25, 987, 1904.
17. Lange, P. W., Concerning the nature and distribution of the lignins in spruce wood (trans.,) *Sven. Papperstidn.,* 47, 262, 1977.
17a. Lange, P. W., Ultraviolet absorption and dichroism of lignin in wood, *Sven. Papperstidn.,* 48, 241, 1945.
18. Freudenberg, K., Lignin: its constitution and formation from *p*-hydroxycinnamyl alcohols, *Science,* 148, 595, 1965.
19. Adler, E., Lignin chemistry — past, present, and future, *Wood Sci. Technol.,* 11, 169, 1977.
20. Freudenberg, K., Grion, G., and Harkin, J. M., Detection of quinone methides in the enzymic formation of lignins, *Angew. Chem.,* 70, 743, 1958. (in Ger.).
21. Björkman, A., Isolation of lignin from finely divided wood with neutral solvents, *Nature (London),* 174, 1057, 1954.
22. Freudenberg, K., Sohns, F., Dürr, W., and Niemann, C., Lignin, coniferyl alcohol and saligenin, *Cellul. Chem.,* 12, 263, 1931.
23. Fergus, B. J., Procter, A. R., Scott, J. A. N., and Goring, D. A. I., The distribution of lignin in spruce wood as determined by ultraviolet microscopy, *Wood Sci. Technol.,* 3, 117, 1969.
24. Goring, D. A. I., Polymer properties in lignins, in *Lignins: Occurrence, Formation, Structure and Reactions,* Sarkanen, K. V. and Ludwig, C. H., Eds., Wiley-Interscience, New York, 1971, 698.
25. Lowe, K. E., Mid-1978 status report on U.S. paper industry, *Pulp Pap.,* 52, No. 7, 1978.
25a. Hass, L., Kalish, J., and Bayliss, M., 1977 Paper and pulp production for 142 countries, *Pulp Pap.,* 52, No. 7, 103, 1978.
26. Salvesen, J. R., Brink, D. L., Diddaus, D. G., and Owzarski, P., Process for making vanillin, U.S. Patent 2,434,626, 1948.
27. Craig, D. G. and Logan, C. D., Vanillin and other products from sulfite waste liquor, Canadian Patent 615,553, 1961.
28. Tomlinson, G. H. and Hibbert, H., The formation of vanillin from waste sulfite liquor, *J. Am. Chem. Soc.,* 58, 345, 1936.
29. Freudenberg, K., Lautsch, W., and Engler, K., Formation of vanillin from spruce lignin, (trans.,) *Chem. Ber.,* 73B, 167, 1940.
30. Freudenberg, K. and Lautsch, W., The constitution of pine lignin, (trans.), *Naturwissenschaften,* 27, 227, 1939.
31. Lautsch, W., Plankenhorn, E., and Klink, F., The formation of vanillin from the wood, lignin and sulfite liquor from fir trees, (trans.), *Angew. Chem.,* 53, 450, 1940.
32. Goheen, D. W., Low molecular weight chemicals in lignins, in *Lignins: Occurrence, Formation, Structure and Reactions,* Sarkanen, K. V. and Ludwig, C. H., Eds., Wiley-Interscience, New York, 1971, 803.
33. Gangaard, D. H., Manufacture of cellulosic products, U.S. Patent 2,928,868, 1960.
34. Brink, D. L. and Schaleger, L. L., Chemical production by oxidative hydrolysis of lignocellulose, *Tappi,* 61(4), 65, 1978.
35. Hägglund, E. and Enkvist, T., Method of improving the yield of methyl sulfide, U.S. Patent 2,711,430, 1955; U.S. Patent Reissue 24,293, 1957.
36. Hearon, W. M., MacGregor, W. S., and Goheen, D. W., Sulfur chemicals from lignin, *Tappi,* 45, No. 1, 28A, 1962.
37. Goheen, D. W., Chemicals from lignin by nucleophilic demethylation, *For. Prod. J.,* 12, 471, 1962.
38. Enkvist, T., Turunen, J., and Ashorn, T., The demethylation and degradation of lignin or spent liquors by heating with alkaline reagents, *Tappi,* 45, 128, 1962.
39. Enkvist, T. and Lindfors, T., Manufacture of organic chemicals from spent pulping liquors by alkali treatment, *Pap. Puu,* 48(11), 639, 1966.
40. Enkvist, T., Finnish Patent 37,402, 1969; as cited in *Chem. Abstr.,* 72, 33465V, 1970.
41. Enkvist, T., Phenolics and other organic chemicals from kraft blacking liquors by disproportionation and cracking reactions, *Applied Polym. Symp.,* 1, 285, 1975.
42. Willstatter, R. and Kalb, L., Concerning the reduction of lignins and carbohydrates with hydrogen iodide and phosphorous, (trans.), *Ber.,* 55, 2637, 1922.
43. Dorée, C. and Hall, L., The lignosulfonic acid obtained by action of sulfurous acid on spruce wood, *J. Soc. Chem. Ind., London.,* 43, 2571, 1924.
44. Harris, E. E. and Adkins, H., Reactions of lignin with hydrogen, *Pap. Trade J.,* 107(20), 38, 1938.
45. Harris, E. E., D'Ianni, J., and Adkins, H., Reaction of hardwood lignin with hydrogen, *J. Am. Chem. Soc.,* 60, 1467, 1938.
46. Adkins, H., Frank, R. L., and Bloom, E. S., The products of the hydrogenation of lignin, *J. Am. Chem. Soc.,* 63, 549, 1941.

47. Cooke, L. M., McCarthy, J. L., and Hibbert, H., Hydrogenation studies on maple ethanolysis products, *J. Am. Chem. Soc.,* 63, 3052, 1941.
48. Lautsch, W., Oxidative and reductive degradation of wood, lignin and sulfur containing waste liquor from spruce, (trans.), *Cellul. Chem.,* 19, 69, 1941.
49. Lautsch, W. and Freudenberg, K., Phenol or its derivatives from lignin or lignaceous materials, German Patent 741,686, 1943.
50. Lautsch, W. and Piazolo, G., Concerning the hydrogenation of lignin and lignin-containing materials with hydrogen yielding media, especially alcohols, (trans.), *Ber.,* 76, 486, 1943.
51. Swiss Patent 305,712, 1955.
52. Giesen, J., Process for cleavage of lignin to produce phenols, U.S. Patent 2,991,314, 1961.
53. Oshima, M., Maeda, Y., and Kashima, K., Process for liquefaction of lignin, Canadian Patents, 700,209 and 700,210, 1964.
54. Goheen, D. W., Hydrogenation of lignin by the Noguchi process, in *Lignin, Structure and reactions,* Marton, J., Ed., American Chemical Society, Washington, D.C., 1966, 205.
55. Goheen, D. W., Chemicals from lignin, paper presented to the 8th World Forestry Congress, Jakarta, Indonesia, 1978.
56. Heuser, E. and Skiöldebraud, C., Destructive distillation of lignin, (trans.), *Z. Angew. Chem.,* 321, 41, 1919.
57. Fischer, F. and Schrader, H., The dry distillation of lignin and cellulose, *Ges. Abh. Kennt. Kohle,* 5, 106, 1920, (in Ger).
58. Tropsch, H., concerning the dry distillation of lignin in a vacuum (trans.), *Ges. Abh. Kennt. Kohle,* 6, 293, 1921.
59. Fletcher, T. L. and Harris, E. E., Destructive distillation of Douglas fir lignin, *J. Am. Chem. Soc.,* 69, 3144, 1947.
60. Greudenberg, K. and Adam, K., The low-temperature carbonization of lignin in a stream of hydrogen, (trans.), *Ber.,* 74, 387, 1941.
61. Goheen, D. W., Orle, J. V., and Wither, R. P., Indirect pyrolysis of Kraft black liquors, in *Thermal Uses and Properties of Carbohydrates and Lignin,* Shafizadeh, F., Sarkanen, K. V., and Tillman, D. A., Eds., Academic Press, New York, 1976, 227.
62. Goheen, D. W. and Henderson, J. T., The preparation of unsaturated hydrocarbons from lignocellulose material, *Cellul. Chem. Technol.,* 12, 363, 1978.
63. Goheen, D. W. and Hoyt, C. H., Polymeric products in *Lignins: Occurrence, Formation, Structure and Reactions,* Sarkanen, K. V. and Ludwig, C. H., Eds., Wiley-Interscience, New York, 1971, 833.
64. Zeigerson, E. and Block, M. R., Novel flame retardants, U.S. Patent 3,962,208, 1976.
65. Raffael, E. and Rauch, W., Use of sulfite liquor in combination with alkaline phenolic resins as particleboard binder, *Holzforschung,* 27(6), 214, 1973.
66. Raffael, E., Rauch, W., and Beyer, S., Lignin containing phenolic-formaldehyde resins as adhesives for gluing veneers, Part I, *Holz. Roh. Werkst.,* 32(6), 225, 1974.
66a. Raffael, E., Rauch, W., and Beyer, S., Lignin containing phenolic-formaldehyde resins as adhesives for gluing veneers, Part II, *Holz. Roh. Werkst.,* 32(12), 469, 1974.
67. Plywood adhesives from sulfite pulp spent liquor, Japanese Kokai Patent 7,401,642, 1974; as cited in *Chem. Abstr.,* 80(26), 147,112, 1974.
68. Falkehag, S. I., Modified lignin surfactants, U.S. Patent 3,865,803, 1975.
69. Glasser, W. G. and Hsu, O. H., Polyurethane adhesives and coatings from modified lignin, *Wood Sci.,* 9(2), 97, 1976.
70. Goldstein, I. S., Potential for converting wood into plastics, *Science,* 189, 847, 1975.
71. Sarkanen, K. V., Renewable resources for the production of fuels and chemicals, *Science,* 191, 773, 1976.

Chapter 9

TURPENTINE, ROSIN, AND FATTY ACIDS FROM CONIFERS*

Duane F. Zinkel

TABLE OF CONTENTS

---

* This chapter was written in late 1978; therefore, some of the facts (i.e., prices) may be out-of-date today.

I. INTRODUCTION

Rosin and turpentine, the traditional products of the naval stores industry, are established commodity chemical raw materials from forest conifers, primarily pines. Both rosin and turpentine, as well as the degraded products of rosin, tars, pitches, and oils, are derived from the oleoresin in the wood (an oleoresin consists of a resin and an essential oil; in pine oleoresin these are rosin and turpentine, respectively). As methods of producing naval stores have changed, so have the quality and nature of the products. One result of changing methods is the emergence of (tall oil) fatty acids as a new commodity from this industry. Current annual production of rosin, turpentine, and tall oil fatty acids in the U.S. is over 1 billion lb with a value of $200 to 300 million.

II. SOURCES OF NAVAL STORES

A. Historical

"Make yourself an ark of gopherwood
put various compartments in it and
cover it inside and out with pitch."

In these instructions to Noah (Genesis 6:14),[1] are chronicled the earliest known reference to the use of "naval stores." (The term first appeared in 17th century English records for naval-oriented commodities including pitch and tar.) As late as the 19th century, pitch was used for caulking ships to keep them seaworthy. Even 100 years after the demise of commercial wooden ships, naval stores remains the generic name for the commodities of rosin and turpentine.

Theophrastus, in his *Enquiry into Plants*[2] (approximately 300 B.C.), gives detailed descriptions of various techniques for gathering oleoresins from the terebinth *(Pistacia terebinthus)* and from several pines. Pitch was an important commodity in the ancient centers of power of insular Greece, Macedonia, Asia Minor, and Egypt. Pitches and volatile oils were produced by cooking conifer exudates in open pots. A sheepskin stretched across the pot would catch the vapor and the turpentine then could be wrung out by hand. The residue remaining in the pot was the pitch. The term colophony, used for centuries to denote crude pine exudate or rosin, was derived from Colophon, the ancient Greek name for the general coastal region and islands of western Asia Minor.

Theophrastas also chronicled[2] a Macedonian method for producing pitch by fire, an ancient approach to obtaining chemicals from biomass. An updated version of this tar-burning process is depicted in the vignette of a medicine stamp, circa 1870 (Figure 1). The process involved the controlled, air-deficient burning of a pile of resinous wood stacked in the shape of a hive and covered with earth. Pine tar was collected at the base of the hive; the higher viscosity pitch was obtained by burning the tar in open kettles. This method and the collection of pine exudates continued to be the primary production methods for naval stores into the 20th century.

B. American Naval Stores Industry

Because of the importance of pitch in keeping a wooden navy afloat, any seafaring nation of significance needed a controlled and dependable source of supply. Thus, England's desire to eliminate her dependency on Scandinavian sources of naval stores contributed to the settlement of North America, as did the great potential for producing naval stores from the pine forest of the New World. Colonial settlers, particularly

FIGURE 1.   Ancient process of making pine tar by fire is depicted in this medicine revenue stamp of the 1870s.

in Virginia and North Carolina, were charged to produce tar and pitch as payment for their charters.

As the forests of pines suitable for naval stores production disappeared in the East, the industry moved into the Southeast. Today this area continues to be the center of U.S. naval stores production.

1. Gum Naval Stores

Wounding pines and collecting the exudate is an operation referred to as turpentining. The collected oleoresin, more commonly called pine gum, is then processed to rosin and turpentine. Early wounding techniques, such as boxing, were highly detrimental to the trees. Extensive tree mortality resulted not only as a direct consequence of the wounding, but also by the effects of insects, fire, and wind on the weakened trees. Vast improvements have been made in turpentining operations due to the pioneering efforts of Dr. Eloise Gerry[3] of the Forest Products Laboratory and the continuing work at the Southeastern Forest Experiment Station. Whereas the timber was by and large wasted after being subjected to the old turpentining methods, modern methods leave the trees saleable for wood products.

Processing of oleoresin remained essentially unchanged for centuries. Replacement of cast iron retorts by copper stills improved products considerably. However, most important was the development and widespread use of the Olustee process[4] that resulted in improved yields of higher quality turpentine and rosin. The process consists of these steps:

1. Dissolve crude oleoresin in turpentine to lower the density to less than that of water.
2. Filter to remove trash
3. Wash with dilute oxalic acid to remove iron contaminants that promote oxidation and color formation.
4. Distill the oleoresin solution with steam sparge to separate turpentine and rosin.

In this country, the gum naval stores industry (now located in Georgia) is based on two southern pine species, slash (*Pinus elliottii* Engelm.) and longleaf (*P. palustris* Mill.).

There is no American equivalent of gathering fossil resins such as the copals or the more valuable gem quality ambers found in the Baltic region. However, an era existed

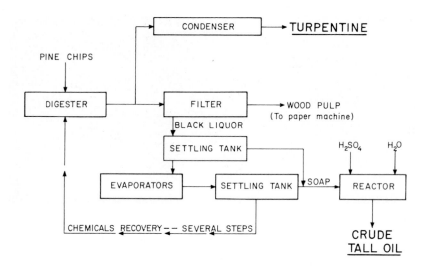

FIGURE 2.   Recovery of naval stores by-products from the kraft pulping process.

in the U.S. during which rosin was mined from lakes and streams of the Carolinas. These deposits were from an earlier period in which turpentine was in demand, but rosin had little value and was discarded into adjacent waters. Some of this rosin mining was still being done in the early 1920s, by which time at least 10 million lb had been recovered.

2. Wood Naval Stores

An outgrowth of the production of naval stores by tar burning was the destructive distillation process. Resinous wood was heated in large retorts, producing charcoal in addition to wood turpentine, various pine oils, and tars. With the demise of wooden ships, the need for pine tar diminished greatly. However, the industrialization of the nation created new demands for raw materials including rosin. These new demands not only required an increased supply, but also required higher quality.

In the early 1900s, a new method for producing naval stores was developed that supplanted the tar-burning and destructive distillation methods. The new method utilizes the Yaryan process, a process originally designed to extract flaxseed with petroleum hydrocarbons. The raw material for the process is virgin pine stumps, from which the sapwood has rotted away. The remaining resinous heartwood that contains about 25% extractives, is chipped, shredded, and extracted with various petroleum solvents (methyl isobutyl ketone is also used). The extract is distilled to recover the solvent, turpentine, pine oil, and a crude resin. This dark red resin is further purified by selective absorption on fuller's earth or by extraction with furfural. A decolorized wood rosin and a dark pitchlike product are obtained.

3. Sulfate Naval Stores

By-product streams from the kraft pulping process are the source of sulfate naval stores (Figure 2). As pine chips are cooked to produce pulp, the volatilized gases are condensed to yield sulfate turpentine. During the pulping, the alkaline pulping liquor also saponifies the fats and converts fatty and resin acids to the sodium salts. In recovering the pulping chemicals, the aptly named black liquor is concentrated in multiple-effect evaporators. Depending upon the design of the system, the soaps can be skimmed from the surface of the black liquor at various points in this operation. The skimmed soap is acidified to yield a material known as crude tall oil.

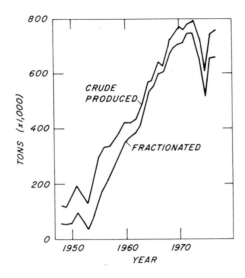

FIGURE 3.   Tall oil production and fractionation in the U.S.

The term "tall oil" is derived from the Swedish word "tallolja" that translates as pine oil. However, such literal translation would have caused confusion with the essential oil known as pine oil. Thus, the simple transliteration to tall oil.

Crude southern tall oil contains 40 to 60% resin acids (rosin), 40 to 55% fatty acids, and 5 to 10% neutral constituents. During the early years of the recovery process, most tall oil was burned for its fuel value. Although distillation units for tall oil purification provided commercial products from the mid-1930s, commercial-scale production of high-purity fatty acids and rosin was not demonstrated until 1949. After 1953, the rate of tall oil production and fractionation increased rapidly (Figure 3).

By far, the major portion of tall oil is now being fractionated into tall oil rosin and fatty acids. Fractionation (Figure 4) of one ton of tall oil yields approximately 600 lb of fatty acids, 700 lb of rosin, and 700 lb of intermediate (distilled tall oil), head, and pitch fractions. Fractionating capacity in the U.S. is over 1 million tons/year, nearly three times that of the rest of the world.

4. Current Production

The production of naval stores since 1930 is presented for rosin and turpentine in Figures 5 and 6. Whereas gum naval stores accounted for over 80% of both rosin and turpentine production in 1930, the contribution from this source has shown a steep, nearly linear decline to about 4% of present naval stores production. (In contrast to this 4% of U.S. production, about half of the world production of rosin and turpentine is of the gum type.) Since 1950, sulfate rosin and turpentine have filled the void created by decreased production of gum and wood naval stores. Pulping by-products now account for over 80% of the turpentine and 60% of the rosin produced in the U.S.

Escalating labor costs have been the prime factor in decreased production from turpentining. Neither improved practices and chemical stimulation[5] nor development of high gum-yielding progeny[6] have been able to stop the rapid decline in gum naval stores production. Labor costs and shortage of raw material have caused major decreases in rosin and turpentine production from virgin pine stumps. This stumpwood supply is finite and is not a renewable resource. Second-growth stumps are not satisfactory because there is not sufficient resin-rich heartwood for economical processing. Although

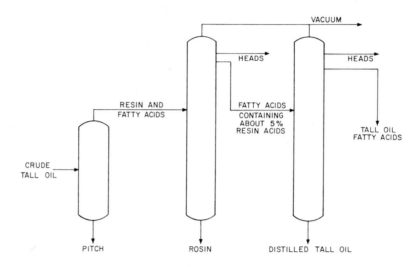

FIGURE 4.   Fractionation of crude tall oil.

there is a large supply of ponderosa pine stumps[7] and commercial processing is imminent,[8] the extract has a high content of neutrals that affects either the product quality or cost of processing.

While recovery of kraft pulping by-products has increased dramatically, it has not been able to compensate completely for the loss in production from the other sources. Output of these by-products is correlated, obviously, with production of kraft pulp from pine. However, output is further influenced by wood sources (e.g., decreased tall oil potential in younger trees of shorter rotation forest management and inclusion of hardwoods in the furnish), the use of purchased chips and the storage of wood inventory as chips, pulping conditions, recovery efficiency, and by-product processing. Although improvements in sulfate by-product recovery are possible, such as the use of stand-by storage[9] to minimize the losses of turpentine and tall oil precursors from chips, drastic changes in the output of naval stores as pulping by-products do not appear likely in the years immediately ahead. Unless economics stimulate greater production of gum oleoresin, new sources of pine silvichemicals will be needed.

5. Potential New Sources

Several sources can be considered as possibilities for naval stores: by-products from solvent drying of lumber, conifer foliage oleoresin (extractives) from logging slash, and induced lightwood.

a. By-Products From Solvent Drying

Solvent drying[10] of conifer lumber with water-miscible solvents such as acetone has some potential; however, even with credit for extractives from by-product recovery, the economics remain unfavorable because of solvent losses and energy requirements. Some hardwood railroad ties are being dried with water-immiscible solvents as an inventory control method, even though this is more expensive than the usual air-drying. Hardwoods, however, are not a source of naval stores.

b. Conifer Foliage

The greatest potential for large volume use of conifer foliage is in animal feed supplements.[11] Pine needles have about 40% available carbohydrate and are 40% digesti-

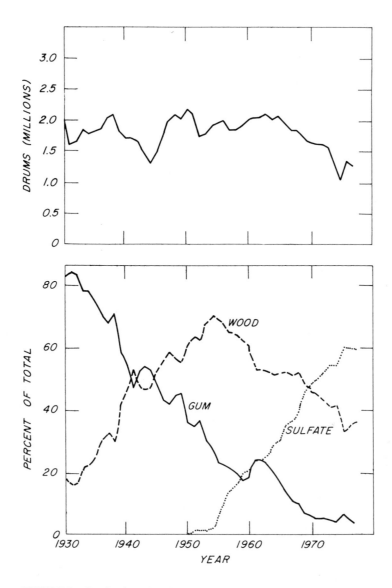

FIGURE 5. Production of rosin in the U.S. Top, in drums (520 lb), bottom, by source.

ble in in vitro ruminant digestibility assays.[12] A limited quantity of a high-quality pine needle oil is presently distilled from some southern European species. Scots pine has also been considered a source of oil.[13] The needle oils of most American pine species, however, too closely resemble turpentines to command any special value. Although the rosins that could be derived from needles of many pines are essentially similar to the xylem rosins, the needle rosins from other species are of unusual composition.[14] These rosins could be the source of specialty fine chemicals, but would not have any significant impact on rosin as a commodity material.

c. Induced Lightwood

The finite supply and nonrenewability of virgin pine stumpwood fostered an interest in producing artificially induced resinous wood as a future raw material for the wood naval stores industry. For some time, any attempts were unsuccessful. Recently, how-

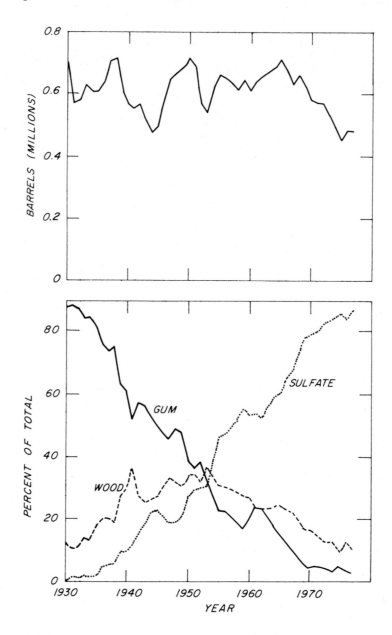

FIGURE 6.   Production of turpentine in the U.S. Top, in barrels (50 gal), bottom, by source.

ever, research at the U.S. Forest Service Naval Stores and Timber Production Laboratory (Olustee, Fla.) to develop new stimulants to improve gum (oleoresin) yields produced an unexpected dividend. Scientists found that treatment of pines with certain dipyridyl herbicides (paraquat, diquat) stimulate extensive oleoresin formation and subsequent diffusion into the wood substance.[15] Extractives contents as high as 40% have been found in some wood samples.[16] Such oleoresin-soaked wood is known as lightwood (despite its relatively high density) because it kindles and burns readily and at one time, served as a source of light (candlewood). Not only does the oleoresin-soaked wood occur in the immediate area of treatment (Figure 7) as in turpentining,

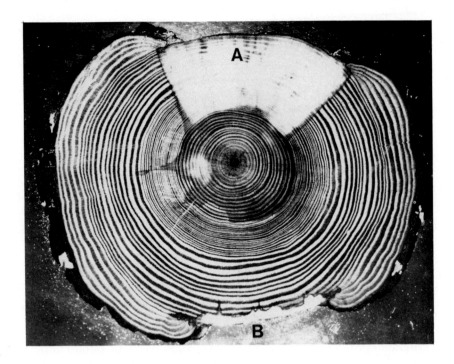

FIGURE 7. Induced lightwood formation on treatment of longleaf pine: (A) with the herbicide paraquat, (B) compared with limited resin-soaking during normal turpentining.

but the zone of lightwood formation extends to the pith and many feet along the trunk above the point of treatment. Among the conifers, only the pines give an adequate lightering response for commercial consideration.

Research activity on lightwood has been intensive since discovery of the phenomenon in 1973. The results of most of this research have been published yearly as the *Proceedings of the Lightwood Research Coordinating Council Annual Meeting.**

Label registration has been approved, permitting commercial use of paraquat to induce lightwood for naval stores production. However, this does not imply that all problems in the technology have been solved. Differences of opinion exist on the impact of a treatment-related insect problem, as well as the overall economics of naval stores from lightwood. Nevertheless, improved treatment methods and alternative lightwood-inducing chemicals, could offer solutions to the problems.

Even with all problems solved, the current excess world inventory of rosin precludes immediate implementation of the lightwood technology. Unfortunately, this situation also obscures the view of the longer-term implications of lightwood naval stores. A primary effect of the implementation of lightwood technology should be to stabilize rosin and turpentine production against the chronic shortages seen in the past, and thus, secure current markets. This would be beneficial to the entire naval stores industry. With further implementation of the technology, sufficient supplies of naval stores would be available for new large-volume markets. Doubling U.S. and world rosin and turpentine production should be readily achievable.

* Information as to the availability of the proceedings can be obtained from the Pulp Chemicals Association, 60 East 42nd St., New York, N.Y. 10017.

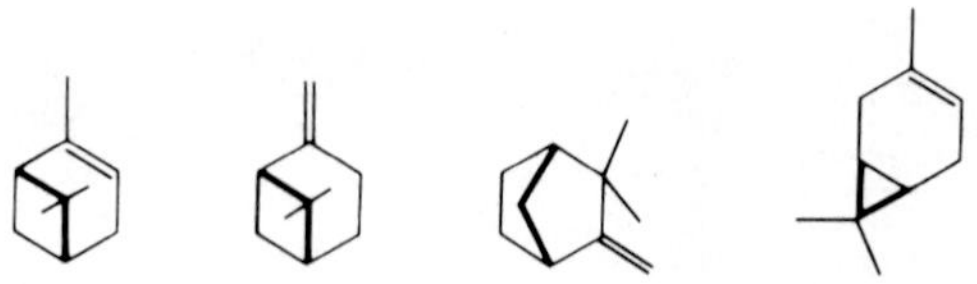

α-Pinene   β-Pinene   Camphene   Δ³-Carene

FIGURE 8.   Typical composition of some commercial turpentines.

III. TURPENTINE

The major use for turpentines once was as a solvent in paints, but in the past 15 years turpentines have been used increasingly as chemical raw materials. As a result, the price of turpentine has increased from $0.20/gal ($0.03/lb) in 1962 to over $1.00/gal (about $0.15/lb) in recent years.

A. Composition

Components of turpentines are important to end use. Structures of the major components and typical compositions of various turpentines are given in Figure 8. The composition of sulfate turpentine can be considerably different from that indicated in the figure because of the number of species (pines and other conifers) pulped, and the large geographical range from which they are taken. Western sulfate turpentine has appreciable amounts of 3-carene (as do Asian and European turpentines), but this component has only limited uses other than as a solvent. Wood turpentines are relatively consistent in composition because the stump raw material is mostly from longleaf pine. The β-pinene of commercial turpentines is always the *l*-isomer, but the α-pinene varies. Among the southern pines, only slash pine has *l*-α-pinene. The α-pinene of other important species is either *d* or *dl*. The α-pinene from sulfate turpentine usually has about 30% optical activity (dextrorotatory).[17]

	Gum (%)	Wood (%)	Sulfate (%)
α-Pinene	60—65	75—80	60—70
β-Pinene	25—35	0—2	20—25
Camphene	—	4—8	—
Other terpenes	5—8	15—20	6—12

B. Current Uses

Gum turpentine is valued at 1.50 to $1.60/gal. Most of this turpentine is repackaged for the retail market or is used as a raw material for small-volume specialty products. Wood turpentine is priced at $1.25/gal. Crude sulfate turpentine (0.50 to $0.60/gal) is refined by fractional distillation and chemical treatment, primarily to remove organic sulfur contaminants. Although most of the refined sulfate turpentine is used captively, it can be assigned a comparative value of 1.00 to $1.10/gal.

The primary use of turpentine (Figure 9) involves the hydration of α-pinene by aqueous mineral acids to synthetic pine oil whose primary constituent is α-terpineol. The process can be controlled to favor further hydration

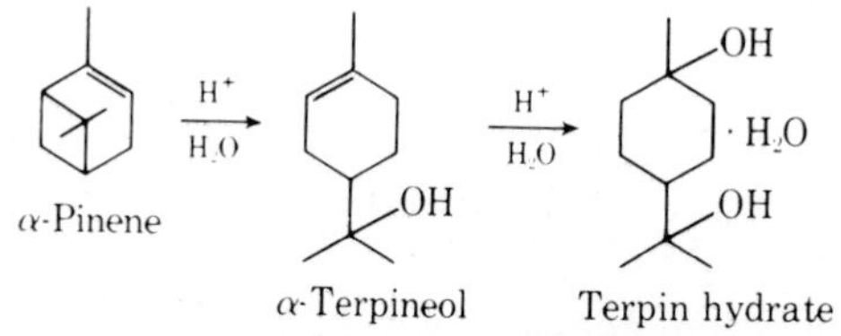

α-Pinene   α-Terpineol   Terpin hydrate

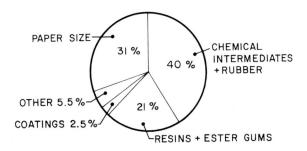

ROSIN

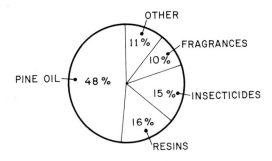

TURPENTINE

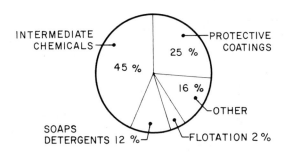

FATTY ACIDS

FIGURE 9. Uses of naval stores materials.

to form the well-known expectorant, terpin hydrate. Variations in the reaction conditions, fractionation, and blending give pine oils with different compositions and properties. These pine oils find use in mineral flotation, in processing textiles, and as solvents, odorants, and bactericides. Over 80% of the pine oil produced is synthetic. The remainder is the so-called natural pine oil that is obtained on fractionation of the stumpwood extractives.

The second largest and growing use of turpentine is in the production of polyterpene resins. A variety of low molecular-weight (600 to 1500 daltons) polymers are produced

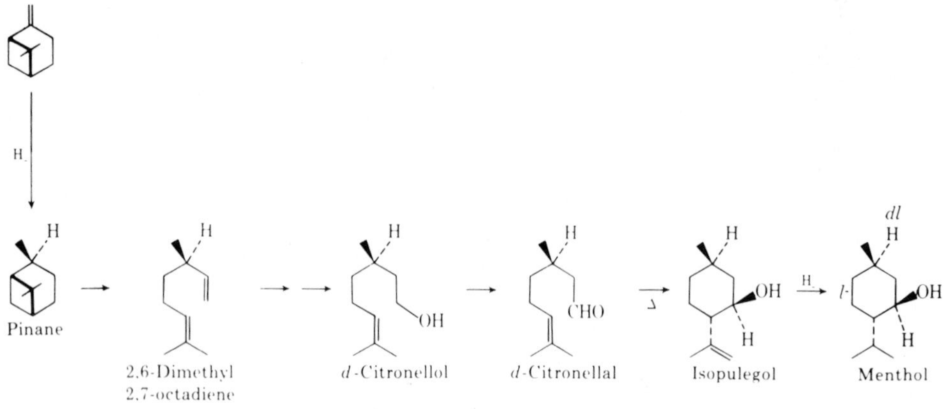

FIGURE 10.   Synthesis of menthol.

from β-pinene, α-pinene, *d*-limonene (from orange and lemon peels), and dipentene, and mixtures of monomers including terpene-styrene. Cationic polymerization of the different monoterpenes yields polymers having differing repeating units. Chain termination has been postulated as a nonpropagating reaction involving camphenic carbenium ion.[18] The resultant polyterpene resins have unique tack and thermoplastic properties and find use in pressure-sensitive adhesives, coatings, sizing, rubber compounding, and other industrial applications.

The terpene insecticides are based on camphene that is obtained either from wood turpentine or by isomerization of α-pinene. The toxaphene-type insecticides are prepared by chlorinating camphene or camphene-α-pinene mixtures to form products having an approximate empirical formula of $C_{10}H_{10}Cl_8$. The future of these chlorinated insecticides, however, is uncertain.

Perhaps the most interesting area of turpentine utilization is in the small, but growing production of flavor and fragrance chemicals.[17] Synthesized products such as lemon, lime, peppermint, spearmint, and nutmeg essential oils are being marketed along with a wide range of fine chemicals for the flavor and fragrance industry.

β-Pinene is the primary starting material in the synthesis of these fine chemicals. Tonnage quantities of *l*-menthol are produced from *l*-β-pinene by synthesis through a sequence of reactions involving *d*-citronellol as shown in Figure 10. Racemic menthol is produced as a by-product of this process and also from optically inactive materials such as 3-methene and *dl*-citronellol. Menthol is used for its cooling effect (the *l*-isomer is most active) in cigarettes and cosmetics and in flavoring, particularly peppermint.

The synthesis of other fine chemicals (Figure 11) does not require optically active β-pinene, as the first step involves pyrolysis of β-pinene to myrcene. The demand for β-pinene for these fine chemicals and for polyterpene resins has resulted in the development of a commercial process for preparing β-pinene by isomerization of α-pinene.

The addition product of myrcene and HCl is reacted with sodium acetate to form the acetates of linalool, geraniol, and nerol that can then be hydrolyzed to the corresponding alcohols. The reaction conditions can be controlled to yield either linalool or geraniol-nerol products. Linalool and its esters have lilac-like fragrances. Geraniol and nerol and their esters are rose-like as is their hydrogenation product, citronellol. Rearrangement of geraniol-nerol over copper catalyst produces either citronellal or a mixture of citrals (geranial and neral), depending on the conditions used. Citronellal finds use as an odorant and can be reduced to citronellol or hydrated (via the bisulfite adduct to prevent cyclization) to hydroxycitronellal which has a lily-of-the-valley fra-

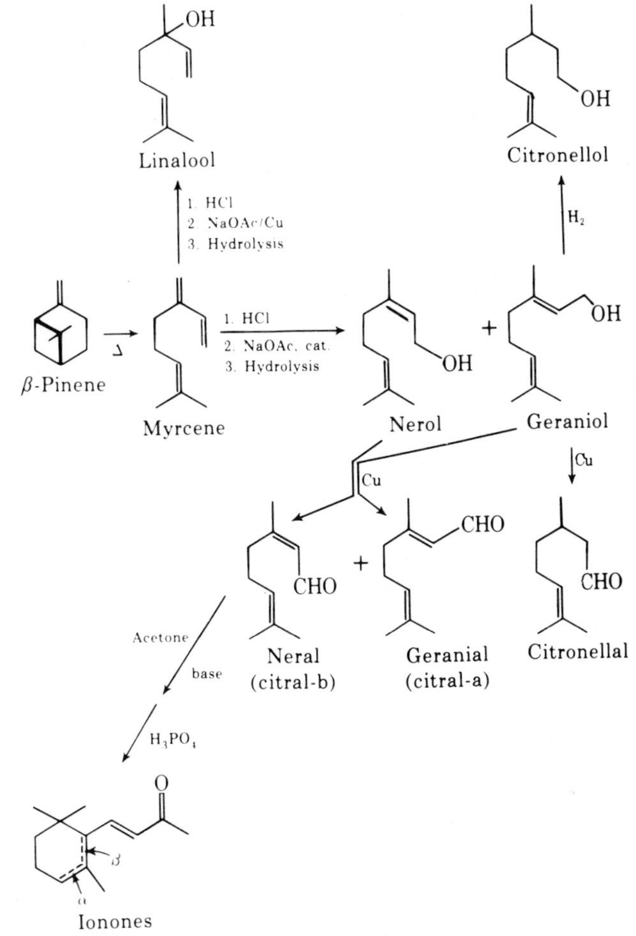

FIGURE 11. Synthesis of some flavor and fragrance chemicals from β-pinene.

grance. Citral is used for its intense lemon character in the preparation of citrus flavors and fragrances. Condensation of citral with acetone followed by isomerization produces a mixture of the violet fragrance ionones. However, the major use for citral, a use requiring tonnage quantities, is in the manufacture of vitamins A and E.

In addition to the use for insecticides, camphene from α-pinene is also used to prepare isobornyl acetate, a product having the characteristic pine

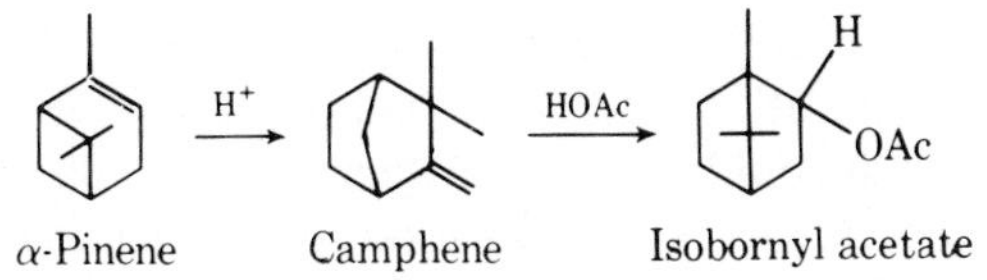

needle scent. Saponification of isobornyl acetate and oxidation of the isoborneol to dl-camphor is a commercial process in several countries, but not in the U.S. today.

The above examples show the use of synthesized flavor and fragrance chemicals to replace diminishing supplies of natural materials. However, other nonnatural products

are used as specialty perfumery ingredients. The condensation of β-pinene with formaldehyde yields nopol, and alcohol with a woody note. Lyral, the condensation product of myrcenal and acrolein, has a fragrance similar to hydroxycitronellol.

Nopol   Lyral

    A number of other monoterpene products are obtained by esterification, fractiona-
tion of turpentine or the reaction products described above, and by other processes.
One such product is ρ-menthane hydroperoxide, which is used as an initiator in low-
temperature polymerization of synthetic rubber.
    In addition to terpenes, turpentine is a source of phenolic-type flavor and fragrance
materials. Estragole (methyl chavicol) and anethole are both isolated during fractionation. The estragole can be isomerized to anethole, which is well-known for its licorice-
anise flavor. Oxidation of anethole produces anisaldehyde, which has wide use in perfumery because of its hawthorne or coumarin-type fragrance.

Methyl chavicol   Anethole   Anisaldehyde

C. Future Utilization
    Current markets for turpentine encompass a wide variety of commodity and fine
chemicals valued from 0.15 to $10.00/lb (Table 1). The markets for many of these
materials will grow and prosper. Pine oils are established products that do not depend
only on fads in promotions of household cleaning formulations. The growing demand
for hot melt and pressure-sensitive adhesives provides an expanding market for the
polyterpene resins. The world-wide demand for synthetic flavors and fragrances will
escalate as improvements in the standard of living decrease the availability of labor-
intensive natural products. Even if a ban or stringent control is placed on use of chlorinated terpene insecticides, developments in present turpentine markets should bring
supply and demand back into good balance in relatively short order.
    Although increased availability of turpentines by commercialization of the induced
lightwood technology would be gradual, the quantity of available turpentine could be
doubled in a few years. That could translate to an additional 100,000 tons of turpentine
each year. The method of recovery of lightwood turpentine will be important to the
quality and price of the material. Extraction or isolation by presteaming prior to kraft
pulping of the wood substance would yield a high-quality turpentine having greater
flexibility for end uses. Because of an increased proportion of β-pinene, lightwood
turpentine has an additional appeal.[19]
    Because the basic and applied chemistry of monoterpenes has been studied exten-
sively, significant large-volume uses should be feasible with off-the-shelf technology.

Table 1
COMPARATIVE PRICE[a] ($/lb) FOR SOME TURPENTINE PRODUCTS

Commodity Chemicals

Product	Price	Product	Price
α-Pinene (96%)	0.13	α-pinene resins	0.60
β-Pinene (85%)	0.30	β-pinene resins	0.45
Dipentene	0.10—0.13	Dipentene resins	0.38
Camphene	0.40	Toxaphene (90%)	0.42
Terpineol	0.46	Pine oil natural	0.25
p-Methane hydroperoxide (55%)	0.82	synthetic	0.28—0.35

Fine Chemicals and Organoleptics

Product	Price	Product	Price
α-Pinene	1.65	Citronellol	2.50-3.00
β-Pinene	2.25	Citronellal	2.75
α-Terpineol	1.45	Hydroxycitronellal	6.00
Menthol (and esters)	7.00—10.00	Dihydrolinalool	3.50
Geraniol	2.80	Dihydroterpineol	1.25
Nerol	4.00	Pseudoionone	2.70
Linalool	2.40	Nopol	1.75
Acetate	2.55	Anethole	4.00
Citral	2.00	Methyl chavicol	11.00

Synthetic Essential Oils

Product	Price	Product	Price
Bergamot	3.75	Siberian pine needle	3.25
Lemon	3.00—6.00	Bois de rose	4.60
Lime	3.50—8.00	Nutmeg	3.80—5.00
Peppermint	10.00—11.00	Spearmint	8.25

[a] Approximate prices as of September 1978 and do not represent quotes. The prices are not corrected for contracts, shipping, etc.

For example, turpentine components can be converted to isoprene for polymerization to polyisoprene-type synthetic rubber for use in tires (petrochemical polyisoprene precursors already are in short supply). Pinenes can be converted by well-known rearrangements to dipentene and then to p-cymene that can be oxidized to terephthalic acid for use in the diverse and large-volume area of polyester products. The p-cymene can also be oxidized

p-Cymene   Terephthalic acid

Pinonic acid

to pure *p*-cresol, currently selling for about $0.80/lb. Turpentine can be oxidized to pinonic acids (the esters are good plasticizers) and related derivatives.[20]

Turpentine has been used for its energy value in lamps, and more recently has been considered for blending in gasoline. However, economics would require an increase in the price of gasoline to well above $1.00/gal for this to be feasible.

The availability of a renewable chemical raw material as a substitute for petroleum is appealing. However, comparing even optimistic estimates of potential turpentine production will show that turpentine alone or with rosin could make only a small contribution in replacing petrochemical feedstocks. Nevertheless, the naval stores industry would benefit by having large volumes of turpentine to serve as feedstocks for integrating into production of materials such as polyisoprene or polyesters. The head-on competition with petroleum feedstocks for production of common products would help maintain a stable supply of turpentine at dependable prices for all users, a situation unusual to the naval stores industry.

IV. ROSIN

A. Composition

Rosins from the wood of pines consist primarily of diterpene resin acids of the abietic type (abietic, neoabietic, palustric, and dehydroabietic) and pimaric type (pimaric, isopimaric, and sandaracopimaric) as shown in Figure 12. Although levopimaric acid is present in substantial amounts in southern pine, it is readily isomerized to the other abietadienoic acids and does not survive processing.

The small amounts of neutral materials such as anhydrides, phenolics, sterols (mainly sitosterol), and diterpene aldehydes and alcohols present can affect the properties and end uses of commercial rosins. More important however, is the composition of the resin acids. Typical resin acid compositions for rosins are given in Table 2. Quality rosins from throughout the world are similar in general composition with only minor variations such as the presence of the dicarboxylic mercusic acid[21] in Indonesian or Philippean gum rosin from *Pinus mercusii* and incidental inclusion in tall oil rosins of

Mercusic (dihydraagathic) acid

Lambertionic acid

Anticopalic acid

anticopalic acid[22] (from western white pine, P. *monticola*) or lambertianic acid (from sugar pine,[23] P. *lambertiana,* or from P. *koriensis*[24] in the U.S.S.R.). High proportions of the conjugated abietadienoic acids are best for preparing the maleic modified (i.e., the maleic anhydride Diels-Alder) product. Predominance of an individual resin acid in rosin results in increased tendency to crystallize, thus causing handling problems in preparing the rosin for further use. For many purposes, rosins and rosin derivatives are interchangeable, regardless of source.

B. Current Uses

Rosins are used mostly in some chemically modified form: hydrogenated, disproportionated (the process is primarily a dehydrogenation), esterified, polymerized, as

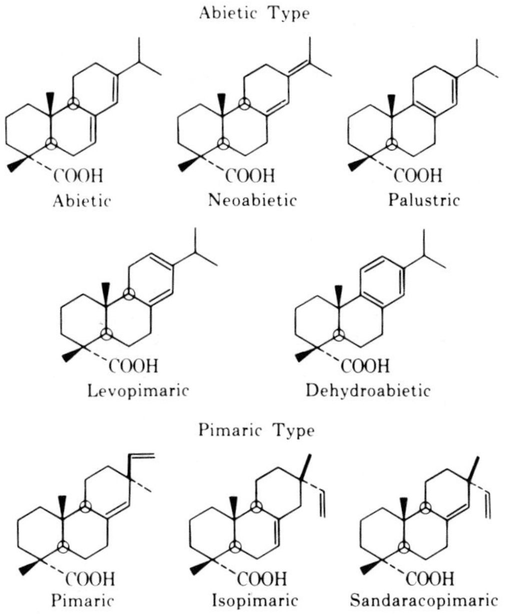

Abietic Type

FIGURE 12. Common resin acids of pine. (To simplify structure representations, the Linstead convention
is used to indicate relative configuration of angular hy-
drogens: ● βH and ● αH.)

Table 2

TYPICAL DISTRIBUTION OF PRINCIPAL RESIN ACIDS IN
ROSINS

Type	Abietic (%)	Neoabietic (%)	Palustric (%)	Dehydroabietic (%)	Pimaric (%)	Isopi-maric (%)
Gum	22	20	25	6	5	17
Wood	50	5	10	13	6	13
Tall oil	32	4	10	30	4	10

salts, or reacted with maleic anhydride or formaldehyde, for example. The largest single use (Figure 9) is in the sizing of paper to control water absorptivity.[25] A rosin soap
or emulsion is added to the pulp and is precipitated onto the paper fibers with aluminum sulfate. This major use of rosin as paper size has been declining, partially because
of more efficient use of rosin and more efficient, chemically modified sizes.[25]

Although rosin found considerable use at one time in the old yellow bar laundry
soaps (38% of rosin was used for this purpose in 1938), this use is almost negligible
now. Disproportionated rosin soaps, however, find important use as emulsifying and
tackifying agents in synthetic rubber manufacture. The use of rosin in the production
of synthetic rubber began during World War II, and now accounts for about 200 million lb of disproportionated rosin per year, especially for SBR (Styrene-butadiene rubber) polymers.[26] Other uses for rosin are in adhesives, surface coatings, printing inks,
and chewing gum.

Comparative price data on rosin and rosin derivatives are shown in Table 3. Ameri-

Table 3
COMPARATIVE PRICES[a] FOR ROSIN AND SOME
DERIVATIVES

Product	¢/lb	Product	¢/lb
Gum rosin (WG)	28	Rosin dimer	37—50
Wood rosin	22—24	Rosin esters	—
Tall oil rosin	16—20	Glycerol (ester gum)	37—50
Distilled tall oil	19	Pentaerythritol	35—50
Pitch (tall oil)	3—5[b]	Methyl	50
Hydrogenated rosin	35	Methyl of hydrogenated	58
Disproportionated rosin	27—37	Dehydroabietyl alcohol	76
Rosin size (maleic anhydride modified)	23	Dehydroabietyl amine	99

[a] Approximate prices as of September 1978 and do not represent quotes. The prices are not corrected for contracts, shipping, etc.
[b] Fuel equivalent is about $0.05 (crude oil at $13.00/bbl).

can gum rosins have a variety of small-volume specialty uses. In some of the uses, the characteristics of gum rosin are of major importance, but in other uses, top-quality gum rosins command premium prices because of limited availability and because reformulations using tall oil and wood rosin are not justified by rosin price differentials.

C. Future Utilization

The key to expansion of rosin as a prime chemical raw material is the development of a large-scale, dependable new source such as offered by the induced lightwood technology. Doubling domestic production by this technology would yield an additional 700 million lb of rosin/year. A further 200 million lb/year currently entering the export market (the U.S. historically has been a net exporter of rosin) could be reclaimed for domestic use. Thus, the amount of domestic rosin available as a raw material for organic chemicals could be increased to about 1.5 billion lb/year.

The resin acid composition of the oleoresins produced during lightwood formation shows no change from that of the normal oleoresins in the wood. Selection of the pine species to be treated can be used to modify rosin composition when processed from mixed species, or rosins can be tailor-made if single species processing can be achieved. The resin acid composition of several oleoresins, rosins, and extractives is given in Table 4. For example, loblolly pine resin acids consist of over 80% conjugated abietadienoic-type acids, whereas slash pine resin acids consist of 30% of the nonconjugated (exocyclic vinylic) pimaric-type acids.

Processing methods will affect the composition of resin acids in rosin from lightwood just as from the usual pine wood. To avoid losses in available turpentine in lightwood, there is an economic incentive to process lightered trees rapidly after cutting. This will also improve the quality of the corresponding tall oil rosin, e.g., the tall oil rosin will have a lower content of dehydroabietic acid and higher content of abietadienoic acids.

Efficient and economical extraction of lightwood requires a method to separate the lightered and normal wood of treated trees. It has been reported that a proprietary process has been developed to effect this separation. Improvements in extraction technology such as using low-boiling solvents and removing by-products (e.g., stronger acids that cause double bond isomerization as well as degradation of the wood and, thus, pulp quality) would give extractive oleoresin. The primary value of the extractive oleoresin, as with gum oleoresin, is as a source for levopimaric acid.

Table 4

COMPOSITION (IN PERCENT) OF THE RESIN ACIDS IN SOME PINE EXTRACTIVES, OLEORESINS, AND GUM ROSINS

Acid

Sample	Pimaric	Sandaraco-pimaric	Palustric	Average of palustric/ levopimaric	Levopimaric	Iso-primaric	Abietic	Dehy-droabietic	Neoabietic	Total abieta-dienoic type	Total pimaric type	Ref.
By Species												
Pinus taeda (loblolly)												
Extractives	8.0	1.8		59.5		1.6	12.6	8.4	8.1	80	11	27
Oleoresin	8.7	2.2		64	—	tr	8.6	6.3	9.5	83	11	28
Rosin	5.4	0.9	10		—	0.9	69	8.1	4.7	84	7	28
P. elliottii (slash)[a]												
Extractives	7.5	2.0	9.7		19.6	23.9	12.1	4.8	18.6	65	33	29
Oleoresin	5.1	1.8		37		21	9.7	3.7	16	65	28	28
Rosin	5.5	1.8	25		—	23	19	7.2	16	67	30	28
P. palustris (longleaf)												
Extractives	6.6	2.0	27.8		19.0	8.5	13.5	10.5	12.1	72	17	29
Oleoresin	5.4	1.1		52		10	9.4	8.3	13	74	17	28
Rosin	4.8	1.6	35		—	16	18	8.6	15	68	21	28
P. echinata (shortleaf)												
Extractives	6.0	1.8	27.9		15.2	5.0	15.1	15.3	13.7	72	13	29
Commercial Gum Rosins												
American	6.2	2.3	20.1		—	14.3	30.6	6.9	16.5	67	23	29
Chinese	8.5	1.5	26.7		—	—	42.6	3.2	17	86	10	29
Spanish	8.7	1.5	27		—	0	36	1.9	24	87	10	28

a   Also contains 2 to 4% of the labdane, *trans*-communic (elliotinoic) acid.

Perhaps proprietary processes for utilization have been developed that will require large volumes of rosin, but there is no tangible evidence that such on-the-shelf technology exists. Major efforts have been made in the past, however, in both industrial and other laboratories, particularly at the now-closed U.S. Department of Agriculture-ARS Naval Stores Laboratory. Reexamination of the past research results is timely.

Much of the past development research was directed at modifying the brittle resin properties of commercial rosins. The free carboxylic group and the unsaturation of the resin acids have been the obvious functional targets in devising chemical transformations to effect desired physical properties of products. A more direct approach to rearrange the tricyclic skeleton such as cleaving the 9,10-bond could produce a highly versatile chemical product having significantly different characteristics from other rosins. Products of this skeletal rearrangement have been seen in tall oil,[30,31] possibly as products of base reactions of levopimaric acid.[32] Treatment of rosins with organic sulfur compounds have been shown recently to give good yields of the 9,10-secodehydroabietic acids.[33,34]

A Secodehydroabietic acid

Using naval stores to replace some uses of petrochemicals was considered in the CORRIM (Committee on Renewable Resources for Industrial Materials) report.[35] One of the first objectives would be to recover past markets lost to petrochemicals. This can be accomplished by demonstrating the reemergence of naval stores as a dependable source of rosin at a dependable and competitive price (about $0.20/lb). Attempts to enter stable, mature markets new to rosin will be a difficult way to approach large-scale utilization of rosin. Growing markets, such as polyurethanes with a projected annual growth rate of 10% for the next several years,[36] offer greater opportunity for success. Rosin products based on secodehydroabietic acid and derivatives could well provide the necessary characteristics to participate in polyurethane markets, a situation that has eluded rosin. The growing area of synthetic lubricants has been suggested[37] as another market for rosin with the challenge, "The technical details of such developments are limited only by the imagination and ability of the organic chemist".

Another conceptual approach would concentrate on reactivity of the exocyclic-vinyl group of the pimaric-type acids in rosins, especially those derived from slash pine, for the preparation of intermediate chemicals. These intermediate chemicals could either be separated from the unreacted rosin components or the crude intermediate reacted further with other materials to provide new rosin products (polymerics). Most of the uses of rosin involving chemical modifications rely on using a modified rosin as an entity.

The development of specialty chemicals based on separated individual resin acids is a possibility such as was involved in the research at the Naval Stores Laboratory on levopimaric acid from oleoresin.[38] As noted above, recovery by a gentle extraction method could provide a large-scale source for levopimaric acid. Levopimaric acid is readily convertible in high yield to relatively pure abietic acid.[39]

All the foregoing general approaches consider using rosin components without deliberate degradation to much smaller molecules. The potential use of rosin as a source

Table 5

COMPARATIVE PRICES[a] FOR TALL OIL
FATTY ACIDS AND SOME DERIVATIVES

Product	¢/lb	Product	¢/lb
Crude tall oil	8—9	Oleic acid	
Tall oil fatty acids		From crystallization	40
92—95%	26	From disproportionation	29
96—98%	28	Linoleic acid	43
Tall oil heads	8	Conjugated	50
Fatty acid dimer	48	C_{21} Dicarboxylic acid	45

[a] Approximate prices as of September 1978 and do not represent
quotes. The prices are not corrected for contracts, shipping, etc.

for mellitic acids has been suggested,[37] but this seems highly impractical considering the potential availability of competing aromatic chemicals from coal.

V. TALL OIL FATTY ACIDS

Rosin and turpentine are common products of all naval stores sources. The nature of rosin recovery as a by-product of kraft pulping involves coisolation with fatty acids. Thus, it is appropriate to consider tall oil fatty acids as a naval stores product, even though its geneological credentials as such are lacking, because of the importance of kraft by-products in U.S. production of naval stores.

A. Composition and Current Uses

Fatty acids from the distillation of tall oil are available with resin acid contents from about 10 to 40% (distilled tall oil) down to less than 0.5%. These fatty acids consist primarily of C_{18} acids with the monoenoic (oleic) and the dienoic (linoleic) acids predominating. Only small amounts of trienoic and saturated acids are present. The linoleic fraction is mostly the *cis,cis*-9,12-isomer with some isomerization products due to processing. A 5,9-linoleic acid isomer has been found in small amounts in tall oil and pine extractives.

Specialty tall oil fatty acid products are also available. One such product obtained from tall oil heads contains predominately saturated fatty acids, of which 55% is palmitic. High-quality oleic and linoleic acids are produced by crystallization of tall oil fatty acid soaps. Another product consisting of about equal amounts of oleic acid and its *trans*-isomer, elaidic acid, is prepared by disproportionation of tall oil fatty acids.

Of the 1 billion lb of fatty acids produced in this country, about 35% of the total, and over one half of the unsaturated fatty acids portion, comes from tall oil. Although only a cheap substitute for other fatty acids in the past, tall oil fatty acids are increasingly being used as chemical raw materials. Major uses (Figure 9) are in protective coatings, intermediate chemicals (a catchall category, of which the major uses are for epoxy tallates and for dimer acids that are used in polyamide resins for inks, adhesives, and coatings), soaps and detergents, and as flotation agents.

One of the most recent developments is the marketing of the C_{21} dicarboxylic acid product from the Diels-Alder reaction of conjugated linoleic acid with acrylic acid. The C_{21} acid and its derivatives have found a number of diverse uses including polyamide resins, detergent coupling agents, fabric softeners, and rust inhibitors.[40] Comparative prices of tall oil fatty acids and derivatives are given in Table 5.

B. Future

Tall oil and tall oil fatty acid output are correlated, obviously, with the production of kraft pulp, which is influenced by the extent of paper recycling and general economic factors affecting demand. Tall oil production also depends on pulping conditions, along with the kind and quality of mill furnish. For example, losses in tall oil precursors occur during handling and particularly during storage of chips because of cell metabolism, oxidation, and microorganisms.[41] Decreased availability of tall oil fatty acids also result from using younger trees (lower fatty acid content) and incorporating hardwoods or mixing of pine and hardwood black liquors (decrease tall oil recovery efficiency). Significant improvement can be achieved in tall oil production by better wood inventory control (particularly in chip storage[9]), separation of hardwood and pine black liquors, and using improved recovery techniques and equipment for recovery of soap from the black liquors. Special processing to remove nonsaponifiables by solvent extraction[42] and new techniques such as membrane technology[43] can result in better yields of tall oil and reduce fatty acid losses due to esterification during handling and distillation.

No major new sources of tall oil fatty acids appear imminent. Although many reports on induced lightwood refer to increased tall oil yield, only the rosin portion increases. Total fatty acid content in the wood does not appear to change on lightering.[19] However, implimentation of lightwood technology would probably result in a decline in the absolute production of tall oil fatty acids because of tbe decreased ratio of fatty acids in tall oil from lightwood, a constant proportion of fatty acids left in intermediate and rosin fractions on distillation, and somewhat greater losses of fatty acids by esterification (because of increased relative amounts of diterpene alcohols).

No promising ideas have come forth on stimulating fatty acid accumulation in trees in analogy to inducing lightwood. However, a clue may be present in our observation that fatty acid content of southern pine sapwood (15- to 20-year-old, plantation-grown trees) increases towards the centers of the trees.[29] Expanded knowledge of the mechanism of heartwood formation and aging in pines may well provide the key to devising a method for inducing fatty acid accumulation. Barring such a development, tall oil fatty acid demand will continue to exceed the supply. "Thus, where at one time tall oil fatty acids were merely used as cheap substitutes for other source fatty acids and oils, we are now witnessing the opposite phenomenon, attempts to substitute for tall oil fatty acids, the ultimate compliment to the high regard that tall oil fatty acids have achieved."[44]

VI. NONSAPONIFIABLES

Tall oil soaps from southern pines normally have nonsaponifiable contents of 5 to 10%. However, nonsaponifiables contents of 25% or more are found in tall oil soaps from mixed conifers that include firs and spruces (such as from the Pacific Northwest, Canada, and the U.S.S.R.), and in Scandinavian soaps that are derived from furnish rich in birchwood. Because of the inferior yields and quality of tall oil fatty acids and rosin from the tall oil soaps containing high percentages of nonsaponifiables, processes[42] (solvent extraction) have been developed to remove the major portion of the nonsaponifiables. Some Scandinavian soaps are now being refined commercially to produce low nonsaponifiables tall oil.

The major portion of nonsaponifiables in fresh pine tall oil consists of sterols (of which sitosterol is the predominant component) and diterpenes (primarily alcohols and aldehydes).[45] Approximately 20,000 tons of sitosterol are potentially available annually in the U.S.[46] In addition to recovery of sitosterol directly from the nonsaponifiables removed by solvent extraction of tall oil soaps, sitosterol can be recovered from tall oil still bottoms according to a number of publications and patents.

Isolation of sitosterol will be more complicated from sources other than southern U.S. tall oils. For example, the nonsaponifiables in western U.S. tall oil have higher proportions of triterpene alcohols due to the use of Douglas-fir[29] and other conifer species in the pulpmill furnish. Scandinavian tall oil contains large amounts of birch-wood polyprenols and pentacyclic triterpene alcohols.[47] Sitosterol has promise as a substrate in fermentation

SITOSTEROL   Fermentation ———→   4-ANDROSTEN-3,17-DIONE

to steroid precursors such as androsta-1,4-diene-3,17-dione and androst-4-ene-3,17-dione.[46] These intermediates can be used for synthesis of spironolactone (blood pressure depressant), hydrocortisones (antiinflammatory agents), and steroid hormones (reproductive and geriatric therapy).

The relative proportion of neutral diterpenes in the nonsaponifiable fraction increases in lightwood naval stores and, thus, the sterols are diluted. The ready esterification with the primary hydroxyl of diterpene alcohols would result in increased losses of fatty acids during processing. Preventing such increased losses of fatty acids may provide sufficient incentive for removal of the neutral fraction from the crude lightwood naval stores, whether recovered as extractives or tall oil. The diterpene-rich neutral fraction left from sitosterol recovery could be oxidized to acids and recycled and processed in crude tall oil.

VII. SUMMARY

The naval stores commodities rosin, turpentine, and tall oil fatty acids already are sources of chemical raw materials from plants. Various constraints limit the production of these commodities by the traditional methods of turpentining, stumpwood extraction, and the relatively recent recovery as by-products of kraft pulping. Commercialization of lightwood technology, however, could lead to a new era for naval stores in offering controllable and dependable production of rosin and turpentine.

REFERENCES

1. *The New American Bible,* Catholic Book Publishing, New York, 1970.
2. Theophrastus, *Enquiry into Plants,* Hort, A., (transl.), Vol. 2, G. P. Putnam's Sons, New York, 1916, 223.
3. Gerry, E., A naval stores handbook dealing with the production of pine gum or oleoresins, *U.S. Dep. Agric. Misc. Publ.,* 209, 1935.

4. **McConnell, N. C.,** Operating instructions for Olustee process for cleaning and steam distillation of pine gum, U.S. Department of Agriculture, Agric. Res. Serv., (SURD), unnumbered report, Olustee, Fla.
5. **Clements, R. W.,** Modern gum naval stores methods, U.S. Department of Agriculture For. Serv. Gen. Tech. Rep. SE-7, Asheville, N.C., 1974.
6. **Squillace, A. E. and Harrington, T. A.,** Olustee's high-yielder produces 487 bbls pine gum per crop for four straight years, *Nav. Stores Rev.,* 77(12), 4, 1968.
7. **Joye, N. M., Jr., Proveaux, A. T., Lawrence, R. V., and Barger, R. L.,** Naval Stores from ponderosa pine stumps, *Ind. Eng. Chem. Prod. Res. Dev.,* 8, 297, 1969.
8. **Anon.,** Summer '80 target date for new plant to process Arizona ponderosa stumpage, *Nav. Stores Rev.,* 88(5), 9, 1978.
9. **Springer, E. L.,** Losses during storage of southern pine chips. The case for standby storage, *Tappi,* 61(5), 69, 1978.
10. **Anderson, A. B.,** Solvent drying pays dividends with ponderosa pine, *For. Prod. J.,* 16(12), 31, 1966.
11. **Keays, J. L.,** Foliage. I. Practical utilization of foliage, *Appl. Polym. Symp.,* 28, 445, 1976.
12. **Zinkel, D. F.,** unpublished data, 1974.
13. **Hannus, K. and Pensar, G.,** Silvichemicals in technical foliage. I. Water steam distilled oil from pine material, *Pap. Puu,* 55, 509, 1973.
14. **Zinkel, D. F.,** Pine resin acids as chemotaxonomic and genetic indicators, in Tappi Conference Papers Forest Biology Wood Chemistry Conference, Technical Association of the Pulp and Paper Industry, Atlanta, 1977, 53.
15. **Roberts, D. R.,** Inducing lightwood in pine trees by paraquat treatment, U.S. Department of Agriculture For. Serv. Res. Note SE-191, Asheville, N.C., 1973.
16. **Roberts, D. R., Joye, N. M., Jr., Proveaux, A. T., Peters, W. J., and Lawrence, R. V.,** *Nav. Stores Rev.,* 85(5), 4, 1973.
17. **Derfer, J. M.,** Turpentine as a source of perfume and flavor materials, *Perfumer Flavorist,* 3(1), 45, 1978.
18. **Ruckel, E. R., Arlt, H. G., Jr., and Wojcik, R. T.,** The chemistry of tackifying terpene resins, in *Adhesion Science and Technology,* Vol. 9A, Lee, L. H., Ed., Plenum Press, New York, 1975, 395.
19. **Zinkel, D. F. and McKibben, C. R.,** Chemistry of naval stores from pine lightwood—a critical review, in *Proc. Lightwood Res. Coord. Counc. 5th Annual Meeting,* Esser, M. J., Ed., Southeastern Forest Experiment Station, Asheville, N.C., 1978, 133.
20. **Hedrick, F. W. and Lawrence, R. V.,** From the piney woods—pinonic acid—a promising chemical raw material, *Ind. Eng. Chem.,* 52, 853, 1960.
21. **Weissmann, G.,** Mercusic acid, a dibasic acid of the oleoresin from *Pinus merkussi, Holzforschung,* 28, 186, 1974.
22. **Zinkel, D. F., Toda, J. K., and Rowe, J. W.,** Occurrence of anticopalic acid in *Pinus monticola, Phytochemistry,* 10, 1161, 1971.
23. **Dauben, W. G. and German, V. F.,** The structure of lambertianic acid. A new diterpenic acid, *Tetrahedron,* 22, 679, 1966.
24. **Raldugin, V. A., Kashtanova, N. K., and Pentegova, V. A.,** Neutral part of *Pinus koraiensis* oleoresin. I. Main diterpenoid components, *Khim. Prir. Soedin.,* 6, 481, 1970.
25. **Davison, R. W.,** The sizing of paper, *Tappi,* 58(3), 48, 1975.
26. **Booth, J. W.,** Tall oil and rosin, *Rubber Age,* 104(7), 47, 1972.
27. **Zinkel, D. F.,** Tall oil precursors of loblolly pine, *Tappi,* 58(2), 118, 1975.
28. **Joye, N. M., Jr. and Lawrence, R. V.,** Resin acid composition of pine oleoresins, *J. Chem. Eng. Data,* 12, 279, 1967.
29. **Zinkel, D. F. and Foster, D. O.,** unpublished data, 1978.
30. **Zinkel, D. F., Rowe, J. W., Zank, L. C., Gaddie, D. W., and Ruckel, E. R.,** Unusual resin acids in tall oil, *J. Am. Oil Chem. Soc.,* 46, 633, 1969.
31. **Holmbom, B.,** Improved gas chromatographic analysis of fatty and resin acid mixtures with special reference to tall oil, *J. Am. Oil Chem. Soc.,* 54, 289, 1977.
32. **Takeda, H., Schuller, W. H., Lawrence, R. V., and Rosebrook, D.,** Novel ring openings of levopimaric acid salts, *J. Org. Chem.,* 34, 1459, 1969.
33. **Inoue, Y. and Ishigami, M.,** Disproportionation and cleavage reaction of abietic acid in the presence of *bis*(2-hydroxy-1-naphthyl) disulfide, *Yukagaku,* 26, 176, 1977.
34. **Thorpe, S. D., Davis, C. B., and Whellus, C. G.,** U.S. Patent 3,872,073, 1975.
35. **Anon.,** Extractives as a renewable resource for industrial materials, in *Renewable Resources for Industrial Materials,* National Academy of Sciences, Washington, D. C., 1976, chap. 12.
36. **Anon.,** Isocyanate needs price hike to spur expansion, *Chem. Eng. News,* 55(19), 6, 1977.
37. **Collier, T. J.,** Feasibility of petrochemical substitution by oleoresin, in *Proc. Lightwood Res. Coord. Counc. 4th Annual Meeting,* Esser, M. H., Ed., Southeastern Forest Experiment Station, Asheville, N.C., 1977, 172.

38. Hedrick, G. W., Lawrence, R. V., Summers, H. B., Jr., and Mazzeno, L. W., Jr., Levopimaric acid. A product for industry, U.S. Department of Agriculture Agric. Res. Serv. Rep. 72-35, New Orleans, 1965.
39. Schuller, W. H., Takeda, H., and Lawrence, R. V., Levopimaric acid as a ready source of abietic acid, *J. Chem. Eng. Data,* 12, 283, 1967.
40. Ward, B. F., Jr., Force, C. G., Bills, A. M., and Woodward, F. E., Industrial utilization of C_{21} dicarboxylic acid, *J. Am. Oil Chem. Soc.,* 52, 219, 1975.
41. Feist, W. C., Springer, E. L., and Hajny, G. J., Spontaneous heating in piled wood chips-contribution of bacteria, *Tappi,* 56(4), 148, 1973.
42. Gronfors, H. and Holmbom, B., Process for upgrading hardwood and softwood soaps, in *New Horizons Chem. Eng. Pulp Pap. Technol.,* Vol. 72, Prahacs, S., Ed., AICHE Symp. Ser. 157, American Institute of Chemical Engineers, New York, 1976, 89.
43. Cooper, W. W., IV, De Filippi, R. P., and Timmins, R. S., U.S. Patent 3,519,558, 1970.
44. Zachary, L. G., Tall oil fatty acid marketing, *J. Am. Oil Chem. Soc.,* 54, 533, 1977.
45. Conner, A. H. and Rowe, J. W., Neutrals in southern pine tall oil, *J. Am. Oil Chem. Soc.,* 52, 334, 1975.
46. Conner, A. H., Nagaoka, M., Rowe, J. W., and Perlman, D., Microbial conversion of tall oil sterols to C_{19} steroids, *Appl. Environ. Microbiol.,* 32, 310, 1976.
47. Holmbom, B. and Avela, E., Studies on tall oil from pine and birch. II. Unsaponifiable constituents in sulfate soaps and in crude tall oils, *Acta Acad. Abo. Ser. B,* 31(16), 1, 1971.

GENERAL

Cooke, D. W., Innovation Diffusion and the Changing Spatial Patterns of Pine Chemical Production in the United States, Ph.D. dissertation, University of Tennessee, Knoxville, 1975.
Sandermann, W., *Naturharze- Terpentinol- Tall Oil, Chemie und Technologie,* Springer-Verlag, Berlin, 1960.
Gamble, T., Ed., *Naval Stores, History, Production, Distribution and Consumption,* Review Publishing and Printing, Savannah, 1921.
Faler, F., Ed., *Naval Stores Review International Yearbook 1974,* Naval Stores Review, Metairie, La., 1975.
Enos, H. I., Jr., Harris, G. C., and Hedrick, G. W., Rosin and rosin derivatives, in *Kirk-Othmer Encyclopedia of Chemical Technology,* Vol. 17, 2nd ed., John Wiley & Sons, New York, 1968, 475.
Stonecipher, W. D. and Turner, R. W., Rosin and rosin derivatives, in *Encyclopedia of Polymer Science and Technology,* Vol. 12, Bikales, N. M. and Conrad, J., Eds., Wiley-Interscience, New York, 1970, 139.
Chiang, T. I., Burrows, W. H., Howard, W. C., and Woodard, G. D., Jr., Study of the problems and potentials of the gum naval stores industry, Project A-1111, Engineering Experiment Station, Georgia Institute of Technology, Atlanta, 1971.
Drew, J., Russell, J., and Bajak, H. W., Eds., *Sulfate Turpentine Recovery,* Pulp Chemicals Association, New York, 1971.
Anon., Summary of annual and monthly statistics 1955—74, Naval Stores, *U.S. Department Agriculture Stat., Bull. 567;* and yearly annual summaries SpCr 3, Washington, D.C., 1977.
Tate, D. C., Tall oil, in *Kirk-Othmer Encyclopedia of Chemical Technology,* Vol. 19, 2nd ed., Standen, A. and Dukes, E. P., Eds., Interscience, New York, 1969, 614.
Weiner, J., Tall Oil, Bibliographic Series 133—135, 3rd ed. 1959; Weiner, J. and Byrne, J., Suppl. 1, 1965; Weiner, J. and Roth, L., Suppl. 2, 1971; Pollock, V., Suppl. 3, 1976, Institute of Paper Chemistry, Appleton, Wis.
Zachary, L. G., Bajak, H. W., and Eveline, F. J., Eds., *Tall Oil and its Uses,* Pulp Chemicals Association, New York, 1965.
Drew, J., Ed., *Tall Oil,* Pulp Chemicals Association, New York, 1981.
Gonzenbach, C. T., Jordan, M. A., and Yunick, R. P., Terpene resins, in *Encyclopedia of Polymer Science and Technology,* Vol. 13, Interscience, New York, 1970, 575.
Stonecipher, W. D., Turpentine, in *Kirk-Othmer Encyclopedia of Chemical Technology,* Vol. 20, Interscience, New York, 1969, 748.
Weiner, J. and Pollock, V., *Turpentine,* Bibliographic Series 260, Institute of Paper Chemistry, Appleton, Wis., 1974.

Chapter 10

BARK: ITS CHEMISTRY AND PROSPECTS FOR CHEMICAL UTILIZATION

R. W. Hemingway

TABLE OF CONTENTS

I. INTRODUCTION

The bark of forest trees is a large renewable resource that is seldom used except as a fuel. At present fossil fuel prices, the gross fuel value of bark is generally about 1.5¢/dry pound. When appropriate costs of handling and installation of bark burning equipment are considered, returns on investment are low. Nevertheless, direct burning of bark for fuel is the most economical use of bark for most of the forest products industry. Efficient systems specially designed for burning bark are being designed since it is a clean and readily available fuel.

In hopes of obtaining higher returns, the forest products industry has been interested in developing specialty chemical products from bark. Bark is especially promising as a raw material base because it is harvested and transported to centralized locations in obtaining wood, is rich in compounds that have potential as specialty chemicals, and no alternative uses provide high returns. That portion which is not used for chemical manufacture can generally be used as fuel. However, few efforts have resulted in new commercial ventures. Successes include wood adhesives formulated from wattle tannins, a variety of specialty polymers derived from western hemlock polyflavonoids, and wax products from Douglas-fir bark.

One of the difficulties that has hampered product development is that the chemistry of bark is far more complicated than that of wood. Extraneous materials that offer opportunities for the development of new products are generally present in much higher proportions in bark than in wood. However, despite the substantial effort that has been devoted to understanding the chemistry of these compounds, our knowledge of their structure and particularly of their properties remains inadequate. This chapter summarizes what is known of the chemistry of tree barks, points out areas where our knowledge is incomplete, and summarizes prospects for the development of chemical products from tree barks as of 1979.

II. FORMATION AND STRUCTURE OF BARK

To understand the reasons for the considerable chemical differences among wood, inner bark, and outer bark, it may be helpful to review briefly the formation and structure of bark. Tree barks are generally composed of a thin inner bark (phloem) and a layer of outer bark (rhytidome) which are separated by a thin layer called the periderm.[1,2] The thickness of the outer bark varies greatly among tree species and with the age of the bark. The bark of many hardwoods and some conifers, such as *Abies, Picea,* or *Pinus contorta,* may be composed largely of inner bark (phloem) with comparatively little outer bark (rhytidome). In other species, such as redwood (*Sequoia sempervirens*) or Douglas-fir (*Pseudotsuga menziesii*), the outer bark may approach 8 to 10 in. in thickness when old-growth trees are harvested, and in these instances, the bark will be almost entirely rhytidome. However, where bark supplies are obtained from young trees, tops, and branches of these species, the phloem constitutes a significant part of the material. Because the percentage of inner and outer bark varies so widely, differences in their chemistry are particularly important.

The phloem is formed by the vascular cambium elaborating cells outward which differentiate into sieve cells and ray or longitudinal parenchyma. The sieve cells of conifers are long thin-walled cells that conduct carbohydrates and other nutrients from the leaves to the vascular cambium. As the sieve cells age and stop functioning, they become clogged with callose and finally collapse. At the same time the parenchyma cells are enlarging greatly. These thin-walled cells contain starch, fats, and tannins together with other materials. The pits of the longitudinal parenchyma are paired with those of the ray parenchyma but do not appear to be associated with sieve cells.

The phloem ray cells are contiguous with the xylem ray cells. A group of specialized cells, called albuminose cells in the conifers and companion cells in hardwoods, form the margins of the rays and are paired with the sieve cells. These regulate the translocation of material from the sieve cells to the ray parenchyma and thus to the vascular cambium. Bark of some conifers contains fusiform rays that have resin ducts contiguous with the horizontal resin canals of the xylem. Pockets of oleoresin (cortical oleoresin) are encapsulated by the periderm in some conifer species such as the true firs (*Abies* sp.). In many instances the oleoresin found in the bark is the result of a wounding and is of a composition similar to the wood oleoresin.

The inner bark of many hardwood species contains fibers. In *Eucalyptus obliqua* for example, the fibers of the phloem constitute a large portion of its volume. These cells are about the same length as the wood fibers, but the cell walls are so thick little lumen volume is present. These fibers contribute to the high cellulose content of this bark. They are also lignified, and the chemistry of the inner bark lignin appears to be similar to that of wood lignin. Similar fibers are found in other barks such as the oaks.

The periderm is composed of three layers of cell types. The phelloderm is a layer of parenchymatous cells formed to the inside of the phellogen (cork cambium). The phellum (cork cells) are formed to the outside of the phellogen. Bands of cork of varying width isolate zones of old phloem and phelloderm. Cork cells are made up of a thin, randomly oriented cellulosic primary wall; a secondary wall of suberin; and a thin wall layer of neutral solvent-soluble wax at the lumen surface. The bands of cork seal off the phloem from loss of water or exposure to air. The proportion and thickness of the cork layers are variable among barks of different species (i.e., large cork layers of Douglas-fir or oaks compared with the thin cork layers found in the pines). The amount and distribution of cork layers in tree barks are particularly important to the usefulness of a bark for isolation of wax or suberin-based compounds.

In some species, sclereids may also be formed by the phellogen. Sclereids appear in many shapes (i.e., the sprocket-wheel shaped cells in southern pines and the spindle-shaped bast fibers in Douglas-fir). These sclereids have exceptionally thick cellulosic walls and are lignified. The bast fibers of Douglas-fir bark find use as reinforcing fibers for plastics.

Differences in the phenolic compounds in the inner bark compared with those found in wood suggest that they are formed in the longitudinal or ray parenchyma of the inner bark. Additional secondary compounds are found after the transformation from inner to outer bark. These compounds may be formed by the phellogen or by the parenchyma cells of the phloem and phelloderm when they are dying. The process of suberization (i.e., formation of cork layers and isolaton of the phloem), together with the formation of additional secondary compounds in the parenchyma during the transformation from inner to outer bark, results in a vastly different chemistry of the outer bark compared with inner bark.

It is particularly important that analytical results be reported for both inner and outer bark tissues because the proportions of these tissues vary widely depending upon the tree species and age of the bark and because inner and outer bark are so different chemically. Where studies are made of whole bark, definition of the proportion of each tissue should be provided. Unfortunately, many of the analytical data available today are obtained from whole bark and the percentages of inner and outer bark are not given. Therefore, considerable caution must be exercised in making comparisons among species.

III. CARBOHYDRATES

A. Total Carbohydrates

Compared with wood, most tree barks contain exceptionally high extractive contents.[3] Therefore, the appropriate analytical figure for consideration of the potential of bark as a source of carbohydrate is probably the yield of carbohydrate as a percentage of the unextracted bark weight.[4,5] When expressed on the basis of unextracted weight, total sugar yields after hydrolysis are, without exception, much lower from bark than from corresponding wood. Some reducing sugar yields obtained from unextracted bark, from bark extracted with neutral solvents, and from alkali-extracted bark expressed as a percentage of unextracted bark weight are summarized in Tables 1 and 2. Yields of sugars obtained by extraction can be determined by difference.

The sugars from the barks of many tree species contain exceptionally high proportions of pentosans.[4-8] However, because of the low total carbohydrate content, the pentosan contents of unextracted barks only range from 8 to 14%, values much lower than those obtained from hardwoods (i.e., 14 to 25%).[6-8*]

B. Water-Soluble Carbohydrates

Extraction of most tree barks with hot water does not remove significant quantities of carbohydrates. Important exceptions include the barks of spruce. Hot water extraction[4,5] removes 20 to 24% of the total carbohydrate present. The material extracted contains a high proportion of glucan, which is also soluble in ethanol, indicating high concentrations of phenolic glucosides. These materials were isolated from methanol or water extracts of whole *Picea glauca* bark and studied intensively by Bishop[9] and Harpham.[10] Despite considerable study of different approaches to fractionation and elucidation of the structure of Bishop's "fraction 21", its nature remains obscure. Results suggest a mixture or complex of glucan with a polyflavonoid.

Painter and Purves,[11] studying the carbohydrates in the inner bark of *Picea glauca,* isolated starch, a polygalacturonic acid containing low proportions of arabinose and galactose moieties, a highly branched galactan, and an arabinan from water extracts. Becker and Kurth[12] isolated an unusual water-soluble gum from *Abies shastensis* bark which gave glucuronic acid, glucuronone, and an aldobiuronic acid together with galactose and arabinose on hydrolysis. Laver et al.[13] found starch and arabinogalactan in water extracts from the inner bark of *Pseudotsuga menziesii.* Hot-water extraction of *Pinus echinata* outer bark removes 5% of the unextracted bark weight as an ethanol-insoluble, carbohydrate-rich polymer which gives mainly arabinose and galactose after hydrolysis.[14]

Of the hardwood barks, sizable quantites of carbohydrates are extracted with hot water from *Ulmus rubra (fulva),*[15,16] *Fagus sylvatica,*[5] and *Quercus robur.*[5] The mucilage of *U. rubra* bark has been intensively studied by Beveridge et al.[15,16] who showed that it is made up of an alternating α(4-2) galacturonic acid-rhamnose backbone with side chains of 3-*O*-methyl-D-galactose units attached to the C-4 of the L-rhamnose and some of the 3-*O*-methyl-D-galactose units attached as nonreducing end groups. The hot-water-soluble carbohydrates from *F. sylvatica* bark are composed of over 84% glucans, and like those obtained from spruce bark, are largely soluble in ethanol, suggesting high concentrations of phenolic glucosides.[5] Hot water extraction of *Q. robur* bark removes large quantities of arabinose moieties (2.4% of the unextracted weight) but the pentosan content of the water extract is only 15.9%[5]

---

*   For another recent review of the chemistry of bark carbohydrates see Dietrichs, H. H., Polysaccharides der Ringen, *Holz als Roh- und Werkstoff*, 33, 13, 1975.

C. Holocellulose

Holocellulose yields provide an estimate of the amounts of water-insoluble carbohydrates present in bark. Isolation of holocellulose from outer bark is especially difficult because suberin is not degraded by the usual delignification reagents, and it is so hydrophobic that it limits access of these reagents to the lignin in bark. Timell[17] emphasizes the requirement of a repetitive acid-chlorite delignification to obtain a holocellulose pure enough to permit fractionation of individual bark polysaccharides. In some instances, yields of holocellulose obtained in this way are much lower than expected from yields of reducing sugars obtained by hydrolysis.

Substantial quantities of carbohydrates are often in the acid-chlorite liquor.[11,13,17] Painter and Purves[11] isolated starch, an arabinan, and a galactan from the acid-chlorite liquor when preparing holocellulose from *Picea glauca*. Laver et al.[13] found uronic acids, and after hydrolysis, glucose, arabinose, galactose, mannose, xylose, and rhamnose in proportions of 59, 12, 4, 4, 1, and 1 in the acid-chlorite liquor obtained in holocellulose preparations from *Pseudotsuga menziesii* inner bark. Total yields of carbohydrates soluble in the acid-chlorite liquor amounted to 12.4% of the original bark sample. Therefore, caution must be used in interpreting yields of holocellulose obtained from tree barks. Holocellulose yields should be compared with the yield of reducing sugars obtained by hydrolysis of the extractive-free bark. Pavlovova et al.[18] addressed this problem in studies of the carbohydrates in *Salix alba* bark and recommended delignification and washing of the holocellulose with 80% ethanol, rather than water, to preserve water-soluble carbohydrates in the holocellulose isolate. When expressed as a percentage of extractive-free bark, holocellulose yields vary widely among species, ranging from 35 to 75% of the extractive-free bark weight (Table 3). Although holocellulose yields are of questionable analytical value, the preparation of a pure holocellulose isolate is essential to successful fractionation of the polysaccharides in tree barks.[17]

D. Cellulose

Isolation and purification of cellulose from tree barks has proven to be especially difficult, and as a result, questions remain as to whether the cellulose in bark is identical to that from wood. Cellulose yields, like holocellulose yields, vary widely among the barks of different tree species and are generally about 20 to 40% of the extractive-free bark weight (Table 3).

The repetitive acid-chlorite delignification required to obtain a holocellulose of sufficiently low lignin content causes severe degradation of the cellulose. The degree of polymerization (DP) of cellulose isolates prepared in this way are very low (67 to 1000).[19-21] However, when isolated by direct nitration, cellulose nitrates are obtained in yields comparable with those obtained from holocellulose, and the DPs are similar to those of wood cellulose (7000 to 10,300 by light scattering or 4000 to 6900 by viscometry).[19,20] Recent work by Garves[21] demonstrates the severe degradation of bark cellulose when prepared by a number of methods. Even when prepared by direct nitration, bark cellulose DPs were lower than those of wood (i.e., *Picea excelsa* bark cellulose 2050 to 2800 compared with wood cellulose 4200 by viscometry).[21]

α-Cellulose isolates from conifer barks frequently contain significantly higher proportions of mannose impurities than are found in wood cellulose isolates. Gardner[26] found 3.2% mannose in an α-cellulose preparation from *Picea glauca* inner bark. Kiefer and Kurth[24] isolated an α-cellulose preparation from *Pseudotsuga menziesii* bast fiber that contained 1.04% mannose. Chen[27] also found unusually high proportions of mannose residues which were resistant to hydrolysis with formic acid in an α-cellulose isolated from Douglas-fir outer bark.

Table 1
REDUCING SUGARS OBTAINED AFTER HYDROLYSIS[a] (CONIFER BARKS)

Species	Extraction	Total	Glucose	Mannose	Galactose	Arabinose	Xylose	Pentosan
				% of unextracted bark weight				
Abies amabilis	None	46.6						
	Neutral	45.3	29.0	5.4	2.3	4.1	3.2	7.3
	NaOH	32.9						
Larix occidentalis	None	46.6						
	Neutral	46.0	31.7	5.1	1.8	2.8	4.1	6.9
	NaOH	38.0	28.1	3.4	1.5	1.5	3.4	4.9
Picea abies	None	49.1	32.6	3.4	2.4	4.8	4.5	9.3
	Neutral	37.2	23.2	3.3	1.7	3.6	4.4	8.0
	NaOH	27.2	19.9	2.0	0.6	0.9	3.8	4.7
P. engelmanni	None	42.9						
	Neutral	34.3	20.9	3.1	1.7	4.5	3.1	7.6
	NaOH	24.2						
P. mariana	None	47.9						
	Neutral	44.3	28.4	3.1	2.7	4.9	4.0	8.9
	NaOH	32.3						
Pinus banksiana	None	30.6						
	Neutral	28.8	18.4	1.7	2.0	2.9	3.2	6.1
	NaOH	21.1						
P. contorta	None	28.3						
	Neutral	32.9	16.5	2.0	2.3	8.6	2.6	11.2
	NaOH	19.2						

	Treatment							
P. elliottii	None	29.7	18.8	2.1	2.1	2.1	4.5	6.6
	Neutral	29.8	17.7	1.8	1.6	1.3	4.0	5.3
	NaOH	26.4						
P. lambertiana	None	22.1						
	Neutral	19.8	13.7	1.6	1.2	1.4	1.8	3.2
	NaOH	16.1						
P. sylvestris	None	27.6	16.3	2.0	2.6	2.2	4.1	6.3
	Neutral	24.0	15.7	1.8	1.7	0.8	3.8	4.6
	NaOH	20.2	15.2	1.2	0.9	0.8	2.2	3.0
Tsuga canadensis	None	34.9	22.3	4.3	1.0	2.7	2.3	5.0
	Neutral	33.3	22.1	2.9	0.6	1.5	2.0	3.5
	NaOH	29.1						

[a] Data obtained from Chang and Mitchell[4] and Dietrichs et al.[5]

Table 2
REDUCING SUGARS OBTAINED AFTER HYDROLYSIS[a] (HARDWOOD BARKS)

Species	Extraction	Total	Glucose	Mannose	Galactose	Arabinose	Xylose	Pentosan
				% of unextracted bark weight				
Acer saccharum	None	35.4						
	Neutral	34.3	21.6	0.3	1.0	2.1	8.6	10.7
	NaOH	31.1	21.5	0.3	0.6	1.2	7.5	8.7
Alnus rubra	None	38.6						
	Neutral	38.0	20.5	0.4	1.1	2.3	12.9	15.2
	NaOH	30.3						
Betula papyrifera	None	32.2						
	Neutral	30.1	16.0	0.3	0.6	1.8	10.8	12.6
	NaOH	21.8	12.0	0.2	0.4	1.1	8.1	9.2
Fagus sylvatica	None	41.2	22.4	0.8	2.6	5.7	8.5	14.2
	Neutral	34.1	16.7	0.7	2.3	5.4	8.1	13.5
	NaOH	20.9	11.5	0.3	1.0	1.6	6.2	7.8
Liquidambar styr-iciflua	None	35.6						
	Neutral	33.5	20.1	1.0	1.0	3.7	6.7	10.4
	NaOH	26.4	18.7	0.3	0.5	1.3	5.5	.8
Platanus occiden-talis	None	40.9						
	Neutral	39.0	23.0	0.4	1.6	1.6	11.7	13.3
	NaOH	31.1						

Table 2 (continued)

REDUCING SUGARS OBTAINED AFTER HYDROLYSIS[a] (HARDWOOD BARKS)

Species	Extraction	Total	Glucose	Mannose	Galactose	Arabinose	Xylose	Pentosan
				% of unextracted bark weight				
Populus tremuloides	None	41.4						
	Neutral	39.0	23.0	0.4	0.8	2.0	11.9	13.9
	NaOH	34.9						
Quercus alba	None	27.8						
	Neutral	28.2	16.9	0.6	1.1	2.0	6.8	8.8
	NaOH	21.2						
Q. rubra	None	32.4						
	Neutral	31.7	16.8	0.3	1.0	1.9	11.1	13.0
	NaOH	28.3						
Q. robur	None	48.7	30.8	1.1	2.3	3.3	10.2	13.5
	Neutral	40.6	28.1	0.8	1.2	0.8	9.6	10.4
	NaOH	28.1	21.1	tr	0.4	0.3	6.4	6.7

[a] Data obtained from Chang and Mitchell[4] and Dietrichs et al.[5]

Table 3
YIELD OF HOLOCELLULOSE AND α-CELLULOSE
FROM TREE BARKS

Species	Holocellulose	α-Cellulose	Reference
	% of extractive-free bark		
Conifers			
Abies amabilis	57.9	38.1	17, 19
Ginkgo biloba	56.5	37.6	17, 19
Picea abies	71.3	50.8	5
P. engelmanni	58.2	30.9	17, 19
P. excelsa	44.3	30.5	21
P. glauca (inner)		48.8	11
Pinus contorta	63.8	30.4	17, 19
P. ponderosa	37.4	22.1	22
P. sylvestris (inner)	65.0	16.5	23
P. sylvestris	40.2	24.9	5
Pseudotsuga menziesii (bast)	56.4	36.2	24
P. menziesii (inner)	38.2	19.2	13, 27
Hardwoods			
Acer pseudoplatinus (inner)	70.8	45.0	22
A. pseudoplatinus (outer)	50.3	29.0	22
Betula papyrifera	75.0	28.4	20
B. platyphylla	41.5	22.3	22
Fagus sylvatica	53.2	33.3	21
	52.3	29.6	5
Fraxinus elatior	58.7	28.5	22
Populus tremuloides	66.0	29.0	25
Quercus robur	61.7	40.1	5

Gardner[26] found X-ray diffraction patterns of cellulose isolated from *Picea glauca* inner bark that were essentially the same as those from wood celluloses. Mian and Timell[20] and Timell[19] report typical cellulose II diffraction patterns from the celluloses isolated from *Betula papyrifera*, *A. amabilis*, *Picea engelmanni*, *Pinus contorta*, and *Ginkgo biloba* barks. Binotta et al.[28] isolated celluloses from *Q. primus* wood, phloem, and rhytidome and found crystallinity indexes of 51.07, 43.37, and 43.09, respectively.

E. Hemicelluloses

The carbohydrates of *Picea glauca* bark were the subject of research directed by Purves at McGill University from 1947 to 1960. A series of studies of material obtained from whole *Picea glauca* bark by successive extraction with methanol,[29] hot water,[9,10] liquid ammonia,[30] subsequent extraction with water[31] followed by 2% NaOH,[32] and finally characterization of the exhaustively extracted bark residue[26] established an understanding of the basic features of bark carbohydrates.* These studies showed the presence of starch, pectin, a water-soluble arabinan, cellulose, and complex hemicelluloses that could not be resolved into distinct polysaccharides despite many approaches to fractionation.

Painter and Purves[11] were probably the first to report the isolation of bark polysaccharides pure enough to warrant detailed examination of their structures. The four

* The efforts of Dr. T. E. Timell and Dr. A. S. Perlin in making copies of these theses available are greatly appreciated.

relatively pure polysaccharides obtained from the inner bark of *P. glauca* included an arabinan containing a low proportion of galactose, and a highly branched galactan containing a small proportion of arabinose. These polysaccharides were apparently part of the water-soluble pectic triad as they were associated with a polygalacturonic acid that contained only small portions of galactose and arabinose moieties. Besides these water-soluble polysaccharides, fractionation of the acid-chlorite holocellulose provided a glucomannan containing very small proportions of galactose units, and a xylan that contained unusually high proportions of glucan with low proportions of 4-*O*-methylglucuronic acid and arabinose residues.

Timell[17] succeeded in isolating several polysaccharides from acid-chlorite holocellulose preparations of the barks of *A. amabilis*, *Pinus contorta*, *Picea engelmanni*, and *G. biloba* in more highly purified states than had been obtained previously. From *A. amabilis* bark, Timell[33] isolated and characterized a water-soluble galactoglucomannan, an arabino-4-*O*-methyl-glucuronoxylan[34] and an alkali-soluble glucomannan consisting of a linear β (1-4) linked glucomannan to which a few galactose residues were linked as side chains.[35]

In further studies of the hemicelluloses of *Picea engelmanni* bark, Ragmalingam and Timell[36] isolated and characterized an arabino-4-*O*-methyl-glucuronoxylan. A notable feature of this xylan was the low proportion of uronic acid groups in comparison with similar xylans isolated from *Picea engelmanni* wood. An alkali-soluble galactoglucomannan consisted of a β (1-4) linked D-glucopyranose and D-mannopyranose backbone to which a few (1-6) linked D-galactopyranose side chains were attached. Properties of this polysaccharide obtained from bark were essentially identical to those of a polysaccharide isolated from *Picea engelmanni* wood. An incompletely characterized glucan appeared to be made-up of β (1-4) linked glucose residues with branch points at C-6. Partial hydrolysis gave a neutral fraction that contained galactose and glucose residues in a ratio of 1:2 and a galactobiose of undetermined structure. Two acidic fractions were also isolated, one of which was composed of glucuronic acid, galactose, and xylose at the reducing end, and the other a glucuronosylgalactose fragment. It is important to note that glucan polysaccharides of this type are not found in wood hemicelluloses. Small amounts of a water-soluble galactoglucomannan were also present.

Studies of the hemicelluloses in *Pinus sylvestris* bark[37] revealed an alkali-soluble galactoglucomannan which, like others isolated from wood and barks, consisted of a linear chain of β (1-4) linked D-glucopyranose and D-mannopyranose units in which about 3% of the chain moieties were substituted at C-6 with α-D-galactose residues. The properties of the bark polysaccharide were similar to a corresponding one isolated from wood. In addition, an impure arabinan was isolated by extraction of holocellulose with sodium carbonate, enzymatic hydrolysis with α-amylase, and extraction into 70% aqueous methanol.[38] This crude arabinan contained residues of galactose, glucose, arabinose, and xylose.

Although callose has been observed as a major carbohydrate that clogs the sieve cells of phloem as they die or stop functioning, there are few instances in which enough pure callose has been isolated to permit detailed structural studies. Fu et al.[23] isolated callose from acid-chlorite liquors of the holocellulose preparations from *Pinus sylvestris* by treatment with α-amylase and repetitive extraction with hot water to remove water-soluble hemicelluloses and residual starch. The callose obtained in this way contained low proportions of galacturonic acid impurities. Partial hydrolysis and methylation showed a β (1-3) linked D-glucan.

Laver et al.[13] found polysaccharides of the arabinoglucuronoxylan, galactoglucomannan, and glucomannan class in *Pseudotsuga menziesii* inner bark. The xylan was characterized as a backbone chain of 90 β (1-4) linked D-xylose residues with arabinose side chains. The proportion of arabinose units was much higher than found in comparable polysaccharides obtained from wood or in arabino-4-O-methylglucuronoxylans isolated from barks of other species. Fernandez[39] investigated the properties of the glucomannan obtained from *P. menziesii* inner bark holocellulose by extraction with 15% aqueous sodium hydroxide. Analytical data indicated a backbone chain of β (1-4) linked D-mannopyranose and D-glucopyranose units with small proportions of (1-6) linked D-galactopyranose units as side chains similar to the corresponding polysaccharides isolated from other barks and from wood.

Less work has been directed to the hemicelluloses in hardwood barks. Mian and Timell[40] isolated and characterized a 4-O-methylglucuronoxylan from the inner bark of *Betula papyrifera*. It consisted of a linear framework of β (1-4) linked D-xylopyranose units to about 1/10th of which were (1-2) linked 4-O-methyl-D-glucuronic acid units, which is similar to the xylan found in wood. Jiang and Timell[25] isolated a similar 4-O-methylglucuronoxylan from the inner bark of *Populus tremuloides* which contained two to three branch points in the main xylan chain and 4-O-methyl-glucuronic acid residues on one of 12 xylopyranose units. From the same bark they isolated and characterized a galactoglucomannan that consisted of a nearly linear chain of β (1-4) linked D-glucopyranose and D-mannopyranose units with α-D-galactopyranose residues attached at C-6 of both the glucose and mannose residues.[41] This polysaccharide differed from analogous glucomannans isolated from wood because of the presence of galactose units in the bark polysaccharide. An arabinan was obtained by direct extraction of the bark with 70% methanol.[42] It was a highly branched α (1-5) linked arabinofuranose polymer with branch points at C-2 or C-3 and in some instances both C-2 and C-3.

Toman[43] isolated a 4-O-methylglucuronoxylan from *S. alba* which consisted of a linear β (1-4) linked D-xylopyranose chain to which one of nine xylose residues had a 4-O-methyl-α-D-glucuronic acid unit attached to C-2. In addition, Toman et al.[44] isolated two water-soluble galactans in nearly equal proportions that had similar composition and electrophoretic mobilities. They were combined to give an isolate with a β (1-4) linked galactopyranose chain and about 4% of the residues linked (1-6). Karacsonyi et al.[45] described a water-soluble arabinan which was a highly branched polymer.

F. Pectins

Inner bark contains far higher proportions of pectin than does wood. Pectins are found in the primary cell walls as a matrix material in which randomly oriented cellulosic microfibrils are embedded. They are α-(1-4) linked galacturonic acid polymers which are generally associated with arabinan and galactan polymers. Crude pectins have been isolated from barks of a number of gymnosperms and angiosperms.[11,18,23,46] However, few pectins have been isolated from tree barks in sufficient purity to permit detailed structural analyses.

Battacherjee and Timell[47] isolated a pectin from *A. amabilis* bark which contained galacturonic acid, galactose, and arabinose in ratios of 85:4:11 together with traces of rhamnose. This was separated into an acidic fraction that gave only galacturonic acid on hydrolysis, and a second fraction that contained galacturonic acid, galactose, and arabinose units in ratios of 74:7:19. Methylation and hydrolysis of the polygalacturonic acid fraction showed that it was a linear α (1-4) linked D-galacturonic acid polymer of at least 450 units. The second isolate was made up of a similar α (1 to 4) linked galacturonic acid framework, but it was branched at C-2 and C-3 with L-arabinofuranose side chains terminated with either D-galactopyranose or L-arabinofuranose units.

Timell and Mian[48] isolated a pectin from the inner bark of *B. papyrifera*. This pectin isolate gave galacturonic acid, galactose, and arabinose in ratios of 66:7:27. Electrophoresis separated the isolate into three fractions. An α (1-4) linked D-galacturonic acid polymer containing seven to eight branch points together with very small proportions of arabinose units was obtained as a major fraction. A second fraction was made up of galacturonic acid, galactose, and arabinose residues, while a third fraction was composed only of neutral sugars. Toman et al.[49] have begun detailed studies of the structure of pectin in *S. alba* bark.

G. Prospects for Use

Because the concentrations of carbohydrates in the barks of mature forest trees are so low (Tables 1 and 2), little opportunity exists for development of chemical products based on bark carbohydrates. However, if whole-tree chipping of young trees (as envisioned for energy plantations) becomes a reality, then chips could be fractionated to remove quality wood fiber for paper or composition board products; the residual chips would contain high proportions of inner bark. In some species, such a furnish would contain relatively high concentrations of water- or mild alkali-soluble carbohydrates[4,5] together with potentially valuable phenolic compounds. It is possible that these materials could be removed by extraction before burning for fuel.

Polysaccharide impurities in tannin preparations for uses such as wood adhesives are important to the quality of the tannin isolates. The water resistance of tannin based adhesives is highly dependent upon the carbohydrate content of the tannin extract used in the formulation.[50] When water-soluble carbohydrates are a substantial part of the extracts, development of products from the condensed tannins may depend on finding inexpensive processes for separation of the carbohydrate fraction. Since water-soluble carbohydrates from many species are rich in pentosans, they could be combined with utilization processes for hardwood prehydrolysis liquors to produce furfural for example. Undoubtedly, however, most use of bark carbohydrates will be through their use in paper products derived from pulps produced from furnish containing some whole tree chips.

IV. LIGNINS

A. Conifer Bark Lignins

Many questions persist about the lignins in conifer barks. Early observations on conifer bark lignins[51-56] focused on the difficulty in isolating bark lignins that were not contaminated with phenolic impurities (in conifers most notably the polyflavonoids). The sulfuric acid-insoluble suberins also complicate analysis of the lignin in outer bark just as they do the determination of carbohydrates. For these reasons, when analytical values for the amounts of lignins in bark are determined by procedures developed for wood, the values are generally high. Sarkanen and Hergert[57] have critically reviewed the contamination of bark lignin isolates by polyflavonoids and consequent difficulties in obtaining accurate determinations of the amounts of lignin in bark tissues. Although much information has been published, none is compiled here because of this problem. The contamination of bark lignin isolates by polyflavonoids and suberins also complicates the elucidation of the structure of bark lignin. However, bark lignins similar to wood lignins can be isolated from alkali-extracted bark by treatment with dioxane-HCl,[54-56,58] ball milling,[59-60] or phenolysis.[51] Up to half of the "true lignin" in bark resists solubilization by treatment with dioxane-HCl and other delignifying reagents.

Several studies of the accessible conifer bark lignins suggest that they are generally similar to corresponding wood lignins. Sarkanen and Hergert[57] compared spectral properties and functional group proportions of the dioxane-HCl-soluble bark and wood lignins of *Abies amabilis* and, except for a 1% lower methoxyl content in the bark lignin, found no significant differences between the wood and bark lignins. Sogo et al.[59-66] also found great similarity in the spectral properties, functional group proportions, and permanganate oxidation products of ethylated Bjorkman lignins obtained from the bark and wood of *Pinus densiflora* and *Picea jezoensis*. Sarkanen and Hergert[57] reported that the phenolic products obtained from treatment of *Pinus palustris, A. amabilis, Tsuga heterophylla,* and *Picea sitchensis* barks with dioxane-HCl at high temperatures were typical of wood lignin degradation products. In addition, treatment of the extractive-free barks of *Pinus palustris* and *A. amabilis* with dioxane-water at 160 to 180°C for about 40 min also yielded phenols typical of wood lignin degradation.[57] Higuchi et al.,[67] studying *Pinus thunbergii, A. firma,* and *Cryptomeria japonica* barks, and Sogo and Kawahara,[68] studying *Pinus densiflora* bark ethanolysis products, found *p*-hydroxyphenyl- and guaiacyl-type phenylpropane derivatives typical of those obtained from the ethanolysis of wood lignins. Hata et al.[69] and Nord and Hata[70] found phenols typical of those obtained by fungal degradation of wood lignins in the decomposition of *Pinus monticola* Bjorkman lignin and whole bark with *Fomes annosus* or *F. fomentarius*. Particularly significant was the finding of coniferyl alcohol, guaiacylglycerol, and guaiacylglycerol-β-coniferyl ether. Swan,[71] studying the H-NMR spectra of acetates of the dioxane-HCl lignins from *Thuja plicata* wood, inner bark, and outer bark, found essentially identical spectral properties for the wood and inner bark lignins. The lignin obtained from the outer bark contained fewer methoxyl groups, fewer aromatic protons, and higher proportions of aliphatic protons together with some catechol groups that may have originated from suberin and phenolic extractive impurities.[57]

Other studies suggest that accessible bark lignins differ somewhat from wood lignins. Nitrobenzene oxidation products obtained from Bjorkman lignin of *Pinus monticola* contained over twice as much *p*-hydroxybenzaldehyde as did the oxidation products from milled wood lignins.[69] Nitrobenzene oxidation of *Pinus thunbergii* bark also gave comparatively high ratios of *p*-hydroxybenzaldehyde to vanillin,[67] and Sogo and Hata[61] found high proportions of 4-ethoxybenzoic acid derivatives compared with 4-ethoxy-3-methoxybenzoic acids after permanganate oxidation of ethylated Bjorkman lignins obtained from *Pinus densiflora* and *Picea jezoensis* bark.

Differences between the dioxane-HCl lignins obtained from the bark and wood of *Pinus densiflora* were also indicated by the yields of phenols obtained by hydrogenolysis of dioxane-HCl lignins over a copper-chromite catalyst.[65] The products obtained from bark lignin contained comparatively higher proportions of phenol, *p*-hydroxybenzoic acid, guaiacol, vanillic acid, and *p*-ethylguaiacol, but significantly lower yields of dihydroconiferyl alcohol than were obtained from a similar treatment of wood dioxane-HCl lignins (Table 4). Andersson et al.[72] examined the oxidation products obtained from methylated Kraft lignins of *Picea abies* and *Taxus baccata* wood and bark. Acids typical of those obtained from methylated wood Kraft lignins were obtained in a much lower yield from the methylated bark Kraft lignins. The relative proportion of *p*-methoxybenzoic acid to 3,4-dimethoxybenzoic acid was over four to five times higher in the bark than in the wood lignin oxidation products. The bark lignins gave particularly high proportions of dicarboxylic acids and lower proportions of 3,4-dimethoxybenzoic acid, diarylether, and biphenyl dimers than were obtained from the wood lignins. In addition, studies of the properties of lignosulfonates obtained by acid sulfite delignification of *Pinus densiflora* bark indicated a much lower maximum molecular weight solubilized from the bark (mol wt 540 to 9700) than obtained from wood (mol wt 920 to 27,000) when pulped under similar conditions.[64] The lignosulfonates obtained from the bark gave over twice as much *p*-hydroxybenzaldehyde as was obtained from nitrobenzene oxidation of the wood lignosulfonates.[63]

Table 4

HYDROGENOLYSIS PRODUCTS FROM WOOD AND BARK LIGNINS OF *PINUS DENSIFLORA*[65]

	Wood dioxane-HCl soluble lignins	Bark dioxane-HCl soluble lignins	Bark resistant lignins
Yield of H_2O-dioxane solubles	63.6	49.0	13.9
Phenols obtained			
phenol	—	1	2
p-hydroxybenzoic acid	1	2	2
p-Propylphenol	1	1	—
Guaiacol	1	2	2
Vanillic acid	2	3	2
p-Cresol	1	1	—
o-Cresol	1	1	—
p-Ethylguaiacol	—	2	—
Acetonvanillone	1	1	1
Dihydroeugenol	1	1	—
Dihydroconiferyl alcohol	4	2	—

Presence of a lignin component in conifer barks which is highly resistant to solubilization by usual wood delignifying reagents is well documented.[51-56] Several investigations of the resistant bark lignins suggest that they differ markedly from the structure of conifer wood lignins. Hata and Sogo[58] isolated the resistant lignin from dioxane-HCl extracted bark residues by digestion with 41% HCl (Willstatter lignin), a procedure which should preserve the basic structure of the resistant lignin. The resistant Willstatter lignin contained a low methoxyl content (6.36%) as well as low C and H compared with dioxane-HCl lignins obtained from *Pinus densiflora*. A series of methylation studies suggested that both aromatic and aliphatic hydroxyl contents were also lower in the resistant Willstater lignin than in the dioxane-HCl lignin. Hydrogenolysis[65] of the resistant Willstatter lignin isolate from *Pinus densiflora* over copperchromite gave very low yields of low molecular weight phenols compared with the yields obtained from the dioxane-HCl bark lignins or wood lignins (Table 4). Notably no dihydroconiferyl alcohol was found in products obtained from the resistant bark lignins although it is the predominant product obtained by hydrogenolysis of wood lignins. Sogo and Hata[66] also compared permanganate oxidation products obtained from ethylated resistant Willstatter bark lignins and methylated bark residues obtained after extraction with dioxane-HCl with the oxidation products obtained from wood or dioxane-HCl lignins. The permanganate oxidation products obtained from the resistant lignin isolate and from the dioxane-HCl-treated bark residues both were composed of comparatively high concentrations of isophthalic acid derivatives and proportionately lower amounts of benzoic acid derivatives typical of wood lignin degradation. These studies indicate that conifer bark lignins, particularly those which are resistant to dioxane-HCl solubilization, are more highly condensed than are the lignins of conifer woods.

B. Hardwood Bark Lignins

Many of the same differences between conifer bark and wood lignins are also found between hardwood bark and wood lignins. Assay of the amount of lignin in hardwood barks is difficult because of interference of suberin and polyphenolic extractives (the condensed and hydrolysable tannins). Smelstorius and Stewart,[73] who studied the wood and bark of *Acacia penninervis,* demonstrate the complexity involved in attempting to accurately determine the amount of lignin in hardwood barks. As in conifer barks, the basic structural units of hardwood bark lignins are similar to the lignins of the corresponding woods. Nitrobenzene oxidation of the barks of *Salix bakko, Prunus domarvarium, Cinnamomum camphora,* and *Melia azedrach* by Sogo[74] and Hata and Sogo[75] established that the bark lignins of these species were composed of guaiacyl, syringyl, and small proportions of *p*-hydroxyphenyl nuclei equivalent to hardwood lignins. Higuchi et al.,[67] studying nitrobenzene oxidation products of the barks of *Fagus crenata, Magnolia obovata,* and *Quercus crispula,* obtained syringaldehyde, vanillin, and *p*-hydroxybenzaldehyde from extractive-free barks in ratios of 3.02:2.5:0.1; 1.5:2.7: trace; and 1.8:2.3: trace, respectively. Ethanolysis of the same barks gave syringoyl-acetyl to vanilloyl-acetyl ratios of 1.1, 0.4, and 0.5, respectively; proportions similar to those obtained in nitrobenzene oxidation but much lower than those generally obtained from hardwood lignins (syringaldehyde/vanillin ratios of about 2.5). Clermont[76] studied the dioxane-HCl lignins obtained from NaOH-extracted stone cells from the inner bark of *Populus tremuloides* and found that the methoxyl content of the bark lignin was only slightly lower than that obtained from the wood lignins. Despite the similarity in methoxyl content found by the above methods, nitrobenzene oxidation of the stone cells gave syringaldehyde to vanillin ratios of only 1:1.1, compared with a ratio of 2.6:1 for the wood lignins.

Sogo[60] and Sogo and Hata[61] examined permanganate oxidation products from ethylated Bjorkman lignins obtained from the barks of *A. deccurens, Q. serrata,* and *Castanea crenata* barks. The products obtained from the bark lignins were dominated by 4-ethoxy-3-methoxybenzoic acid and 3,5-dimethoxy-4-ethoxybenzoic acids, as would be expected of a typical hardwood lignin. However, in comparison with permanganate oxidation products obtained from ethylated Bjorkman lignins from *C. crenata* wood, the oxidation products obtained from the bark lignins contained proportionately high yields of dicarboxylic acids. Andersson et al.[72] found major differences between wood and bark kraft lignins of *Betula verrucosa, Fraxinus excelsior,* and *Vitis vinifera* by oxidation of the methylated kraft lignins. The major difference between wood and bark lignins was the low ratio of syringyl/guaiacyl nuclei in the bark lignin oxidation products (*B. verrucosa,* syringyl/guaiacyl ratios of 1:0.9 for bark compared with 3.9:1 for the wood lignins). In addition, oxidation products obtained from the methylated hardwood bark kraft lignins contained higher proportions of dicarboxylic acids and lower proportions of the benzoic acid derivatives typically obtained from the wood lignins. In studies of the bark lignins of *Q. crispula,* Hata and Sogo[77] found a dioxane-HCl-resistant lignin fraction in proportions nearly equal to the amounts of the dioxane-HCl-soluble lignin. Isolation of this resistant lignin as Willstatter lignin gave a product with a low methoxyl content (13.2%). Nitrobenzene oxidation showed a low syringaldehyde/vanillin ratio and the proportion of *p*-hydroxybenzaldehyde obtained was higher than that obtained from nitrobenzene oxidation of the dioxane-HCl lignin. The above studies indicate that hardwood bark lignins differ from the corresponding wood lignins in the proportions of syringyl, guaiacyl, and *p*-hydroxyphenyl nuclei and in the much higher degree of condensation in the bark lignins.

Studies by Hergert[78] suggest an explanation for the differences between wood and bark lignins. In the lignification of wood, most of the lignin is deposited near the xylem cambium. However, Hergert has demonstrated that secondary lignins (both Brauns native lignin and insoluble lignin) are deposited near the heartwood boundary. The secondary lignin differs from the lignin formed near the xylem cambium in that it contains a lactonic carbonyl indicative of the incorporation of conidendrin type of structure in the secondary lignin. In the lignification of *Pinus palustris* bark, only a small proportion (approximately 28%) of the lignin found in the outer bark is present in the inner bark. The majority of the lignin found in outer bark tissue is apparently a secondary lignin deposited during the transformation of inner bark to outer bark. Because lignification of bark occurs both physically and physiologically remote from the formation of the majority of wood lignin, differences between these polymers seem acceptable on a biochemical basis.

In studies of the chemistry and pulping characteristics of the fibrous bark of *Eucalyptus obliqua*, Hemingway and Davies[79] found that lignin was present in inner bark and wood in nearly the same amounts (20.7% vs. 22.9%) and the methoxyl contents of the inner bark and wood lignins were similar (22.1% vs. 22.6%) after correction for phenolic extractives. However, both inner and outer bark tissues resisted delignification by kraft or Soda pulping conditions even when much higher amounts of chemicals were applied. In this case, the resistance of the bark to delignification was attributed to a lack of accessibility of the chemical caused by the very thick cell walls of the inner bark fibers.

C. Lignin and the Utilization of Bark

The importance of lignin to the utilization of bark rests primarily with its resistance to solubilization by usual pulping reagents. Pulping whole tree chips has been somewhat successful. However, amounts of barky chips that can be tolerated have been limited for a number of reasons, including poor fiber characteristics, higher chemical demand caused by phenolic extractives, higher bleaching costs, and difficulties with dirt content. In addition to these problems, bark lignins resist usual pulping reagents either because of differences in the degree of condensation or because of their inaccessibility when fibers with exceptionally thick walls or highly suberized bark tissues are encountered. The resistance to delignification will contribute to high kappa numbers even when additional alkali is used to compensate for the higher phenolic extractive content of many barks.

V. WAX AND SUBERIN

A rhytidome containing large amounts of cork cells is formed in the bark of many tree species. Litvay and Krahmer[80] showed that the cork cells of *Pseudotsuga menziesii* are composed of a compound middle lamella made up of randomly oriented cellulosic microfibrils embedded in pectin and phenolics, a thick cell-wall layer composed of alternating zones of waxes and phenolic polymers rather than the usual cellulosic wall, a thin layer of benzene-soluble wax, and a thin layer of cytoplasmic debris and phenolic extractives at the lumen surface. A similar structure of the cell wall layering was observed in *Quercus suber* cork except for a thin cellulosic wall layer at the lumen surface in oak cork.[81] Many commercially important trees including *Pseudotsuga menziesii*, *Abies*, *Quercus*, and *Betula* species have barks that contain large amounts of cork. These barks are potentially valuable sources of waxes, rare fatty acids, and fatty alcohols that might be derived from the suberin of cork. Other tree species that do not have large discrete cork areas (i.e., *Pinus*) are also potentially valuable sources of waxes, although yields are lower because the cork tissues cannot be economically separated from other bark tissues.

The waxes of tree barks, located in the cork cells of the outer bark, are extended polyestolides (esters of fatty acids such as lignoceric acid, behenic acid, dibasic acids, and hydroxyfatty acids with fatty alcohols such as lignoceryl or behenyl alcohol).

Waxes of conifer barks also contain substantial proportions of ferulic acid, and waxes of hardwood barks contain sinapic acid as a significant constituent. Tree barks also contain fats that are found primarily in the parenchyma of the inner bark and are similar in composition to the fats found in sapwood (i.e., predominantly esters of oleic, linoleic, or linolenic acids with glycerol). Therefore, the composition of saponification products obtained from tree barks depends upon the age of the bark or the proportion of inner to outer bark in the sample.

The works of Zellner, who from 1924 to 1934 studied the waxlike constituents of over 20 European hardwood barks, have been reviewed by Segall and Purves.[82] Zellner was able to crystallize and identify a large number of fatty acids, fatty alcohols, and oxygenated fatty compounds from hardwood bark waxes. Study of this work, together with the work of Harwood,[83] Kurth and co-workers,[84-85] Kurth,[86] and Hergert and Kurth[87-89] done from the late 1940s to the early 1960s, highlights the recent lack of attention to these compounds.

A. Conifer Bark Waxes

Interest in the utilization of the wax components of North American conifer barks apparently first developed from study of the hydrolysis lignin obtained from the production of ethanol from barky Douglas-fir chips in the middle 1940s.[90] Clark et al.[91,92] obtained 1.5% of waxlike substances from the hydrolysis lignin and reported the identification of arachidic, behenic, and lignoceric acids; arachidyl and behenyl alcohols; and a series of saturated hydrocarbons.

At about the same time, Harwood[83] was applying Zellner's approach to an examination of the ligroin- and ether-soluble constituents of methanol extracts from *Picea glauca* bark, and Kurth[86] was studying the wax constituents of *Pseudotsuga menziesii* bark. Harwood[83] found fats composed primarily of glyceride esters of oleic, linoleic, and linolenic acids; a complex mixture of resin acids; phytosterols; and substantial proportions of waxes composed mainly of lignoceric and smaller amounts of behenic acids esterified with lignoceryl alcohol as the dominant fatty alcohol component. Kurth and Kiefer[84] isolated the waxes from *P. menziesii* bark by separating the benzene-soluble materials into portions soluble and insoluble in hot hexane. The hexane-soluble wax was composed of 60% lignoceric acid, 20% lignoceryl alcohol, and 20% ferulic acid. The benzene-soluble wax contained large amounts of phlobaphene polymers and a more complex mixture of fatty acids and alcohols. Hergert and Kurth[87] obtained a relatively pure separation of the cork of *P. menziesii* bark and isolated hexane- and benzene-soluble waxes from the cork in yields ranging from 5.6 to 10.4%. The hexane-soluble wax was dominated by lignoceric acid (49.3%), lignoceryl alcohol (27.5%), ferulic acid (9.8%), and small amounts of phytosterols (0.6%), together with undetermined acid and phenolic materials amounting to 11.7%. Saponification of the benzene-soluble wax gave lignoceryl alcohol (2.4%), lignoceric acid (19.2%), a hydroxypalmitic acid (7.3%), unsaturated hydroxyfatty acids, and other unidentified fatty acids (6.4%), and a phlobaphene polymer (27.4%).

In further studies of the benzene-soluble wax, Kurth[86] found major proportions of a hydroxybehenic acid and smaller proportions of hydroxyarachidic acid in addition to the hydroxypalmitic acid found earlier. Kurth[86] also found a C_{20} dicarboxylic acid as a significant constituent of the benzene-soluble wax. Adamovics et al.[93] isolated ferulate esters of lignoceryl and behenyl alcohols in nearly equal proportions from chloroform-soluble wax. Laver et al.[94] also reported esters of ferulic acid with behenyl alcohol and lignoceryl alcohol in hexane-soluble waxes. Loveland and Laver[95,96] have examined the constitution of the fatty acids (monobasic and dibasic acids), the hydroxyfatty acids, and the fatty alcohols obtained from saponification of the hexane-

Table 5
SAPONIFICATION PRODUCTS OF HEXANE-INSOLUBLE, BENZENE-SOLUBLE WAX FROM DOUGLAS-FIR BARK[95,96]

Fatty acids	Yield	Dicarboxylic acids	Yield
C-22	32.3	C-16	36.3
C-24	51.0	C-18	14.6
C-26	12.7	C-18[1]	24.8
		C-20	14.3
		C-22	8.7
		C-24	1.3

Hydroxy fatty acids	Yield	Fatty alcohols	Yield
C-16	26	C-16	trace
C-18	7	C-18	4
C-20	18	C-22	45
C-22	24	C-24	48
C-24	6		

insoluble, benzene-soluble waxes of *Pseudotsuga menziesii* and report relative proportions of individual components of each fraction (Table 5).

Hergert and Kurth[88] obtained a hexane-soluble wax that melted at 56 to 57°C from *Abies concolor* cork in yields of 3.46 to 4.50%. Lignoceryl alcohol (9.37%) was separated from the wax by precipitation from acetone prior to hydrolysis. Free behenic acid (31.1%) and unsaturated acids (5.38%) were obtained by extraction into potassium carbonate. Saponification of the neutrals gave lignoceryl alcohol (25.1%), phytosterols (1.26%), and unsaturated alcohols (1.82%) along with behenic acid (18.93%) and an unidentified fatty acid that was insoluble in hexane (4.47%). After extraction of the cork with hexane, a hard, brown wax with a melting point of 70 to 72°C was obtained in yields of 2%. Saponification of this benzene-soluble wax gave lignoceryl alcohol (5.35%), phytosterols (0.3%), and unsaturated alcohols (0.3%) in the neutral fraction. The acid fraction contained 32.5% of a phenolic acid polymer, behenic acid (3.02%), and two unidentified fatty acids (9.30% and 0.39%). A hexane-insoluble hydroxy-fatty acid identified as 13-hydroxymyristic acid was obtained in 48.8% yield. Further work by Kurth[86] on the hydroxyfatty acid fraction showed a mixture of hydroxypalmitic, hydroxyarachidic, and hydroxybehenic acids together with an incompletely resolved mixture of behenic acid, an unidentified dicarboxylic and unsaturated fatty acids.

Rogers and Grierson[97] extracted *A. grandis* bark with petroleum to recover 2.4% of a brown wax. Neutral constituents, approximately 50% of this wax, were separated into ferulate esters, wax esters, free wax alcohols, free and esterified sterols, hydrocarbons, and triterpene lactones. The largest fraction (50% of the neutrals) was composed of ferulate esters of lignoceryl alcohol (56%) and behenyl alcohol (44%). Small proportions of a white colored wax were obtained by recrystallization from acetone. This wax was composed of lignoceric (60%), behenic (37%), and arachidic (3%) acids esterified with lignoceryl alcohol (66%) and behenyl alcohol (33%). Free wax alcohols were dominated by lignoceryl alcohol, although small amounts of behenyl and ceryl alcohols were also isolated. Fatty acid esters with sitosterol and campesterol were isolated in addition to free β-sitosterol. Hydrocarbons were composed of a homologous series of *n*-alkanes (C_{19} to C_{42}) together with four branched alkanes. Two triterpene lactones cyclograndisolide and epicyclograndisolide were also minor constituents.

Extraction of *Tsuga mertensiana* bark with benzene gave 3.7% of a brown wax that was completely soluble in hot hexane. Saponification of this wax gave large amounts of lignoceric acid and lignoceryl alcohol. The phenolic materials present were not investigated further.[86]

The waxes of *Pinus* species have also received considerable attention, although cork is not present in such large layers as is found in the bark of *Pseudotsuga menziesii* or *Abies* species. Kurth and Hubbard[85] isolated both hexane-soluble (3.4%) and benzene-soluble (1.3%) waxes from *Pinus ponderosa* bark. The yellow colored hexane-soluble wax contained 42% acids; behenic acid and lignoceric acid were dominant, and there were small amounts of unsaturated fatty acids and resin acids. Saponification of the neutrals yielded 30% acids, of which arachidic and behenic acids were the major constituents. The unsaponifiable fraction contained relatively large proportions of a fatty alcohol and small proportions of phytosterols. Hata and Sogo[98] obtained a light-brown wax in a yield of 3.31% by extraction of the outer bark of *Pinus densiflora* with hexane. Further extraction with benzene removed only an additional 0.53% of a cream colored wax. When extracted with benzene only, 3.5% of the bark was recovered as a light-brown wax. Saponification gave lignoceric acid (52.4%), hydroxypalmitic acid (5.4%), lignoceryl alcohol (11.6%), and phytosterols (1.7%). Rowe et al.,[99] studying the waxes of *Pinus banksiana*, separated the fatty acids into free and esterified fractions. The major fatty acid in both the free acid and esterified form was lignoceric acid (54% and 44%) with large proportions of cerotic (23% and 20%), and behenic (13% and 27%) acids; smaller proportions of palmitic, stearic, and arachidic acids; and very small proportions of odd chain length fatty acids. The fatty alcohols obtained after saponification and present as ferulates were also examined. Lignoceryl alcohol (58% and 60%) and behenyl alcohol (30% and 30%) predominated, with smaller amounts of steryl, arachidyl, ceryl, and a C-23 alcohol. A smaller portion of a *n*-paraffin fraction was a complex mixture ranging in carbon numbers from C-21 to C-31.

Hergert,[100] studying the extractives of the four major southern pines (*Pinus taeda, P. elliottii, P. palustris,* and *P. echinata*), found petroleum and benzene solubilities in the range of 1.0 to 2.6% and 0.5 to 1.4%, respectively. Hergert[100] separated the petroleum extractives of *P. palustris* into two fractions, both of which were composed mainly of wax-derived compounds including lignoceryl alcohol, lignoceric acid, and ferulic acid, and a third fraction rich in fatty acids, resin acids, and fats derived from the true fats and oleoresin. Hemingway[101] obtained a similar wax from *P. echinata.* When purified by repetitive precipitation from cold methanol this wax was obtained as nearly white wax in yields of slightly over 1%. Pearl[102-104] found petroleum ether solubilities between 1.5 and 4.8% for *P. taeda* and between 1.8 and 2.9% for *P. elliottii.* Pearl saponified the total petroleum ether-soluble fraction to obtain neutral compounds that ranged from 18 to 36% of the petroleum extracts of *P. taeda* and from 20 to 29% of the petroleum extracts of *P. elliottii.* Of the neutrals, lignoceryl alcohol predominated, with significant amounts of ceryl, behenyl, and arachidyl alcohols also present. Yields of acids obtained after saponification ranged from 57 to 78% of the petroleum extract from *P. taeda* and from 57 to 74% of the petroleum extract from *P. elliottii.* The fatty acid composition of the saponified total petroleum ether extracts was more characteristic of the fatty acids obtained from fats of wood. The major fatty acids present were oleic, linoleic, and a *trans,trans*-9,11-octadecanoic acid. Arachidic, behenic, lignoceric, and cerotic acids, and a *cis,cis,cis*—5,11,14-eicosantrienoic fatty acid were also significant constituents; however, no single fatty acid predominated. All of the above mentioned fatty acids were present in proportions ranging from 5 to 37% of the total fatty acids. Saponification of the diethyl ether soluble extract gave lignoceric, linoleic, behenic, and arachidic acids as the major fatty acid contituents, and lignoceryl alcohol dominated the neutral fraction. There was considerable variation among trees both in terms of the yields of fatty alcohols and fatty acids, but no systematic seasonal variations could be determined.

Hartman and Weenick[105] extracted *P. radiata* bark with petroleum: methanol (9:1) followed by ether. The combined extracts (3.4% of dry bark weight) were saponified. The fatty acids obtained from hydrolysis were compared with those obtained from wood and tall oil. Behenic acid (38%), lignoceric acid (29%), and arachidic acid (12%) predominated in saponified bark extracts; oleic acid and linoleic acid were the predominant fatty acids from wood and tall oil.

Weissman,[106] studying the wax components of *P. sylvestris* bark, obtained 3.7% benzene-solubles, composed of 41% neutrals, 37% free acids, and 22% highly polar compounds that were not eluted from DEAE-Sephadex A-25® with CO_2-saturated ether.[107] The free fatty acids were dominated by behenic acid (55%) with large amounts of lignoceric acid (25%) and arachidic acid (11%). The saponifiable acids were similar in composition and were obtained in yields of 20% of the extract. The neutrals obtained after saponification (17% of the extract) contained behenyl and steryl alcohols as major constituents with small proportions of lignoceryl, arachidyl, and palmityl alcohols. Ferulic acid comprised 22% of the extract.

Weissmann[106] obtained similar results in parallel studies of the benzene-soluble extractives of *Picea abies* bark. This bark contained 4.2% benzene-solubles, of which 34% was free acids. The neutrals after saponification gave 9.6% acids and 22% unsaponifiables. The free and saponifiable fatty acids were similar in composition: lignoceric acid (34% and 31%), behenic acid (23% and 20%), oleic acid (9.0% and 9.5%), and palmitic acid (11.5% and 6%). Behenyl and lignoceryl alcohols were the predominant fatty alcohols in the unsaponfiable fraction. Smaller amounts of steryl, arachidyl, and palmityl alcohols were also present. Ferulic acid constituted 13% of the benzene extract.

B. Hardwood Bark Waxes

Except for the early work of Zellner, the waxes in hardwood barks have not been extensively studied. Segall and Purves,[82] summarizing Zellner's work, show that important differences exist between the wax extracts of hardwood barks and those of the conifers. Whereas the petroleum-soluble waxes from conifer barks are usually comparatively simple in constitution (i.e., dominance of lignoceric acid or lignoceric acid and behenic acid esterified with lignoceryl alcohol, with smaller amounts of ferulic acid present), the constituents of hardwood petroleum extracts are more complex. Zellner's work shows that ceryl alcohol appears more frequently in both the esterified and free alcohol state. The esterified fatty acids are much more complex, with myristic, palmitic, stearic, oleic, linoleic, and arachidic acids frequently found. In addition, Zellner crystallized a large number of oxygentated compounds in amounts sufficient for combustion analysis but insufficient for structure determination.

Hossfield and Hunter[108] examined the saponification products obtained from petroleum extracts of *Populus tremuloides*. They isolated lignoceric acid, linoleic acid, and an unresolved mixture of saturated fatty acids. The unsaponifiable fraction contained a hydrocarbon (melting point 56 to 57°C), β-sitosterol, ceryl alcohol, and glycerol. Abramovitch et al.[109] saponified the neutral fraction obtained from acetone extraction of *Populus balsamifera* and obtained a complex mixture of fatty acids and aliphatic dialcohols. Palmitic, palmitoleic, stearic, oleic, and linoleic acids typical of fatty materials and 11-eicosenoic, behenic, lignoceric, and cerotic acids typical of wax esters were found. The wax alcohols ranged in carbon numbers from C_{18} to C_{28}. Streibl et al.[110-113] studied the petroleum extracts of *Fagus sylvatica*. Saturated fatty acids ranged in carbon numbers from C_{12} to C_{30}, unsaturated monocarboxylic acids of carbon numbers from C_{12} to C_{24}, and dicarboxylic acids in carbon numbers from C_{12} to C_{24}. The aliphatic alcohols ranged in carbon numbers from C_{12} to C_{31}.

Lipophillic extracts obtained from many hardwood barks are often dominated by triterpenes rather than waxes similar to those obtained from conifer barks. Most notable are the high yields of triterpenes obtained from *Betula* species (betulin), *Quercus* species (friedelin and cerin), and *Alnus* species (taraxerol and tararexerone). These compounds are discussed more fully below.

C. Suberin

Hergert, Kurth, and their co-workers[84-89] made remarkable advances in their studies of the wax and suberins of conifer barks. The hydroxyfatty acids were especially difficult to study because of their instability (i.e., estolide formation) and their tendency to crystallize as mixtures of compounds which were difficult to separate.[86]

The extractive-free cork of *Pseudotsuga menziesii* is composed of a large proportion of polyflavonoids, complex hydroxyfatty acids, and glycerol.[87] Hydroxyfatty acids reported include 11-hydroxylauric acid,[87] hydroxypalmitic acid,[87] hydroxyarachidic acid,[86] and a monohydroxyhexadecenoic acid.[87] Lignoceric acid,[86] an unresolved mixture of fatty acids containing unsaturated fatty acids,[87] and a C_{20}-dicarboxylic acid[86] were also reported in significant yields. Kurth[86] reported that the same compounds obtained from saponification of the extractive-free cork were also obtained from saponification of the hexane-insoluble, benzene-soluble wax, although relative yields of individual components varied between the two materials. It is instructive to compare the above results with those of Loveland and Laver,[95,96] who used gas liquid chromatography and mass spectral (GLC-MS) techniques in the identification of the fatty acids obtained after saponification of hexane-insoluble, benzene-soluble waxes (Table 5).

The major constituent of the extractive free cork (42%) was an ether-insoluble phenolic polymer that had properties similar to the polyflavonoids obtained from bast fibers. In studies of how these components were combined, Hergert and Kurth[87] isolated fractions from a partial saponification of the cork. A hexane-soluble fraction gave 11-hydroxylauric acid after further saponification. This finding indicated that some of the hydroxyfatty acids in the suberin were linked as esters (not etholide). In addition, an ether-soluble fraction composed of both hydroxyfatty acids and phenolic materials was isolated. On further saponification this fraction could be separated into hydroxyfatty acids and ether-insoluble phenolic fractions. These results were interpreted as a suberin structure in which the hydroxyfatty acids were linked to each other as well as to the polyflavonoids by ester linkages.

Litvay and Krahmer[80] proposed a different structure for the suberin of *P. menziesii* cork. Presumably based on the term etholide linkage,[87] they suggested that the wax layers in the suberin wall were condensed tail to tail by ether linkages and that these wax layers were of a width comparable with two hydroxy-fatty acid molecules aligned in the width direction of the wax lamella. The phenolic materials obtained after saponification were soluble in acetone, and paper chromatography indicated that the phenolics were "ferulic acid like" compounds together with some unidentified phenolics that gave green fluorescence under ultraviolet light. Based on the intense blue fluorescence of the suberin wall layer and the above results, they proposed that the phenolic lamella were composed of ferulic acid type polymers of approximately 11 molecules in thickness. However, it is difficult to reach such a conclusion from the results obtained by Hergert and Kurth,[87] for the phenolic material that they isolated from saponification of extractive-free cork. Hergert[89] did obtain small amounts of ferulic acid from saponification of extractive-free cork of *P. menziesii*, but the major phenolic product was the ether-insoluble polymer.

Hergert and Kurth[88] saponified extractive-free cork of *A. concolor*, acidified the total product, and fractionated the dried acid-insoluble material by successive extraction with hexane, ether, and alcohol. A water-soluble phenolic acid was shown to be ferulic acid. As in saponification of the extractive-free cork of *P. menziesii*, a major fraction obtained by extraction into ethanol was a phenolic acid polymer that closely resembled the tannins and phlobaphenes obtained from this species, except for a rearranged catechin unit.[114] Extraction of the dried saponification product with hexane and subsequent fractionation of the hexane-soluble fraction gave three hydroxy fatty acids tentatively identified as 13-hydroxymyristic acid, hydroxyarachidic acid, and a compound that might be a dihydroxydicarboxylic acid, as had been isolated from suberin saponification products of *Q. suber*. Kurth[86] later showed that the 13-hydroxymyristic acid obtained from the hexane-insoluble, benzene-soluble was a mixture of hydroxypalmitic and dicarboxylic acid. Hydroxyplamitic, hydroxyarachidic, and hydroxybehenic acids were also obtained from saponification of the hexane-insoluble, benzene-soluble wax.[86] Hergert[89] showed that lignin was present in the residue obtained after saponification of the extractive-free cork. Hergert[89] also exacted the extractive-free cork with dioxane-HCl, and the extract was then soluble in chloroform or benzene. Saponification of this extract gave 72% of hydroxyfatty acids, 6% of ferulic acid plus other water-soluble acids, and 22% of a phenolic acid polymer. The infrared spectra indicated that aliphatic esters were dominant. Application of GLC, mass spectroscopy (MS), and nuclear magnetic resonance (NMR) instrumental methods to the study of the hydroxyfatty acids similar to the work done by Loveland and Laver[95,96] and Kolattukudy[115] may provide more accurate information on the constitution of the fatty acids in the suberins.

The chemistry of the suberins of *Betula* and *Quercus* species was ably reviewed by Jensen et al.[3] Although suberins are dominant constituents of many commercially important hardwoods, little work has been published on these compounds since 1960. Zhuchenko and Cherkasova[116] report suberin contents of 40 to 41% together with 35% betulin in the barks of *Betula pubscens* and *B. verrucosa*. Jensen[117] has published a scanning electron microscopic study of the barks of *B. verrucosa* and *Quercus suber*. Gonzales[118] isolated a glyceride ester fraction from the suberin of *Q. suber* and from this fraction isolated a new hydroxy-fatty acid, 9-hydroxy-1, 18-octadecanedioic acid, as a major saponification product. Seoane et al.[119] isolated and proved the structures of two epoxy fatty acids, *cis*-9,10-epoxy-octadecanedioate and 9,10-epoxy-18-hydroxyoctadecanoate from *Q. suber*. Guillemonat,[120] comparing the saponification products from the corks of *Kielmeyera coriacea* and *Q. suber*, found similar but not identical compositions. In saponification products from the corks of *Quercus, Betula*, or *Kielmeyera*, the fatty acids are complex mixtures, and no individual fatty acid dominates. To summarize, the fatty acids identified are: ω-hydroxybehenic acid,[3] eicosanedicarboxylic acid,[3] 9,10,18-trihydroxystearic acid,[3] 18-hydroxy-9-octadecenoic acid,[3] *cis*-9,10-epoxy-18-hydroxystearic acid,[119] 9-hydroxy-1,18-octadecanedioic acid,[118] *cis*-9,10-epoxyoctadecanedioic acid,[119] 8,9-dihydroxy-1,16-hexadecadioic acid,[3] 8-hexadecene-1,16-dicarboxylic acid,[3] and further unidentified unsaturated and dihydroxy fatty acids.

Hergert,[89] studying the phenolic constituents of the cork of *Q. suber*, found synapic acid together with other water soluble acids in a 5% yield, a polyflavonoid polymer in a 30% yield, and hydroxyfatty acids in 65% yield from the saponification of dioxane-HCl-soluble materials obtained from extractive-free cork.

D. Prospects for Utilization of Wax and Suberin

Early attempts to isolate and market waxes derived from *Pseudotsuga menziesii* failed. However, wax products are presently being manufactured by a process developed by Trocino.[121] Ground bark is extracted counter-currently with a hot mixture of aliphatic and aromatic hydrocarbons to remove a light green-brown wax. Applications for this wax include polishes, internal lubricants for molded plastics, concrete additives, slow-release fertilizers, carbon papers, and after bleaching, use as fruit coatings or in cosmetics.[122] A particularly valuable aspect of Trocino's process is the relative ease with which the extracted bark residues are separated into a granulated cork, a bast fiber that has applications as a reinforcing fiber in plastics, and a fine bark powder that has been marketed as a plywood glue extender.[123,124]

Processes have been patented for the recovery of the fatty or waxy components obtained by saponification of bark, particularly *P. menziesii*. Kurth's early work[125] was followed by research and development aimed at fractionation of the alkaline hydrolysates of bark into fatty components, phenolic materials, and residual alkali-extracted fiber. Zenczak[126,127] employed an alcoholic saponification in which the alkali-soluble materials were dried and extracted with organic solvents to recover neutral components and fatty acid fractions. The potential use of other tree barks as a substrate for an alkaline ethanolysis reaction to recover ethyl esters of wax and suberin fatty acids was examined further by Swan[128] and by Swan and Naylor.[129] Crude ethanolysis yields were 21% from *P. menziesii*, compared with 7 to 8% from *Tsuga* species and only 1.2 to 4.5% from the other species studied. Ethanolysis of the *P. menziesii* bark gave 2.84% of pure ethyl octadecanedioate, 0.88% of ethyl 18-hydroxystearate, and 0.32% of ethyl ferulate. These results differed markedly from the products that might have been expected, based on Loveland and Laver's[95,96] findings on hydroxy- and dibasic-fatty acid composition obtained from the saponification of the benzene-soluble wax (Table 5).

Heritage and Dowd[130] employed an aqueous alkaline hydrolysis to recover fatty acids, fatty alcohols, an alkali-soluble phenolic acid of use in the preparation of drilling mud or adhesive additives, and an alkaline extracted bast fiber of use in reinforcing plastics. This work was followed by that of Brink et al.[131] and Dowd et al.,[132] who studied extraction conditions for fractionation of the alkali-soluble products into fatty acids, hydroxyfatty acids, neutral compounds, and two phenolic fractions. Saponifiable fatty acids from the wax and suberins were recovered from aqueous alkaline solutions by adjusting the pH to the range of 9.5 to 10.5[131] or to about 4.0 forming a suspension,[132] and extracting with higher alcohols such as amyl alcohols.[133] The fatty materials soluble in alcohol solutions at pH 4.0 were recovered by steam distillation. Products obtained were a complex mixture of fatty acids, hydroxyfatty acids, fatty alcohols, and residual neutral fatty acid esters. Despite considerable research of methods to fractionate the alkali-soluble products obtained from *P. menziesii* barks, none of these processes has found commercial viability to date.

Fatty acids and alcohols derived from waxes and suberins are recovered in the tall oil when whole tree chips are used in kraft pulping. Therefore, tall oil products from southern pine pulping would be expected to contain higher proportions of arachidic, lignoceric and behenic acids, and lignoceryl alcohol than the amounts obtained from pulping of wood alone. The change in composition of the tall oil components reflects the percentage of bark added as whole tree chips. Apparently there are no major differences in the amounts and composition of these fatty acids relative to tree species.[102-104] The higher proportion of long chain length fatty acids derived from the bark of whole tree chips may promote increased efficiency of tall oil soap separation when pulping the resin acid rich southern pines.[103]

The waxes and suberins of hardwood barks have received little attention in the development of products, even though suberin is a major constituent. A nonionic surfactant has been made from *Betula* bark by grinding the bark in a 14 to 16% NaOH solution, diluting the mixture with isopropanol, and recovering the NaOH insoluble portion by filtration or centrifugation.[134] Distillation of the alcohol from the soluble portion precipitates betulin, which is recovered by centrifugation. Acidification of the NaOH solution to pH 4 to 5 precipitates the suberin complex, which is recovered by centrifugation. The suberin fraction obtained in this way is reacted with ethylene oxide to obtain a series of nonionic surfactants of varying polyoxyethylene content. Yields of betulin (250 kg) and suberin (300 kg) from 1 ton of bark are reported.[134]

VI. TERPENES

The chemistry and use of terpenoids have been reviewed by Zinkel in this volume, so their occurrence in tree barks will only be briefly considered here. Interest in conifer bark terpenoids has centered on their importance as taxonomic indicators and as a source of new and unusual compounds. Zavarin and co-workers[135-144] have studied the bark terpenoids of *Abies* species, particularly with regard to their genetic and geographical variation. Studies of the terpenes of the cortex blisters of *Abies* species led to the discovery of the new terpenes γ-humulene and cyclosativene[135] as well as 4,4-dimethyl-2-cyclohepten-1-one.[143] Other studies conducted with particular regard to genetic and geographical variations have been made of the volatile terpenes in the cortex of *Pinus monticola*,[145,146] *P. strobus*,[146,147] *P. radiata*,[148,149] *P. sylvestris*,[150,151] *P. elliottii*,[152] *P. palustris*,[153] *Picea glauca*,[154] *P. pungens*,[155] and *P. rubens*.[156] The volatile terpenoids of the cortical oleoresins of *Picea abies*,[157] *Larix decidua*,[158] *Pseudotsuga menziesii*,[159-161] and *P. macrocarpa*[161,162] have also been examined.

Intensive study of the neutral diterpenes of conifer barks has led to the discovery of a number of new compounds (Figure 1). Rowe and Scroggins,[163] studying *Pinus contorta* bark, found the new diterpene alcohol 13-epimanool.[164] Rowe et al.[165,166] proved the structures of three additional diterpenes; agathadiol, isoagatholal, and 13-epitorulosol. Bower and Rowe[167] found the known diterpenes manoyl oxide, torulosol, 13-epitorulosol and agathadiol, together with a new compound (+)-13-epimanoyl oxide in the neutral fraction of *Pinus banksiana* bark extracts. Further studies of this species resulted in identification of the new diterpenes 18-norabieta-8,11,13-trien-4-ol, 19-norabieta-8,11,13-trien-4-ol, and 18-hydroxy-8,11,13-abietatrien-7-one.[168] Three 19-norabietatetraenes were indicated in the bark extracts of *P. monticola*.[168] Norin and Winell[169] isolated 19-norpimara-8(14),15-dien-3-one, 19-norisopimara-8(14),15-dien-3-one, and 19-norisopimara-7,15-dien-3-one from extracts of *P. sylvestris* bark. Other diterpenes found in *P. sylvestris* bark extracts included pimaradiene, isopimaradiene, pimaral, isopimaral, dehydroabietal, pimarol, isopimarol and abienol.[151] Zinkel and Evans[170] found strobol, strobal, manoyl oxide, together with *cis* and *trans* abienols in *P. strobus* bark extracts. The neutral diterpenes of *Picea abies* included abienol, manoyl oxide, epimanoyl oxide, pimaradiene, pimarol, and dehydroabietol.[157] 13-Epimanool was isolated from the neutral terpenoid fraction of *Picea sitchensis* bark.[171] The known terpenes sugiol, xanthoprenol,[172] isopimarol,[173] and three new diterpenes, 4α-hydroxy-18-norisopimara-8-(14),15-diene,[173] 4β-hydroxy-18-norisopimara-8(14),15-diene,[173] and 19-norisopimara-8(14),15-diene-3-one[174] were isolated from extracts of *Thuja plicata* bark. Sugiol and xanthoprenol were also found as principal terpenes of *Chamaecyparis obtusa* bark.[175]

In most instances, the cortical resin acids do not differ substantially from those of the xylem oleoresin.[176] However, many new diterpene resin acids have been described

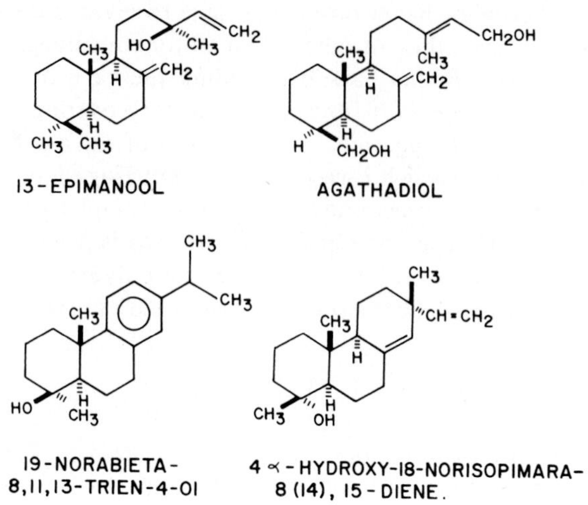

13-EPIMANOOL   AGATHADIOL

19-NORABIETA-   4 α-HYDROXY-18-NORISOPIMARA-
8,11,13-TRIEN-4-OI   8 (14), 15-DIENE.

FIGURE 1.   New neutral diterpenes isolated from bark.

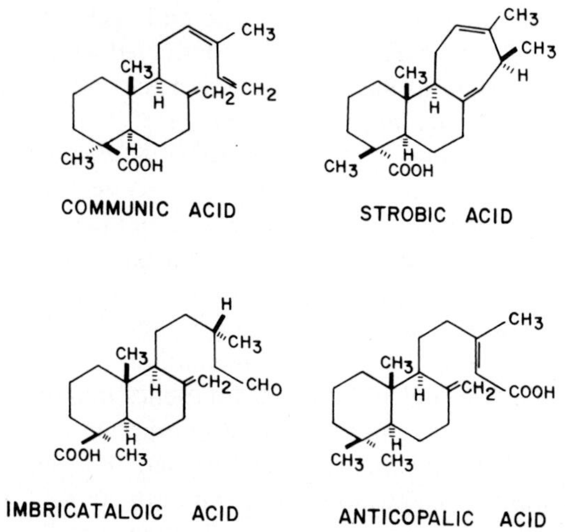

COMMUNIC ACID   STROBIC ACID

IMBRICATALOIC   ACID   ANTICOPALIC ACID

FIGURE 2.   New diterpene resin acids isolated from
bark.

in the examination of leaf or cortex oleoresins (Figure 2). Communic acid was first isolated from the bark of *Juniperous communis*[177] and was subsequently found in a number of *Juniperous* species.[178] Zinkel reports that the cortex oleoresin of *Pinus resinosa* contains significantly high proportions of communic acid.[176] Zinkel isolated and described the new resin acid strobic acid from his studies of the cortical oleoresin of *P. strobus*.[179] This compound is also found in the cortex of *P. aristata* in significantly high concentrations.[176] Anticopalic acid, a resin acid not previously reported in gymnosperms, was first found in the bark and then in the xylem of *P. monticola*[180] and derivatives of imbricatolic acid were found in the leaf and cortical oleoresin of *P. elliottii*.[181] Small amounts of 4-epicommunic and 4-epiimbricatolic acids have also been noted in the cortical oleoresin of *P. densiflora*.[182] Norin and Winell[157] found 15-hydroxydehydroabietic acid, a resin acid not commonly reported, in the cortical oleoresin of *Picea abies*.

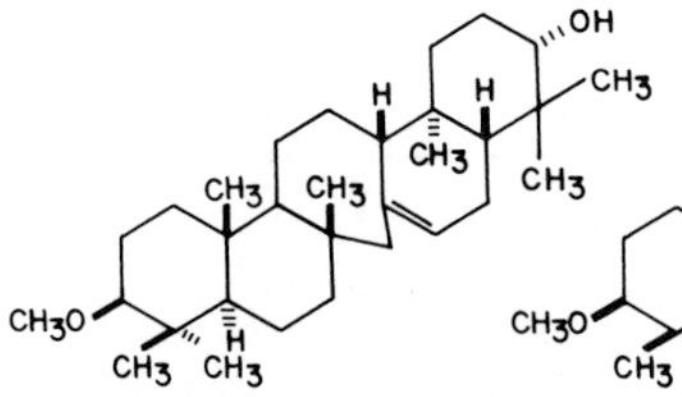

3β-METHOXY-21α-HYDROXY-14-SERRATENE

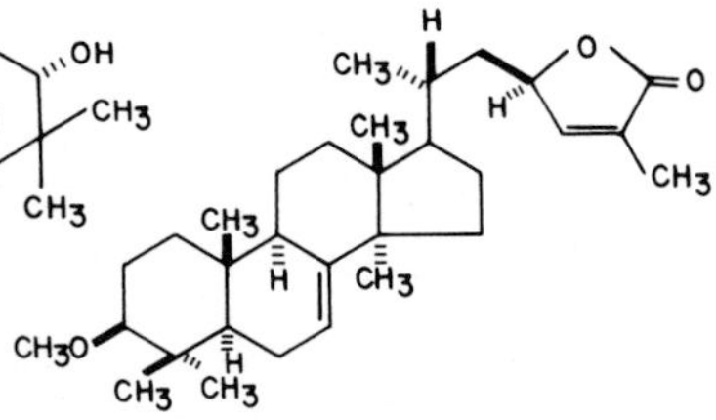

ABIESLACTONE

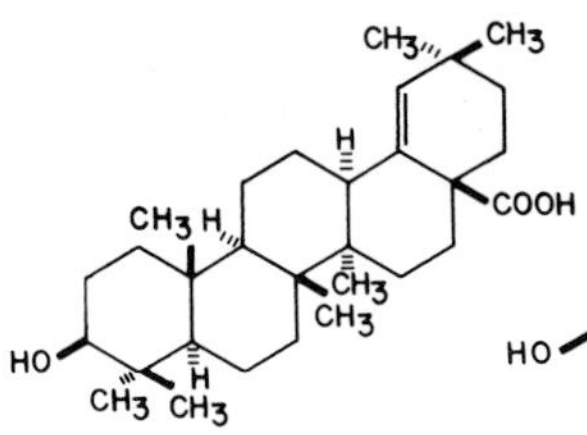

MOROLIC ACID

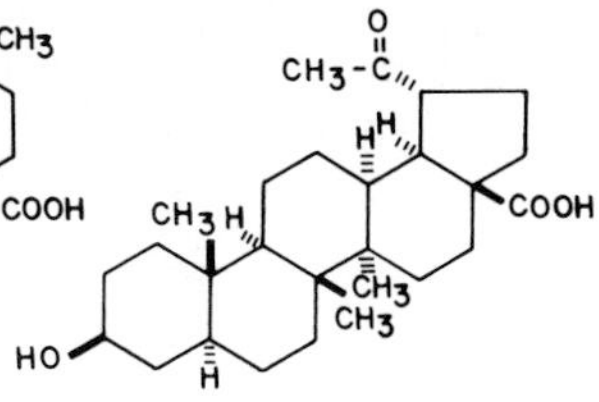

PLATANIC ACID

FIGURE 3. Some triterpenes obtained from bark.

Rowe,[183] Rowe and Bower,[184] and Rowe et al.[185] have identified a series of new triterpenes which were isomers of 3α,21α-dihydroxy-14-serratene, their 3-methyl ethers, 3,21-dimethyl ethers, and 3-methyl-21-keto derivatives in barks of *Pinus* species common to North America (Figure 3). Weston,[186] studying *Pinus radiata*, and Norin and Winell,[151] studying *Pinus sylvestris* bark extracts, also found serratenes as the major triterpenes. Serratenes have been identified in the barks of *Picea sitchensis*[171,187] and *Picea abies*.[157] Barks of several species of *Abies* contain the unusual triterpene abieslactone (Figure 3).[187a] The sterols of the barks of *Pinus* species were dominated by sitosterol with small proportions of campesterol and only trace amounts of other sterols.[188] Sitosterol and campesterol were also reported in extracts from *Pseudotsuga menziesii* bark.[189]

Among the hardwoods, barks of the Betulaceae are best known for their high concentrations of triterpenes; barks of several *Betula* species contain the triterpene betulin in concentrations of 30% or higher.[3,116,190] Lupeol is also commonly reported in *Betula* bark extracts.[190-192] Other triterpenes reported in small proportions include allobetulinol,[192] betulinic acid,[192] betulinaldehyde,[191] betulonic acid,[191] betulonaldehyde,[191] 3β-acetoxy-lupan-20-ol,[193] 3β,28-diacetoxylupan-20-ol,[193] acetyloleanolic acid,[191] 3β-hydroxy-6α-acetoxy-oleanolic acid,[194] and oleanolic acid.[191] Barks of *Alnus* species also contain high concentrations of triterpenes but generally do not rival those of *Betula* species as a source of these compounds. Zellner and Weiss[195] first isolated the triterpene alcohol taraxerol and its corresponding ketone taraxerone from the barks of *Alnus incana* and *A. glutinosa.* Kurth and Becker[196] found the same compounds in the hexane extracts of *A. rubra.* Matyukhina et al.[197] found taraxerol, taraxerone, lupeol, betulin, and betulinic acid in the bark of *A. subcordata.* Other compounds reported in *Alnus* species include glutinone,[198] 3β-hydroxyglutinene,[199] lupenone,[200] alnincanone,[199] and alnuseneone.[199] Friedelin and cerin are present in the barks of *Quercus* species,[3] morolic acid in *Eucalyptus papauana,*[201] and betulin, betulinaldehyde, betulinic acid, betulonic acid, platanic acid and its 3-oxo-derivative are present in *Platanus* species[202] (Figure 3).

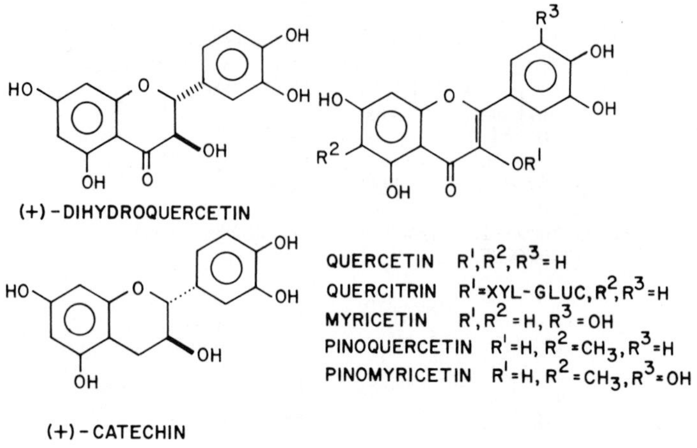

(+) - DIHYDROQUERCETIN

(+) - CATECHIN

QUERCETIN $R^1, R^2, R^3 = H$
QUERCITRIN $R^1 = XYL-GLUC, R^2, R^3 = H$
MYRICETIN $R^1, R^2 = H, R^3 = OH$
PINOQUERCETIN $R^1 = H, R^2 = CH_3, R^3 = H$
PINOMYRICETIN $R^1 = H, R^2 = CH_3, R^3 = OH$

FIGURE 4.   Flavonoids available in substantial quantities from commercially important tree barks.

VII. MONOMERIC POLYPHENOLS AND THEIR GLYCOSIDES

Barks of certain tree species contain polyphenols of low molecular weight in concentrations high enough to attempt commercial production of fine chemicals. Many bark-derived compounds that were important items of commerce in the past have lost their markets because they have been replaced by less expensive substitutes. In addition, the U.S. Food and Drug Administration has ruled that some of these compounds were carcinogenic or were ineffective as pharmaceutical preparations.

Examples include monotropitoside (methylsalicylate-6-xylosylglucoside) obtained from *Betula lenta* bark before being replaced by inexpensive synthesis,[203] and storax (cinnamylcinnamates and related esters) obtained from wound exudates of the bark of *Liquidambar stryaciflua*. Quercitron was historically a valuable yellow dye exported to Europe from America, where it was obtained from the inner bark of *Querus velutina (tinctora)*.[204] Sassafras bark oil (safrol), which was used to flavor root beer and tea, has been banned by the U.S. Food and Drug Administration because metabolic products from it cause liver cancer.[203] Over 200 bioflavonoid preparations sold by prescription or as over-the-counter pharmaceuticals have been removed from the U.S. market because they were deemed to have no demonstrable usefulness by the U.S. Food and Drug Administration.[90,205] The opinion is not shared by many European scientists. Reviews by Hergert,[206] Ryan,[204] and McClure,[205] of the extensive literature on the pharmacological properties of flavonoid compounds suggest that, particularly in treatment of diseases related to capillary fragility, these bioflavonoid preparations have efficacy.

A. Flavonoids

Quercetin and its dihydro-analog taxifolin are the most common flavonoids in conifer barks (Figure 4).[206,207] Quercetin was first produced from a conifer bark by heating water extracts from *Pinus pinaster* barks with sodium bisulfite.[208] Because of the relatively high concentration of dihydroquercetin in Douglas-fir bark (5% of whole bark or 20% of the cork[206]), and the large amounts of this bark harvested by the North American forest products industry, the isolation of dihydroquercetin or quercetin has received considerable study. Considerable work[209-215] has been directed to the development of processes for the extraction and purification of dihydroquercetin and its conversion to quercetin. Gregory et al.[216] summarized the different approaches to this problem and patents authored by Roberts and Roberts and Gregory,[217,218] Esterer and Dowd,[219] Brink,[220] and Esterer[221] resulted from their studies. These extensive research efforts have been summarized by Hergert[206] and Hall.[90]

Hergert[206] emphasizes that to be an economically viable source of flavonoids (or other fine chemical for that matter) the bark of one species must be available in large quantities at a central location. It must contain high concentrations of flavonoids and the flavonoid distribution must be dominated by one compound or an acceptable mixture of compounds that can be easily separated. Therefore, although the barks of many tree species contain substantial amounts of flavonoid compounds, bark from few tree species meets the other requirements of a raw material base for fine chemical production. Possibilities among conifer barks include potential for production of myricetin from *P. contorta* bark[222,223] catechin and epicatechin from *Abies concolor* bark,[224] and pinomyricetin or a mixture of flavonoids from *P. ponderosa* bark.[225] Details of possible processes have been summarized by Hergert.[206]

Among the hardwoods, bark of several species are of interest because of flavonoid compounds. In surveys of the flavonoid patterns in the wood and bark of the genus *Acacia*,[226,227] species of different morphological classification could also be divided into four major classes based on their heartwood or bark flavonoids. Although the bark of *Acacia mearnsii* (mimosa) is highly valued as a source of tannin, the monomeric flavonoids are composed of a complex mixture which comprise only 3% of the total polyphenols in the bark.[228] Further, although the heartwood of *Acacia* species in one group contained an interesting series of 3,4',7,8-tetrahydroxyflavonoids and no condensed tannins, the flavonoids of the barks of this group were dominated by complex mixtures of catechins, gallocatechins, epicatechin-gallates and leucodelphinidins, and leucocyanidins.[226] Therefore, despite their value as a source of flavonoid polymers, the barks of the *Acacia* species do not generally make a good source of monomeric flavonoids.

The barks of *Quercus* species are also well known for their flavonoid constituents. Quercetin could be produced from extracts of *Quercus* species by purification of quercitrin and acid hydrolysis to quercetin and monosaccharides, but this would not rival production of quercetin production from Douglas-fir bark because of the low yields and small supplies of black oak bark.[206]

The flavonoids of the barks of *Ulmus americana* are of interest because the bark of this species contains catechin-7-β-D-xylopyranoside, which stimulates feeding of the European bark beetle (Figure 5).[229,230] The stimulant, deterrent, and toxicity relationships of flavonoids with insects have been reviewed recently by McClure.[205] The sensitivity of these interrelationships is demonstrated by the flavonoid interactions with the silk worm, where quercetin-3-rhamnoside acts as a feeding stimulant, while quercetin-3-glucoside is inactive.[205]

The unusual flavonoids found in the bark of *Uvaria chamae,* chamanetin, and isochamanetin[231,232] are of interest because of their in vivo activity against P-388 leukemia in the mouse and in vivo activity against cells derived from human carcinoma of the nasopharnx. Other flavonoids, i.e., eupatin, eupatoretin, centoureidin, and 6-demethoxy-centraureidin, are also effective against carcinoma of the nasopharnx,[233,234] and quercetin and its glycosides have been shown to be weakly inhibitory to anaerobic respiration of human brain tumor slices.[205] Quercetin pentamethyl ether was very effective and rutin (quercetin-3-rutinose) moderately effective against benzopyrene induced carcinomas.[235]

The flavonoids of the bark (as well as wood) of the Moraceae are of interest because they contain a series of prenylated compounds. The flavonoid composition of *Morus alba*, *M. rubra*, and *Maclura pomifera* all differ markedly. The major flavonoids of *Morus alba* bark are mulberrin, mulberrochromene, cyclomulberrin, cyclomulberro-

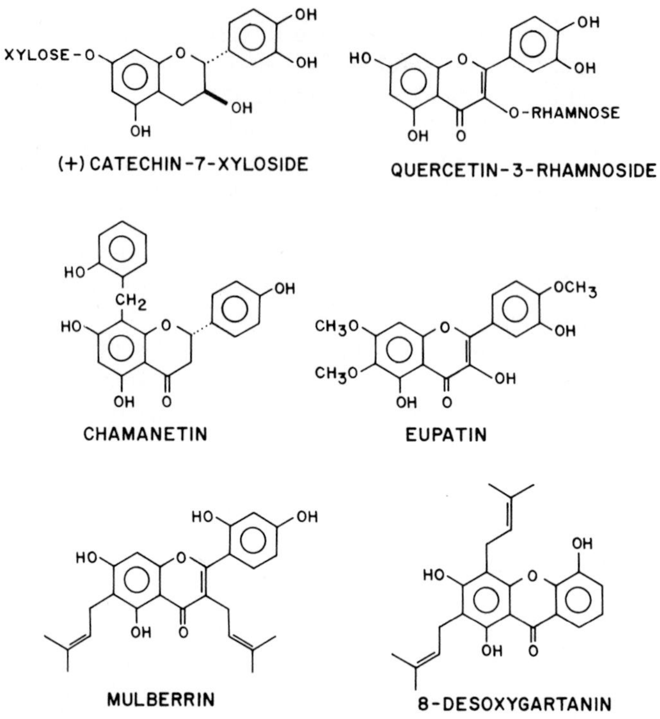

FIGURE 5.   Bark flavonoids with particularly interesting properties.

chromene, mulberranol, albenol, and rubranol. Except for the common occurrence of rubranol, the flavonoids in the bark of *Morus rubra* are entirely different from those in *M. alba*.[236] Such marked differences between species are not evident in the heartwood flavonoids. The polyphenols in the barks of *Maclura pomifera* also differ from those found in the wood, with prenylated xanthones most common in the bark extracts.[203]

The present review of bark flavonoids has dealt only with those that have potential commercial significance or with those that have particularly interesting properties. The two volume series *The Flavonoids* edited by Harbone, Mabry, and Mabry[237] has additional details on the chemistry and occurrence of these compounds. Ryan[204] has prepared an especially thorough review of the chemistry of quercetin (see Figure 5).

B. Salicins

Compounds related to salicin are present in the barks and leaves of trees of the Salicaceae in concentrations high enough to attract interest in using them as pharmaceutical intermediates. Pearl, Darling, and their co-workers have studied these compounds in the commercially important species of *Populus* indigenous to North America, and European species of *Populus* and *Salix* have been studied by Theime and co-workers. Thorough reviews of the progress in the chemistry and occurrence of these compounds in North American *Populus* species have been prepared by Pearl.[238] Table 6 lists the occurrence of these compounds in Populus species. Pearl and Darling[239,240] conclude that salicortin and tremulacin may be the principal compounds formed in vivo and that the other derivatives arise from them by hydrolysis and acyl migration.

C. Hydrolysable Tannins

Although remarkable progress has been made in the chemistry of the hydrolysable

Table 6
OCCURRENCE OF COMPOUNDS RELATED TO SALICIN IN
COMMERCIALLY IMPORTANT SPECIES OF *POPULUS*

Compounds	Species	Ref.	Compounds	Species	Ref.
Salicortin	*deltoides*	241	Deltoidin	*deltoides*	241
	grandidentata	242			
	tremuloides	243	Salicyloyltremuloidin	*grandidentata*	240
				tremula	249
				tremuloides	240, 255, 256
Tremulacin	*grandidentata*	242			
	heterophylla	244			
	tremula	245	Salireposide	*balsamifera*	247
	tremuloides	243		*tremula*	249
				tremuloides	257, 258
Salicin	*balsamifera*	246, 247		*tricocarpa*	251, 252
	deltoides	246			
	grandidentata	248	Tricocarpin	*balsamifera*	247
	heterophylla	246		*tricocarpa*	252, 259
	tremula	249			
	tremuloides	246, 249, 250	Tricoside	*balsamifera*	260
	tricocarpa	251, 252		*tricocarpa*	252, 261
Tremuloidin	*grandidentata*	248	Tricocarposide	*balsamifera*	262
	tremula	249		*deltoides*	241
	tremuloides	246, 249, 253		*tricocarpa*	251, 252, 261, 262
Populin	*grandidentata*	239	Populoside	*balsamifera*	262
	tremuloides	239, 249, 250, 253		*deltoides*	241
				grandidentata	242, 263
Salicyloylsalicin	*grandidentata*	240, 254			
	tremuloides	240, 254	Nigracin	*nigra*	264

tannins in recent years, particularly by Schmidt and Mayer and their co-workers, most of the compounds that have been isolated and had their structures proven have been obtained from fruits, leaves, or galls. Comparatively little is known of the hydrolysable tannins in most hardwoods, and the hydrolysable tannins in tree barks have received very little study. The barks of many commercially important tree species contain gallic or ellagic acids, suggesting the possible occurrence of hydrolysable tannins. Only in a few instances have hydrolysable tannins been isolated and their structures definitely proven from tree barks (Figure 6).

Rowe and Conner[203] report that acertannin was obtained in 0.5% yield from the bark of *Acer spicatum*. This tannin was first isolated from the leaves of *A. ginnale* where it was obtained in yields of 10% of the dried leaf weight.[265] Barks of the white oak group have been said to contain mixtures of hydrolysable and condensed tannin, but no hydrolysable tannin has yet been isolated from the bark of the white oaks. The wood of *Q. alba* contains a complex mixture of largely undefined hydrolysable tannins.[266] One hydrolysable tannin has been obtained from a red oak bark. Mayer et al.[267] proved the presence of hamamelitannin in the bark of *Q. rubra*. Seikel et al.[268] also reported the presence of this compound and of a complex mixture of unidentified ellagitannins in the sapwood and heartwood of *Q. rubra*. Although the inner bark did not contain gallic or ellagic acids, the presence of "tannin esters" was indicated. Hamamelitannin was first isolated from the bark of witch hazel (*Hamamelis virginiana*).[265]

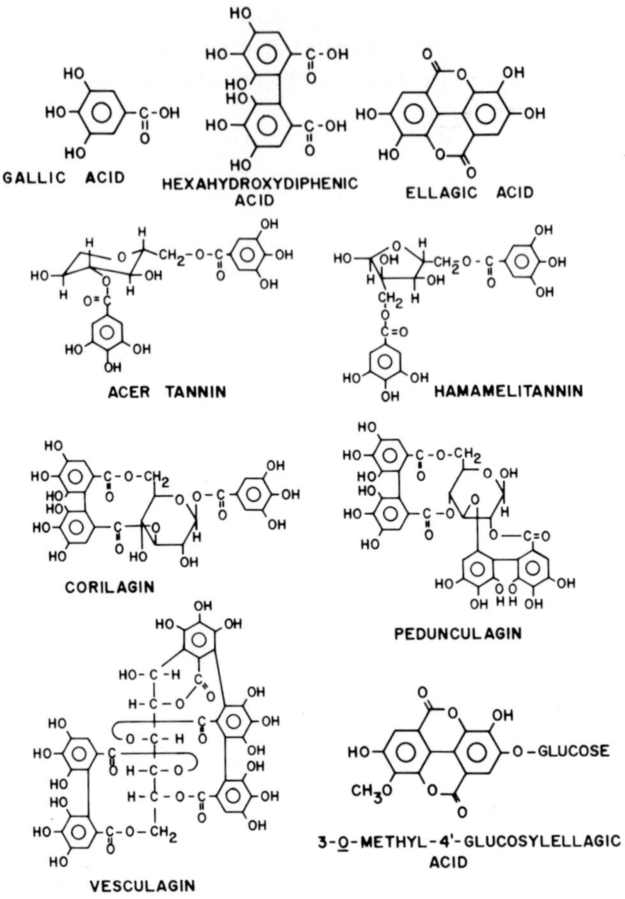

FIGURE 6.   Hydrolysable tannins and ellagic acid derivatives in bark.

The hydrolysable tannins of the wood of *Q. sessiliflora* have been rigorously studied by Mayer and co-workers.[269-274] They isolated the new tannins castalagin, vescalagin, castalin, and vescalin.[269,270] The same compounds also were present in the wood of *Castanea sativa*.[269,270] The structure of castalin obtained from *C. sativa* was rigorously proven[271] as were the structures of castalagin,[272] vescalin,[273] and vescalagin,[274] which were isolated and crystallized from the wood of both *Q. sesseliflora* and *C. sativa*. An unidentified tannin composed of dehydrodigallic acid combined with a sugar also was reported in the leaves, twigs, and bark of *C. vesca*.[265] The tannin pedunculagin has been isolated from the fruit coatings of *Q. sesseliflora* and *Q. pedunculata*.[275] Castalagin and vescalagin,[276] together with the new tannins valolinic acid,[277] valolaginic acid,[277] and their isomers isovalolinic acid,[278] isovalolaginic acid,[278] and castavaloninic acid[279] have been isolated from the fruit cups of *Q. valonea* or from mixtures of *Q. valonia*, *Q. aegilops*, and *Q. macrolepis*. Considering the differences in the chemistry of the wood and bark in other classes of compounds, it would be foolish to project the occurrence of these compounds in fruits or wood to their presence in bark.

Yazaki and Hillis,[280] studying the hydrolysable tannins in wood and bark of *Eucalyptus globulus*, *E. regnans*, and *E. deglupta*, found slight differences in the hydrolysable tannin patterns between the wood and bark of *E. globulus* and *E. regnans*. However, the bark extracts contained more of the unidentified ellagitannin D-13 and

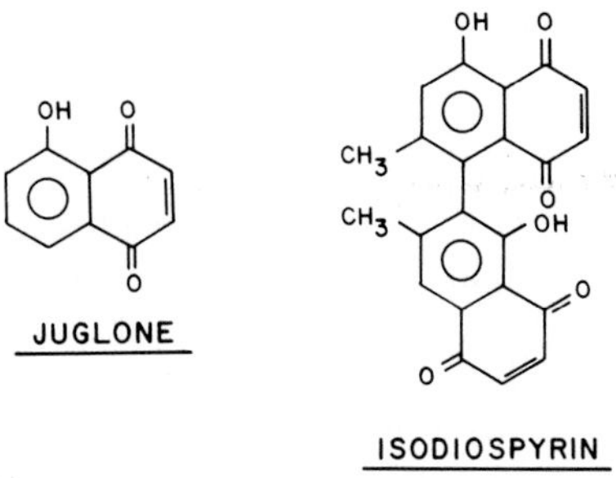

FIGURE 7. Juglone and isodiospyrin, two important naphthoquinones in the bark of the Juglandaceae.

glycosides of methylated ellagitannins than the wood. In the bark of *E. globulus*, 3-*O*-methylellagic acid-4′-rhamnoside[280,281] was the major glycoside present. However, in *E. regnans* bark extracts, substantial amounts of a 3-*O*-methylellagic acid-4′-glucoside were present. The wood and bark of *E. deglupta* contained relatively large amounts of 3,3′-di-*O*-methylellagic acid, 3,3′,4-tri-*O*-methylellagic acid, and 3,3′,4,4′-tetra-*O*-methylellagic acid but hydrolysable tannins were present only in trace amounts. Hemingway and Davies[79] found that the bark of *E. obliqua* contained relatively large amounts of a glycoside of 3-*O*-methylellagic acid, moderate amounts of ellagic acid and gallic acid, but only trace amounts of the hydrolysable tannins characteristic of the wood. Based on their chromatographic properties, corilagin was indicated in the phloem of *E. sieberiana* and juglanin (isomeric with corilagin[265]) was present in the phloem of both *E. sieberiana* and *E. gigantea (delegetensis)*.[282] Corilagin has also been reported in the leaves, twigs, and bark of *Schinopsis* (Quebracho) species.[283]

The ellagitannins of *E. delegetensis* wood were the first of the eucalypt hydrolysable tannins to be isolated in a chromatographically pure state so that attempts could be made to determine their structures.[284] The ellagitannins D-1 (2,3-hexahydroxydiphenoylglucose), D-4 (4,6-hexahydroxydiphenoylglucose), and D-2 most probably peducu-lagin (either 2,3:1,6 or 2,3:4,6-dihexahydroxydiphenoylglucose) were identified. Other unidentified ellagitannins were present.[284,285] Hillis and Yazaki[286] found ellagitannins in wood of *E. polyanthemos*. In addition to these ellagitannins, two gallotannins were indicated by their color reaction with chromogenic spray reagents. Methylated ellagic acid derivatives were also present in substantial proportion in the heartwood. These compounds included 3,3′-di-*O*-methylellagic acid and its glucoside, 3,3′-4-tri-*O*-methylellagic acid and its 4′-glucoside and 3,3′,4,4′-tetra-*O*-methylellagic acid. Heartwood of *E. sideroxylon* is also reported to contain relatively large amounts of 3,3′-di-*O*-methylellagic acid-4′-glucoside.[281]

Except for the tannin juglanin which was isolated from the fruit coating of *Juglans nigra*,[265,287] the hydrolysable tannins possibly present in the wood or bark of other commercially important trees have not been studied in detail. The wood of *J. nigra* contains relatively large amounts of gallic and ellagic acids, but a hydrolysable tannin has not been isolated from either the wood or the bark. The extractives of the wood are characterized by large amounts of a dark violet polymer and the bark contains juglone (Figure 7).[203]

D. Stilbenes

The presence of substantial quantities of stilbenes or their glycosides in *Picea* species was demonstrated in the late 1950s, although structural assignments for these compounds were incorrect. The work of Cunningham et al.,[288] who crystallized and determined the structure of the aglycone of piceatannol (astringenin), and the work of Andrews et al.,[289] who proved the structure of isorhapontin, established the basic structures of the major stilbenes present in barks of the *Picea* species. Manners and Swan,[290] in a survey of the presence of stilbenes in five Canadian *Picea* species, found the *cis* and *trans* isomers of astringin, astringenin, isorhapontin, and isorhapontigenin in the bark from *Picea sitchensis*, *P. engelmanii* and *P. glauca*, while barks of *P. rubens* and *P. mariana* contained much smaller amounts of astringenin. The stilbenes in *P. koraensis* bark were the *trans* isomers of isorhapontin and isorhapontigenin together with small amounts of astringenin and resveratrol;[291] the major stilbenes of *P. ajanensis* were astringin and astringenin together with small amounts of resveratrol and isorhapontigenin. Astringenin was the major stilbene in the outer bark.[292] Further studies of the stilbenes in *P. sitchensis* bark showed that astringin was by far the major stilbene in fresh bark (6.0% of the dry bark), while isorhapontin was only 0.4%; small amounts (0.08%) of piceid were also present. These compounds were fully characterized.[293] Pearson et al.[294] also proved the structures of isorhapontin and astringin in the bark of *P. engelmanni*. Sono and Sakakibara[295] found resveratrol, piceid, astringin, astrigenin, isorhapontin, and isorhapontigenin in the bark of *P. glehnii*.[295] Figure 8 shows some stilbene structures.

The stilbenes present in *Pinus* barks have also been examined, but their concentration is much lower than in bark of *Picea* species. Rowe et al.,[99] studying the benzene extractives of *Pinus banksiana* bark, isolated both *cis* and *trans* pinosylvin dimethylether. Markham and Porter,[296] studying the bark of *P. radiata*, found astringenin and a corresponding glucoside together with smaller amounts of pinosylvin. Compounds of similar chromatographic properties and color reactions were evident in newly formed outer bark of *P. taeda*, *P. echinata*, *P. palustris*, and *P. elliottii*.[297] The freshly prepared phloem did not contain detectable amounts of stilbenes, but after invasion by the blue stain fungus *Ceratocystis minor*, pinosylvin, pinosylvin monomethyl ether, and resveratrol were detected in the early stages of decomposition of the phloem.[298] Yazaki and Hillis[299] isolated and crystallized astringenin from *P. radiata* and proved the structure of the glucoside noted by Markham and Porter. Hergert[207] noted the presence of unidentified stilbenes in the bark extracts of *P. monticola*, *Larix occidentalis*, and *Cedrus atlantica* in addition to the larger amounts of stilbenes in *Picea sitchensis* and *P. engelmannii*.

Although stilbenes are important constituents of the leaves and wood of many commercial hardwood species (cf. *Eucalyptus*[300]) and are possibly important to the durability of woods,[301] stilbenes in hardwood tree barks have not been studied.

E. Other Important Compound Classes

The barks of certain other hardwoods contain classes of compounds of particular interest because of their unusual structures, taxonomic significance, or biological properties. It is not possible to present a thorough review of all these compounds in this chapter. An excellent review of these compounds in trees of the eastern U.S. has been prepared by Rowe and Conner.[203] To present a concise summary of the extractives of the more interesting barks, selected compound classes have been abstracted and summarized in Table 7, and structures of some compounds are presented in Figures 9 and 10. References to research on these compounds are presented by Rowe and Conner,[203] unless otherwise indicated.

F. Prospects for Utilization

It is impossible to present a thorough review of the pharmacological properties of

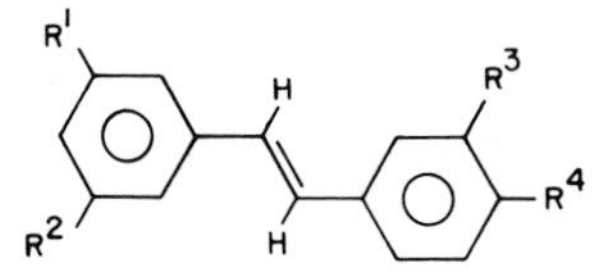

AGLYCONES   GLYCOSIDES

PINOSYLVIN
    R^1, R^2 = OH; R^3, R^4 = H
PINOSYLVIN MONOMETHYL ETHER
    R^1 = OCH$_3$; R^2 = OH; R^3, R^4 = H
PINOSYLVIN DIMETHYL ETHER
    R^1, R^2 = OCH$_3$, R^3, R^4, = H
RESVERATROL   PICEID
    R^1, R^2 = OH; R^3 = H, R^4 = OH R^1 = O - GLUCOSIDE
ASTRINGENIN   ASTRINGEN = PICEATANNOL
    R^1, R^2, R^3, R^4 = OH R^1 = O - GLUCOSIDE
                                             PINUS RADIATA BARK STILBENE
                                                 R^4 = O - GLUCOSIDE
ISORHAPONTIGENIN   ISORHAPONTIN
    R^1, R^2, R^4 = OH, R^3 = OCH$_3$ R^1 = O - GLUCOSIDE

FIGURE 8.   Stilbenes found in the barks of forest trees.

flavonoids here. Readers interested in further details should see Hergert,[206] Ryan,[204] and McClure.[205] The flavonoids which have been shown to be effective in increasing capillary resistance include the flavanones, flavonols, isoflavones, catechin, flavan-3,4-diols, and chalcones.[205] Over 50 diseases associated with capillary fragility respond to bioflavonoid therapy. The flavonoids with free hydroxyls at the 3′,4′ positions derive their activity from inhibition of ascorbate oxidation, inhibition of O-methyl transferase activity that prolongs epinephrine activity, and stimulation of the pituitary-adrenal axis. Flavonoids (particularly those with multiple methoxyl or ethoxyl substitution) are effective inhibitors of blood cell aggregation. Rats kept on a diet severely deficient in flavonoids developed brain edema and subpleural haemorrhages, and these symptoms were significantly reversed after treatment every other day with 50 mg/kg of rutin derivatives or hesperidin.

The action of flavonoids in preventing ascorbate oxidation may also explain their synergistic values in vitamin C therapy. The flavones possess coronary dilating action and those with a 7-O-CH$_2$COOR group are used clinically in Italy as coronary vasodilators since they are nontoxic and are more active than nitroglycerin or khellin.[303] However, it is apparent that much of the research on the pharmacological properties of the flavonoids has been abandoned since the U.S. Food and Drug Administration removed bioflavonoid preparations from use as pharmaceuticals. Recent European research on the biological properties of flavonoids is summarized in a series of papers in *Flavonoids and Bioflavonoids, Current Research Trends.*[304]

The classification of tannins as Category I carcinogens by the U.S. Occupational Safety and Health Administration[305] and reference to catechin from quebracho[306] as a carcinogen have also prompted considerable controversy. In McClure's recent review,[205] only the reference to Morton,[307] who correlated human esophogeal cancer with the consumption of excessive amounts of astringent beverages, is cited as evidence of the carcinogenic activity of flavonoids. Indeed, many of the flavonoids have been

Table 7
SELECTED HARDWOOD BARK EXTRACTIVES NOT
DISCUSSED IN TEXT

Tree species	Compound classes	Common name
Betulaceae	Diarylheptanoid	
Alnus rubra		Oregonin
A. hirsuta		Hirsutanonol
		Hirsutenone
Betula playtyphylla		Platyphyllol
Bignoniaceae	Noriridoside	
Catalpa bignonioides		Catalposide
		Catalpinoside
Cornaceae	Flavonoid	
Cornus stolonifera		Quercetin-3-galactoside
	Iridoid	
C. florida		Cornin (verbenalin)
Ebenaceae	Naphthoquinone (Figure 8)	
Diospyros virginiana		Isodiospyrin
Fagaceae	Lignan (Figure 10)	
Quercus rubra		Lyoniside
Hippocastanaceae	Coumarin (Figure 9)	
Aesculus species		Esculetin
		Esculin
		Scopoletin
		Scopolin
		Fraxetin
		Fraxin
Juglandaceae	Flavonoid	
Carya illinoensis		Quercetin-5-methyl ether
		Quercetin-3,5-dimethyl ether
Juglans nigra		Myricetin
		Myricetin-3-rhamnoside
		Sakuranetin
		Sakuranetin-5-glucoside
C. illinoensis	Naphthoquinone	Juglone
J. nigra		Juglone
		Dihydrojuglone-4-glucoside
Magnoliaceae	Alkaloid	
Liriodendrin tulipifera		Complex mixture
Magnolia species		Complex mixture
L. tulipifera	Lignan	Pinoresinol
		Syringaresinol + glycosides
		Medioresinol
		Lirionol
Magnolia species	Lignan	Magnolol
		Calopiptin
		Galgravin
		Veraquensin
		Accuminatin
Oleaceae	Coumarin	
Fraxinus species		Fraxin
		Fraxetin
		Esculin
		Esculetin
		Pennsylvanol
Rosaceae	Cyanogenic glycosides	
Prunus species		Prunasin
		Amygdalin

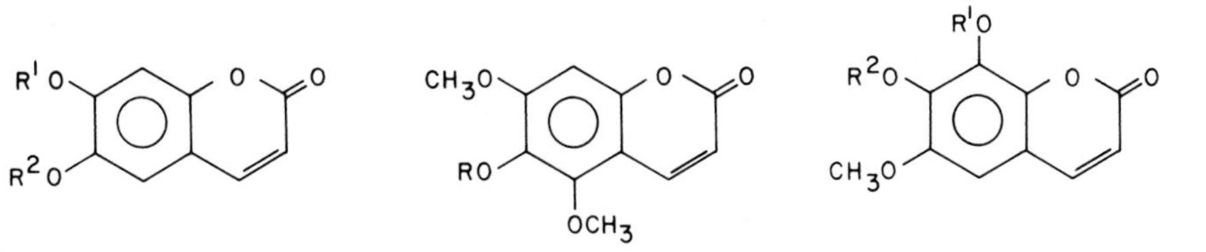

FIGURE 9.   Major coumarins found in the barks of *Fraxinus* species.

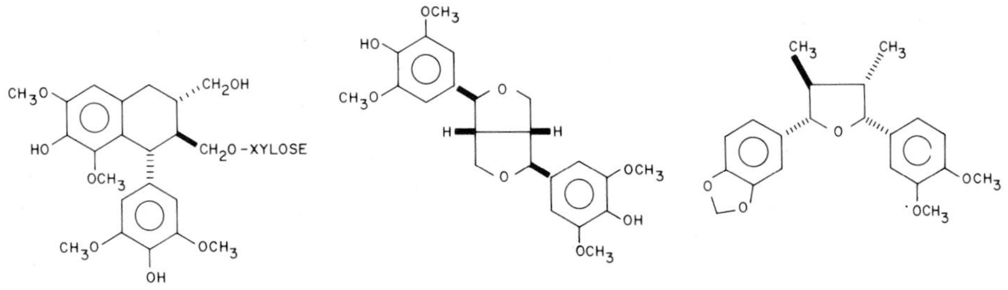

	R_1	R_2
ESCULETIN	H	H
ESCULIN	H	GLUCOSIDE
CHICHORIIN	GLUCOSIDE	H

	R
FRAXINOL	H
MANDSHURIN	GLUCOSIDE

	R_1	R_2
FRAXETIN	H	H
FRAXIN	GLUCOSIDE	H
FRAXIDIN	H	CH_3
ISOFRAXIDIN	CH_3	H
CALYCANTHOSIDE	CH_3	GLUCOSIDE
FRAXIDIN-8-O-B-$\underline{D}$- GLUCOSIDE	GLUCOSIDE	CH_3

FIGURE 10.   Selected lignans obtained from tree barks.

demonstrated to exhibit significant antitumor activity.[231-235] Perhaps classification of these flavonoids as carcinogens will stimulate more thorough research on the biological properties of flavonoid compounds and will not discourage further research. Because of these rulings, prospects for use of bark as a source of flavonoids for pharmaceutical or other fine-chemical applications are limited.

The possible use of the salicin related compounds in *Populus* or *Salix* species as pharmaceutical intermediates has been complicated by the broad mixture of compounds obtained when extracting tree barks or leaves. Although much ellagic acid could readily be obtained from wet process hardboard or certain pulping processes using eucalypts, no commercial products have been developed. Ellagic acid is a sedative and exhibits tumor inhibiting activity.[308,309] Use of the magnesium complex has been explored.[310] Ellagic acid is readily decarboxylated to hexahydroxybiphenyl[311] and use of this compound as an ion-exchange resin after acid-catalysed condensation with formaldehyde, or use of derivatives as topical ointments for treatment of skin diseases have also been briefly explored.[312] Bark constituents such as the coumarins and naphthoquinones are well known for their biological properties, but uses for bark extracts have not been developed.

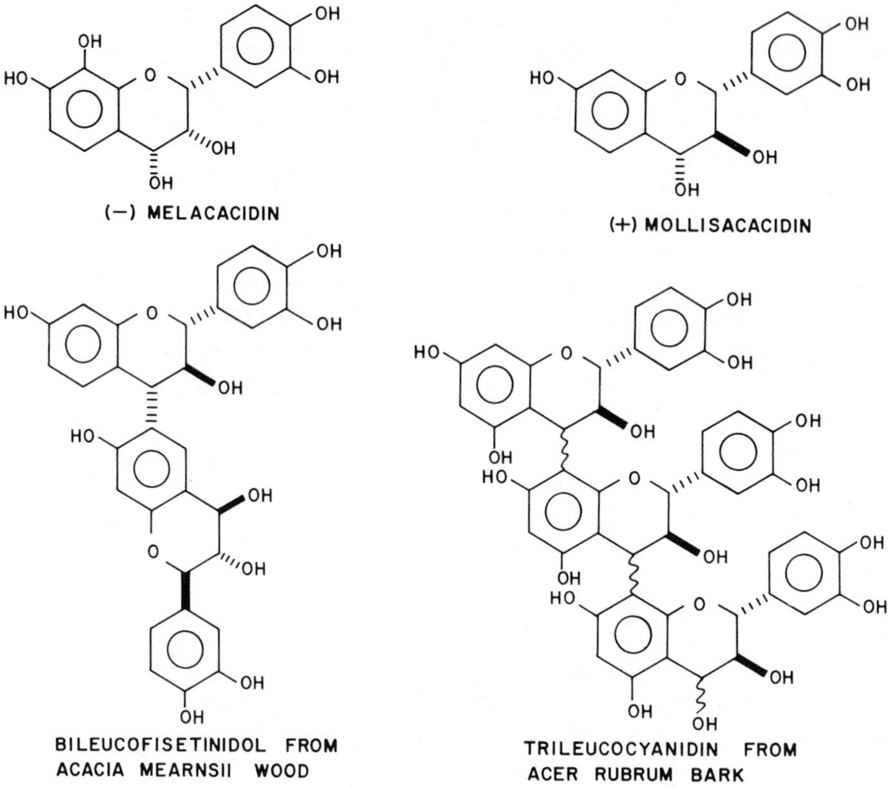

(−) MELACACIDIN

(+) MOLLISACACIDIN

BILEUCOFISETINIDOL FROM
ACACIA MEARNSII WOOD

TRILEUCOCYANIDIN FROM
ACER RUBRUM BARK

FIGURE 11.   Flavan-3,4-diols isolated from wood or bark.[302]

VIII. POLYFLAVONOIDS

A. Leucoanthocyanidins

A discussion of the chemistry and utilization of the polyflavonoids present in tree barks must be preceded by consideration of the leucoanthocyanidins (flavan-3,4-diols) and the proanthocyanidins (oligomeric flavan-3-ols). Although the leucoanthocyanidins and proanthocyanidins are usually not present in commercially significant quantities in tree barks that contain substantial condensed tannins, an understanding of the chemistry of these compounds is essential to development of uses for the higher polymers. Reviews of the chemistry of the leucoanthocyanidins and proanthocyanidins have been prepared by Weinges et al.,[313] Roux,[314] and Haslam.[315]

The report of the isolation of melacacidin (Figure 11) from *Acacia melanoxylon*[316] was received with enthusiasm since the flavan-3,4-diols were logical intermediates in the biosynthesis of the condensed tannins.[206] A number of leucoanthocyanidins have since been isolated, particularly the 5-deoxyflavan-3,4-diols from *Acacia* species by Clark-Lewis and co-workers[317,321] and by Roux's group.[322-325] Seshadri and co-workers,[326-331] Manson,[332] and Hergert[206] also report the presence of flavan-3,4-diols bearing the 5,7-dihydroxy A-ring substitution. Baig et al.,[333] Bokadia et al.,[334] Betts et al.,[335] and Freudenberg and Weinges[336] have synthesized derivatives of the various isomeric 5,7,3′,4′-tetrahydroxyflavan-3,4-diols and none of the compounds isolated from natural sources have physical properties identical with synthesized compounds.

Doubt about the natural occurrence of 5,7,3′,4′-tetrahydroxyflavan-3,4-diols[327] and Roux and Ferreira's[338] isolation of α-hydroxychalcones prompted questions about the

227

biosynthesis of the procyanidins and condensed tannins by reduction of a dihydroflavonol to the flavan-3,4-diols and their corresponding carbocations.[339] Jacques et al.,[337] through studies of the incorporation of specifically ^{3}H and ^{14}C labeled cinnamic acids into the procyanidins, obtained evidence for a flav-3-en-3-ol intermediate that could be derived from an α-hydroxychalcone. Such a mechanism would allow the 4-deoxyflavonoids to be biosynthetically independent of the flavonols and dihydroflavonols. Haslam et al.[339] and Jacques et al.[337] propose that a flav-3-en-3-ol is held by an enzyme permitting stereospecific reduction to either (+) catechin or (−) epicatechin as well as stereospecific protonation to form either the (+) catechin or (−) epicatechin carbocations. This proposition explains the high degree of specificity often found in the flavan-3-ols and procyanidins surveyed by Thompson et al.[340] and also in the flavan-3-ols and polyflavonoids found in the barks of the southern pines (see discussion below).[341-343] However, this proposition has not been proven and there is evidence for the formation of the condensed tannins by a flavan-3,4-diol intermediate in equilibrium with its corresponding carbocation. Wong and Birch[344] found that U-^{3}H-aromadendrin was incorporated into both epicatechin and procyanidin B-4 by *Rubus ideeus* nearly as well as 1-^{14}C-phenylalanine. Thiolysis of the procyanidin B-4 gave epicatechin and the thioether of catechin with approximately the same degree of labeling. In addition, the presence of the three bileucofisetinidols in *Acacia mearnsii* heartwood[345] (Figure 11) and dimeric and trimeric leucocyanidins in the wood and bark of *Acer rubrum*[346] also suggests that the flavan-3,4-diols are important intermediates in the biosynthesis of procyanidins and condensed tannins. The biosynthesis proposal suggested by Haslam and co-workers[337,339] provides an explanation for the differences found in hydroxylation patterns of the flavan-3-ols, procyanidins and condensed tannins compared with the flavonol and dihydroflavonols, and also provides an explanation for the stereospecificity between the flavan-3-ols and the upper and lower units of the condensed tannins.[341-343] A great deal more work is required to verify this hypothesis in conifer bark polyflavonoids.

B. Proanthocyanidins

Drewes et al.[347] found that the bark of *Acacia mearnsii,* in contrast to the wood which contained a series of bileucofisetinidins,[345] contained three all-*trans* proanthocyanidins. The all-*trans* stereochemistry was unusual in comparison with the bileucofiesetinidins isolated from the heartwood and suggested that enzymic control of the condensation of these compounds to form the condensed tannins may exist. Pelter et al.[348] isolated two similar diasteriomeric proanthocyanidin-dimers from the wood of *Jalbernadia globiflora.* DuPreez et al.[349] described a related 8-carboxy derivative together with the above two proanthocyanidins in the heartwood of *A. luederitzii.* Study of the proanthocyanidins in the wood of *A. giraffae*[350] showed the presence of an all-*trans*(−)fisetinidol→(+)catechin dimer.

Investigation of the ^{1}H-NMR spectra of the decamethyl ether-triacetate of a trimeric proanthcyanidin which had been isolated from the wood of *Colophosperman mopane* showed six acetyl signals rather than the expected three, and multiple pairs of methoxyl and C-ring protons were also evident in the spectrum at ambient temperature.[351] These spectral features were attributed to the presence of rotational isomers. Because of these rotamers, the ^{1}H-NMR spectra of proanthocyanidins, and their derivatives are often far more complex than would be expected.

The flavonoids of the Anacardaceae (*Rhus, Schinopsis, Cortinus*) are exceptional in containing the extremely rare 2S,5-deoxyflavan-3,4-diol (−) leucofisetinidin and (+) catechin as major tannin precursors.[352] Ferreira et al.[352] purified a tetrameric proanthocyanidin from *Rhus lancea* which gave a first order ^{1}H-NMR spectrum at ambient temperature. The absence of rotation isomers in this proanthocyanidin permitted assignment of its structure. The isolation and successful characterization of this compound was important because it clearly demonstrated the principle of successive condensation of flavonoid carbonium ions with oligomeric proanthocyanidins in the formation of condensed tannins.

Similar progress has been made in our understanding of the procyanidins [(+) catechin and/or (−) epicatechin oligomers] since Weinges and Freudenberg[353] first described two crystalline procyanidins isolated from cranberries and cola nuts. Further studies of the procyanidins of various fruits[354,355] led to the description of four isomeric procyanidin dimers [(B-1, (Figure 12), (−)epicatechin→(+)catechin, B-2, (−)epicatechin→(−)epicatechin; B-3, (+)catechin→(+)catechin and B-4 (+)catechin→(−)epicatechin]. In addition, Weinges et al.[356,357] synthesized the octamethyl ether-diacetate derivative of procyanidin B-3 by means which established a C_4 to C_8 linkage. Thompson et al.[340] found the same four procyanidins, and four diasteriomeric pairs, as well as a series of higher oligomers in a survey of the distribution of procyanidins in several plants.

Detailed investigation of the ^{13}C and ^1H-NMR spectra of these procyanidins by Fletcher et al.[358] established the absolute stereochemistry at C_4 for the procyanidins B-1, B-2, and B-5 as 4R and for B-3, B-4, and B-6 as 4S. Examination of their ^1H-NMR spectra over a range of temperatures showed differences in rotational energy barriers and differences in populations of rotamer pairs which also provided evidence for their mode of linkage and stereochemistry at C-4. In addition to the dimers, two trimeric procyanidins have been partially described.

Another important class of procyanidins, first crystallized from the seed coats of *Aesculus hippocastanum* by Mayer et al.,[359] was the doubly linked dimer procyanidin A-2. Further investigation of this compound, particularly through ^{13}C-NMR spectral studies of it and related model compounds by Weinges et al.,[355] Schilling et al.,[360] Jacques et al.,[361] and Weinges[362] established that this compound was a C_4-C_8'' and C_2-O-C_7'' doubly linked epicatechin dimer. Two trimeric procyanidins of a procyanidin A-2→(−)epicatechin and procyanidin A-2→(+)catechin which were first isolated from avocado seeds (*Persea gratissima*)[363] and two trimers from *A. hippocastanum* procyanidin A-2→(−)epicatechin and (−)epicatechin→procyanidin A-2 (Figure 12) have been described.[361] Other proanthocyanidins with different hydroxylation patterns are reviewed by Haslam.[315]

Although the procyanidins in the fruits, leaves, etc. of a number of plants have been surveyed,[313-315,340] there have been few investigations of the procyanidins in the barks of commercially important tree species other than Roux's work on the *Acacia* species. Porter[364] and later Yazaki and Hillis[299] found procyanidin B-1, B-3, B-6, and C-2 in the inner bark of *Pinus radiata*. The dominant proanthocyanidin in *P. taeda*[365] inner bark is procyanidin B-1.[365]

C. Polyflavonoid Polymers

Progress in understanding the chemistry of the higher polymers such as the condensed tannins (soluble in water and alcohol), phlobaphenes (soluble in alcohol-insoluble in water), and the phenolic acids (insoluble in neutral solvents-soluble in alkaline solution)[366] has been slow in comparison with the advances made on the proanthocyanidins. The structures of these polymers often are conceptualized as extensions of the molecular weight of the proanthocyanidins since analytical methods and selective degradation reactions which preserve the structural integrity of the basic units are limited. Formation of anthocyanidins by treatment of the polymers with mineral acids in alcohol solution, and alkaline fusion to determine the hydroxylation patterns of the A- and B-ring moieties of the basic structural units remain as important degradation reactions. However no information is obtained relative to the stereochemistry of the polymer, and yields are often low.

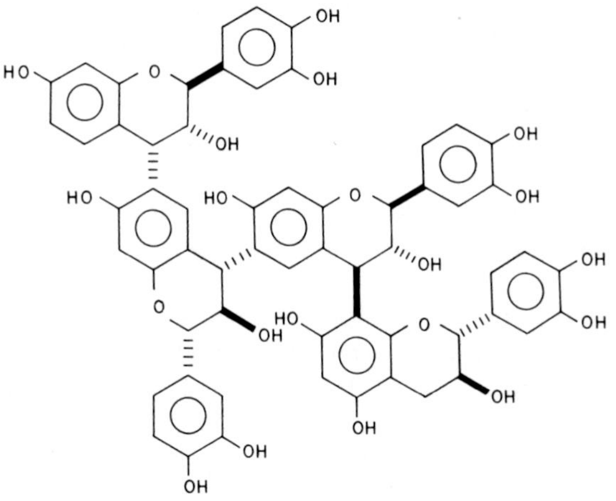

PROANTHOCYANIDIN FROM RHUS LANCEA

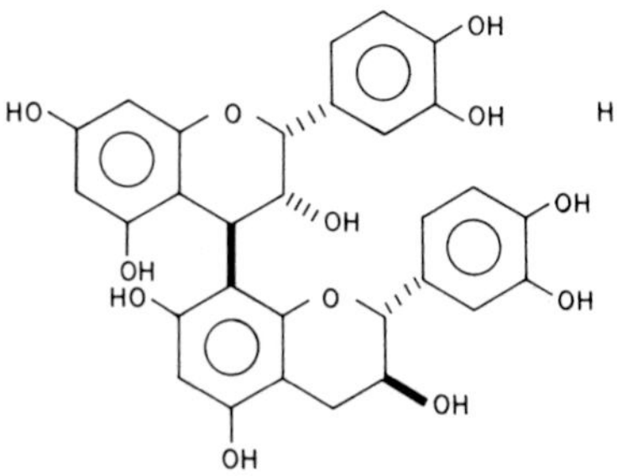

PROCYANIDIN B-1

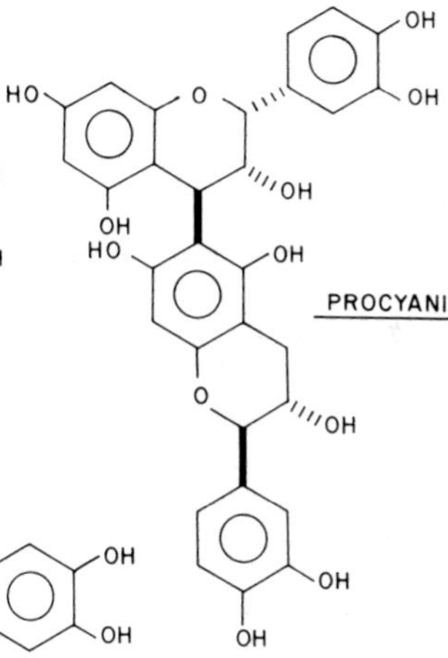

PROCYANIDIN B-7

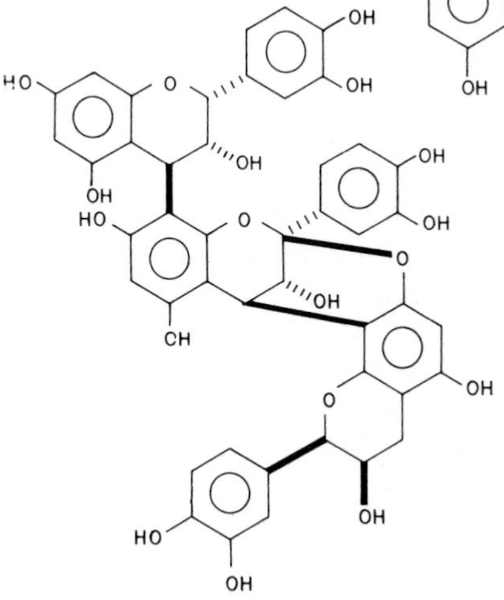

(−) EPICATECHIN→PROCYANIDIN A-2

FIGURE 12.   Selected proanthocyanidins.

A major advance in the study of the 5,7-dihydroxyflavonoid polymers arose from the work of Betts et al.[367] who degraded a tannin from heather (*Calluna vulgaris*) with mercaptoacetic acid. The interflavonoid linkage was considered to be a benzyl ether because this reagent was known to cleave such bonds. However, Sears and Casebier[368] showed that this reagent also cleaved the benzylic C-C bond in synthetic procyanidins. Since then, sulfur nucleophiles other than mercaptoacetic acid[369] including toluene-α-thiol and benzenethiol,[370] have been employed in an effort to improve yields of the thioethers. The principle of this reaction has been used to synthesize procyanidin B dimers using (+)catechin or (−)epicatechin as the nucleophile that captures carbonium ions produced in the reaction of specific condensed tannins or procyanidins with acid.[358]

Sears and Casebier[371] studied the degradation products obtained from polyflavonoid polymers of *Tsuga heterophylla* bark. Reaction of a methylated condensed tannin with mercaptoacetic acid gave equivalent yields of 2,3-*cis* and 2,3-*trans* methyl (3-hydroxy-5,7,3′,4′-tetramethoxyflavan-4-yl-thio)acetates, suggesting that the condensed tannins were composed of nearly equal proportions of catechin and epicatechin units linked either C_4 to C_6 or C_8. Reaction of the extractive free bark with mercaptoacetic acid and subsequent methylation also produced the same thioethers, except that the yield of the epicatechin derivative was about three times that of the catechin derivative. These results indicated that the phenolic acid fraction (i.e., alkali-soluble polyflavonoids) of *T. heterophylla* bark was of a structure similar to that of the condensed tannins and that its insolubility in neutral solvents was probably due to its higher molecular weight and/or inaccessibility.[366] The carbonyl function generally observed in alkali-soluble isolates was shown by Sears et al.[114] to be derived from an alkaline rearrangement of the phloroglucinol A-ring of the catechin unit. Reaction of a phlobaphene fraction with mercaptoacetic acid did not produce recognizable thioethers, thus providing further evidence for the view that the phlobaphenes have a C-ring in a higher oxidation state than do the condensed tannins or alkali-soluble polyflavonoids.[206]

Hemingway and McGraw[341,342] examined thioglycolysis products from methylated condensed tannins and extractive free barks of *Pinus taeda* and *P. echinata*. They found epicatechin and catechin thioglycolates in relative yields of about 5 to 1. The products were ethylated rather than methylated to determine if benzyl ether linkages dominated the polymer structure. No products with aromatic ethoxyl groups were detected, suggesting that the condensed tannins and phenolic acid fractions of *P. taeda* and *P. echinata* were linked C_4 to either C_6 or C_8. The high ratio of epicatechin to catechin thioethers obtained from these polymers was not expected since Porter[364] had previously reported that the procyanidins in *P. radiata* bark were made up largely of (+)(catechin) units in addition to reports of a dominant 2,3-*trans* flavan-3,4-diol in *P. palustris* bark.[206] The epicatechin thioether obtained from degradation of the pine bark tannins could not be crystallized,[341] so its acetanilide derivative was prepared. Its properties were essentially identical[342] to those reported for the (−)epicatechin derivative synthesized by Betts et al.[369] Karchesy and Hemingway[343] isolated three different condensed tannins from the inner bark of *P. taeda*. Thiolysis of each of these tannins with benzene-thiol gave only the epicatechin thioether and a dimeric epicatechin→epicatechin thioether but no catechin isomers were indicated. The flavan-3-ol obtained from the lower terminal unit of these tannins was exclusively catechin: in the monomeric flavan-3-ols, only catechin and no epicatechin has been detected.

Results by Markham and Porter[296] in their investigation of the tannins of *P. radiata* bark suggest that these polymers differ significantly from those obtained from *P. taeda* and *P. echinata*. Both cyanidin-chloride and delphinidin-chloride were obtained on treatment of the tannin with HCl in 2-propanol. Alkaline fusion gave resorcinol, phloroglucinol, pyrogallol, and protocatechuic acids, suggesting that the hydroxylation patterns are far more complex in the tannins of *P. radiata* than in *P. taeda* or *P. echinata*. Hillis and Yazaki[286] isolated a condensed tannin from *P. radiata* bark by chromatography on LH-20 Sephadex® using acetone-dimethylformamide solvents. This tannin isolate gave cyanidin-chloride on treatment with n-butanol-HCl and the IR spectrum of the tannin was nearly identical with the spectra obtained from procyanidin B-1 and B-3.

Karchesy et al.[372] studied the condensed tannins isolated from the bark of *Pseudotsuga menziesii* and *Alnus rubra*. The tannins from both species gave cyanidin chloride on treatment with propan-2-ol and HCl and phloroglucinol and protocatechuic acid were obtained after alkaline fusion. Treatment of the methylated tannins with propan-2-ol and HCl produced 5,7,3′,4′-tetramethylcyanidin chloride, indicating that a benzyl ether bond was not the dominant linkage. The epicatechin thioglycolate was the only product obtained from *A. rubra* bark. Thiolysis of the methylated tannin from *P. menziesii* afforded the epicatechin and catechin thioglycolates in ratios of about 3:1.

The yields of thioglycolates obtained from the tannins of *T. heterophylla*[371] and *A. rubra*[372] were 10% or less and yields obtained from the tannins of *Pinus taeda* and *P. echinata* were also very low.[341,342] Fletcher et al.[358] reported quantitative yields of thiolysis products from the ethyl acetate-soluble tannin that is retained on an LH-20 Sephadex® column eluted with EtOH but which is mobile when eluted with MeOH. Karchesy and Hemingway[343] isolated an ethyl acetate-soluble tannin by chromatography on LH-20 Sephadex®, as described by Fletcher et al.,[358] and two acetone-water soluble tannins that were partitioned on cellulose. Gel permeation of the methyl ether derivatives indicated n = 3 to 4 for the ethyl acetate soluble tannin and n = 6 to 7 and n = 8 to 9 for the two water soluble tannins. Thiolysis of these three tannins with benzenethiol according to the method of Brown and Shaw[370] gave low yields (<15%) of epicatechin thioethers from all three tannins. Although the ^{13}C NMR spectra became increasingly broad with increases in molecular weight, no substantial differences in the spectra of these three tannin isolates could be detected. The spectra could be accounted for as a procyanidin B-type polymers.[343]

Fletcher et al.[358] have determined the preferred conformations of the various procyanidins through extensive ^{13}C- and ^{1}H-NMR studies and from these conclusions have built molecular models for condensed tannins. These procyanidin B type polymers appear to have a thread like structure in which the core is made up of the phloroglucinol A-ring and the heterocyclic C-ring moieties. The catechol B-rings project laterally from this core in a helical arrangement in which (+)catechin polymers are wound in a clockwise direction while a counterclockwise helix is described by the B-rings of a (−)epicatechin polymer. The finding that the phloroglucinol A-ring and the interflavonoid linkage are both buried in the core of the higher polymers is particularly interesting in terms of interpretation of the chemical properties of these polymers. However, use of these structural models must be made with extreme caution. Based on the finding of Fletcher et al.[358] that thiolysis yields from procyanidin B type polymers should be quantitative even when molecular weights are reasonably high, i.e., n = 6-7, polymers that do not give high yields of thioethers may contain other, undefined structural features.

D. Prospects for Utilization

Probably the greatest effort toward utilization of tree barks has centered on the use of their water- or alkali-soluble polyflavonoids as specialty polymers. Excellent reviews of the development of these products from *Acacia mearnsii* by Roux et al.,[373] *Pseudotsuga menziesii* by Hall,[90] and conifer bark polyflavonoids in general (but with particular reference to *T. heterophylla*) by Hergert[206,366] have been prepared.

Because of their comparatively low molecular weight (number average molecular weight = 1250) and resorcinol functionality in reaction with formaldehyde, wattle tannins have found broad application as major constituents of a variety of adhesives. Wattle tannin-based adhesives for manufacture of exterior plywood are described by Plomley,[374] Saayman and Oatley,[375] Scharfetter et al.,[50] and Pizzi.[376] When used for exterior-grade plywood, wattle tannin adhesives are generally fortified with small amounts of phenol-resorcinol-formaldehyde resins. Adhesive formulations with a pot life of about 4 hr at 20°C which cure after pressing for 5 to 6 min at 120°C provide exterior or marine grade bonds. The tannin-urea-formaldehyde adhesive formulated by Pizzi[376] is particularly interesting because of difficulties with formaldehyde release from products made with urea-formaldehyde adhesives, and because it is an inexpensive crosslinking agent in the manufacture of exterior-grade wattle tannin adhesives.

Wattle tannin adhesives also have been used to manufacture a weather resistant particleboard in South Africa. The major problem to be overcome in the formulation of a particleboard adhesive based on tannins is the high viscosity of tannin solutions when concentrated to 45 to 50% solids. Roux et al.[373] indicated that sulfitation, while effective in reducing the viscosity, is detrimental to the water resistance of the bonds. They recommend extraction of the bark with water at lower temperature and the addition of phenol. Saayman and Oatley[375] recommend heating wattle tannin solutions at 45% solids content adjusted to pH 9.5 for 1 hr to achieve a sufficient reduction in the viscosity. After cooling, the pH is adjusted to 7.0 with acetic acid and a 40% formalin solution is added. Boards made with this wattle tannin adhesive had better dimensional stability and resistance to weathering than comparable boards made with phenol-formaldehyde adhesives.

Cold-setting laminating adhesives have also been formulated with wattle tannins. Saayman and Oatley[375] described a laminating adhesive that is prepared by grafting resorcinol to a methylol-tannin intermediate prepared by reaction of the tannin with formaldehyde in 50% methanol solvents. The resin is spray-dried without losing reactivity. Van Der Westhuizen et al.[377] developed a cold-setting adhesive for use in finger-jointing which was designed to permit handling of the glued lumber within 1 hr after gluing. It was a variation of the resin system described by Kreibich.[378] One component of the adhesive was resorcinol-grafted wattle tannin prepared as described above but containing additional paraformaldehyde. The other component was a mixture of wattle tannin and an *m*-aminophenol-formaldehyde novolack. Pizzi[376] also described a laminating adhesive formulated from wattle tannins and a urea-formaldehyde resol fortified with a phenol-resorcinol-formaldehyde resin.

Starch adhesives for corrugated containers have also been formulated with wattle tannin-formaldehyde fortifiers that perform as well as resorcinol-formaldehyde-fortified starch adhesives. These adhesives are used for cartons that must withstand exposure to moisture, such as refrigerated fruit containers.[379,380] In addition, wattle tannins have been used as binders for foundry cores,[373] and after partial benzylation, for the preparation of urethane surface coatings. Benzylated wattle tannin urethanes have high gloss and scratch resistance and are stable when exposed to sunlight.[381]

Use of wattle tannins in drilling mud additives has declined as stability at high temperatures and high salinity has become more important. However, use of chrome salts with wattle tannins has increased their stability enough to permit use in wells as deep as 6000 ft when high salinity is no problem.[373] Wattle extracts have been used to recover fluorspar ore.[382] The condensation product of wattle tannins with ethanolamine is a water-soluble amphotannin that has been used as a flocculant for clay suspensions in municipal water treatment.[373]

Similar uses for conifer bark polyflavonoids have been developed. MacLean and Gardner[383] focused on some of the important properties of conifer bark tannins in their study of the use of *T. heterophylla* bark extracts in plywood adhesives. Conifer bark polyflavonoids exhibit phloroglucinol functionality in reaction with formaldehyde,[384-86] and the molecular weight of conifer bark tannins and particularly the alkali-soluble phenolic acids are higher than those of wattle tannins. The condensation of conifer bark tannins has been too rapid to permit reasonable control,[386] even in the presence of high proportions of alcohols. MacLean and Gardner[383] were unable to obtain acceptable bonds using paraformaldehyde or hexamine as aldehyde donors and recommended the use of methylol-phenols as crosslinking agents. Herrick and Bock[387] made extensive studies of the use of polymethylolphenols as crosslinking agents. To obtain adhesives of satisfactory viscosity, cure rate and stability for plywood manufacture, the molecular weight and methylol content of the methylol-phenol was balanced with the proportion of polyflavonoid used. Brandts and Lichtenberger[388] also employed a polymethylol-phenol as a crosslinking agent but advocated heating alkaline extracts from bark to temperatures of 150°C, precipitation of the heated extract by addition of mineral acid, and washing the precipitate to remove carbohydrates and other impurities before formulation with the methylol-phenol. Hemingway and McGraw[342] found that such heat treatments caused a reduction in the rate of condensation in reaction with formaldehyde as would be expected if rearrangement of the phloroglucinol ring as described by Sears et al.[114] had occurred. Mixtures of heated southern pine bark polyflavonoids and phenol could be cooked with formaldehyde in much the same way as a conventional phenol-formaldehyde resin. Adhesives prepared with these types of resins when used to make southern pine plywood gave wood failure values of only 70% after pressing for 6 min at 140°C. Heated alkali-soluble bark extracts have also been reacted with dimethylolurea to prepare a derivative used to substitute 60% of phenol-formaldehyde resin in plywood adhesives.[389]

Resin formulations for plywood that contain water-soluble tannins, para-formaldehyde, and various fillers, also have been examined. Booth et al.[390] and Herzberg[391] describe plywood adhesives made with *Pinus radiata* bark extracts. To obtain a water-resistant bond, panels were pressed for 15 min at 140 to 145°C. Steiner and Chow,[392] working with *T. heterophylla* tannins, formulated a plywood adhesive containing bark extract and paraformaldehyde. The pH was adjusted to 6.5 to 7.0 just before spreading. Press temperatures of 180°C for 7.5 min provided exterior grade bonds. Plywood adhesives have been made using bark ground to a powder as a reactive extender.[14] The most recent approach was developed by Hartman[393] who compounded 1000 parts of bark powder slurried in NaOH with 1000 parts of a 50% wattle tannin solution and 100 parts of formaldehyde. When used for plywood pressed at 138°C for 6 min, bonds giving 80% wood failure at 190 psi wet shear strength after vacuum-pressure water soaking were obtained.

Particleboard adhesives based on sulfited *P. radiata* bark extracts as described by Hall et al.[394] have been developed from the extracts of the barks of a number of conifer species by Anderson and co-workers.[395-397] In the most recent work, three-layer particleboards (wood-face, bark-core) were bonded with sulfited conifer bark based adhesives.

Herrick and Conca[398] developed a laminating adhesive based on extracts from *T. heterophylla*. The sodium salts of an ammonia extract were combined with either resorcinol or phenol-resorcinol resols in ratios of 1:1 or 1:2, respectively. Considerable experimentation was done to find a solvent system of suitable pot-life, spreading characteristics, and cure rates. Maleic anhydride was added in amounts of 10% of the polyflavonoid to control pH and cure rate. Epoxy and urethane resin or foam products based on conifer bark extracts have also been described.[399-401]

Conifer bark extracts have found application in reducing the viscosity and gel strength of muds used in well drilling.[402,403] Sulfited polyflavonoid preparations are quite effective.[366,403] Examples of the development of drilling mud additives from bark extracts are found in various patents.[404-406] However, sulfonated bark extracts have met with considerable competition from various lignosulfonate preparations.[407] A broad spectrum of dispersants based on sulfited bark extracts has been developed for use in many applications such as ceramic clays, minerals, pigments, dyes, pesticides, boiler water scale, and carbon blacks.[366,403] Other interesting products include water-soluble heavy-metal micronutrient complexes[366] that have been used to correct iron deficiencies in citrus plantations,[408] and a grouting medium used to stop water flow and stabilize soil in earthen structures or building foundations.[409]

Amphoteric flocculants based on the condensation products of the phenolic acid fractions of *Pinus sylvestris* and *Picea abies* barks with dimethylamine and formaldehyde or by etherfication with 2-(NN-diethylamino)-ethyl chloride to form the DEAE ether derivative and reaction of this compound with methyl iodide to form the quaternary ammonia derivative have been recently described by Pulkkinen and Peltonen.[410] Other coagulation aids prepared from bark phenolic acids or have been described by Allan.[411]

In addition to its uses as mulches,[412] decorative landscaping,[412] poultry litter,[413] and other agricultural and horticultural uses, bark ground to small particles has been investigated for many specialty products. Use of ground bark as the polyol for reaction with diisocyanates in the manufacture of urethane foams of particularly good flammability resistance has been studied by Hartman.[414] Ground bark has been studied for its application to solution of a number of pollution problems. It is effective in removing oil spills from water surfaces,[415] as a scavenger for heavy metal ions,[416] and has a high affinity for the odoriferous compounds produced in Kraft pulping.[417] Bark has also been reacted with herbicides to produce a slow release product that is very effective in Douglas-fir plantations.[418] The above examples of the prospects for utilization of bark are by no means a complete list. For further information, readers should see the Institute of Paper Chemistry Bibliographic Series 191.[419-421]

IX. CONCLUSIONS

The reader of this chapter is undoubtedly struck by the large research effort that has been focused on understanding of the structure of bark constituents rather than on product development. Without some understanding of the structure, it is impossible to predict the behavior or to focus on properties that can be exploited for developing specialty chemicals. Therefore, the dedication to structure elucidation seems fully warranted. More work needs to be done to further clarify the chemistry of most of the major classes of compounds discussed in this chapter. However, as argued by Herrick and Hergert,[407] more effort should be directed to specific problems of product development.

Prospects for rapid advances in the chemical utilization of tree barks in the short range are not particularly bright. The major route to utilization of bark carbohydrates will probably continue to be through pulping of whole-tree chips. However, limited amounts of barky chips can be tolerated in many paper making applications. After decades of research, bark waxes are finally being manufactured, but on a small scale. Efforts to broaden the markets for these waxes are needed now. Manufacture of fine chemicals from tree barks has been thwarted by the loss of bioflavonoid pharmaceutical markets and by difficulties in obtaining economic separation of complex mixtures on an industrial scale. Probably the best opportunities for chemical utilization of tree barks in the development of specialty polymers that use the polyflavonoids which comprise 30 to 40% of the weight of many conifer barks. However, a good deal more should be learned of the properties of the conifer bark polyflavonoids. Good progress has been achieved in the utilization of wattle tannins in adhesive formulations. Similar progress has not been achieved in the use of conifer bark polyflavonoids, but results have been encouraging enough to warrant further research. However, North American phenol production is expected to be only approximately 70% of capacity through the mid 1980s. Longer term prospects for the use of bark polyflavonoids depend upon the allocation of oil and developments in coal conversion to phenolic compounds. The recent classification of tannins as Category I carcinogens by OSHA may also restrict utilization.

As fossil bases for organic chemicals become increasingly scarce and expensive, we can expect the forest products industry to assume a larger role as a chemical manufacturer. However, enthusiasm for future prospects must be tempered by the realities of a continuing highly competitive petrochemical industry and longer range production of chemicals from coal. Far more must be learned of how to manipulate the properties of the compounds obtainable from forest trees. The path will be long and arduous, requiring patience and definitive research.

REFERENCES

1. Howard, E. T., Bark structure of the southern pines, *Wood Sci.*, 3, 134, 1971.
2. Howard, E. T., Bark structure of southern upland oaks, *Wood Fiber*, 9, 172, 1977.
3. Jensen, W., Fremer, K. E., Sierila, P., and Wartiovaara, V., The chemistry of bark, in *The Chemistry of Wood*, Browning, B. L., Ed., Interscience, New York, 1963, chap. 12.
4. Chang, Y. -P. and Mitchell, R. L., Chemical composition of common North American pulpwood barks, *Tappi*, 38, 315, 1955.
5. Dietrichs, H. H., Garves, K., Behrensdorf, D., and Sinner, M., Untersuchungen über die Kohlenhydrate der Rinden einheimischer Holzarten, *Holzforschung*, 32, 60, 1978.
6. Browning, B. L., The composition and chemical reactions of wood, in *The Chemistry of Wood*, Browning, B. L., Ed., Interscience, New York, 1963, chap. 3.
7. Hergert, H. L., Sloan, T. H., Gray, J. P., and Sandberg, K. R., The chemical composition of southeast hardwoods, *Wood Fiber*, in press.
8. Smith, L. V. and Zavarin, E., Free mono- and oliogosaccharides of some California conifers, *Tappi*, 43, 218, 1960.
9. Bishop, C. T., The Chemical Investigation of the Aqueous Extract of White Spruce Bark, Ph.D. thesis, McGill University, Montreal, 1949.
10. Harpham, J. A., Further Separations of the Water-Soluble Components of White Spruce Bark, Ph.D. thesis, McGill University, Montreal, 1951.
11. Painter, T. J. and Purves, C. B., Polysaccharides in the inner bark of white spruce, *Tappi*, 43, 729, 1960.
12. Becker, E. S. and Kurth, E. F., The chemical nature of the extractives from the bark of red fir, *Tappi*, 41, 380, 1958.
13. Laver, M. L., Chen, C. -H., Zerrudo, J. V., and Lai, Y. -C. L., Carbohydrates of the inner bark of *Pseudotsuga menziesii*, *Phytochemistry*, 13, 1891, 1974.
14. Hemingway, R. W., Adhesives from southern pine bark: a review of past and current approaches to resin formulation problems, in *Proceedings of Complete-Tree Utilization of Southern Pine*, McMillin, C. W., Ed., Forest Products Research Society, Madison, Wis., 1978, 443.
15. Beveridge, R. J., Stoddart, J. F., Szarek, W. A., and Jones, J. K. N., Some structural features of the mucilage from the bark of *Ulmus fulva* (slippery elm mucilage), *Carbohydr. Res.*, 9, 429, 1969.
16. Beveridge, R. J., Jones, J. K. N., Lowe, R. W., and Szarek, W. A., Structure of slippery elm mucilage *(Ulmus fulva)*, *J. Polym. Sci., Part C*, 36, 461, 1971.
17. Timell, T. E., Isolation of polysaccharides from the bark of gymnosperms, *Sven. Papperstidn.*, 64, 651, 1961.

18. **Pavlovova, E., Rendos, F., and Kovac, P.**, Hemicelluloses from twigs of white willow (*Salix alba* L.). IV. The bark polysaccharides, *Cellul. Chem. Technol.*, 4, 255, 1970.

19. **Timell, T. E.**, Characterization of four celluloses from the bark of gymnosperms, *Sven. Papperstidn.*, 64, 685, 1961.

20. **Mian, A. J. and Timell, T. E.**, Isolation and characterization of a cellulose from the inner bark of white birch *(Betula papyrifera), Can. J. Chem.*, 38, 1191, 1960.

21. **Garves, K.**, Viskosimetrishe Bestimmung der Polymerisationsgrade von Rinden Cellulosen, *Cellul. Chem. Technol.*, 10, 249, 1976.

22. **Thornber, J. P. and Northcote, D. H.**, Changes in the chemical composition of a cambial cell during its differentiation into xylem and phloem tissues in trees. I, *Biochem. J.*, 81, 449, 1961.

23. **Fu, Y. -L., Gutmann, D. S., and Timell, T. E.**, Polysaccharides in the secondary phloem of scots pine (*Pinus sylvestris* L.). I. Isolation and characterization of callose, *Cellul. Chem. Technol.*, 6, 507, 1972.

24. **Kiefer, H. J. and Kurth, E. F.**, The chemical composition of the bast fiber of Douglas-fir bark, *Tappi*, 36, 14, 1953.

25. **Jiang, K. S. and Timell, T. E.**, Polysaccharides in the bark of aspen *(Populus tremuloides).* I. Isolation and constitution of a 4-0-methyl-glucuronoxylan, *Cellul. Chem. Technol.*, 6, 493, 1972.

26. **Gardner, P. E.**, A Study of the Completely Extracted Fibrous Portion of White Spruce Bark, Ph.D. thesis, McGill University, Montreal, 1957.

27. **Chen, C. -H.**, Douglas-fir Bark: Isolation and Characterization of a Holocellulose Fraction, Ph.D. thesis, Oregon State University, Corvallis, 1973.

28. **Binotto, A. P., Murphey, W. K., and Cutter, B. E.**, X-ray diffraction studies of cellulose from bark and wood, *Wood Fiber*, 3, 179, 1971.

29. **Harwood, V. D.**, An Investigation of the Ether and Ligroin Extracts of White Spruce Bark, Ph.D. thesis, McGill University, Montreal, 1949.

30. **Jablonski, W. L.**, Action of Liquid Ammonia on White Spruce Bark Extracted With Methanol and Water, Ph.D. thesis, McGill University, Montreal, 1953.

31. **Sanderson, E. S.**, The Substances Extracted by Water From White Spruce Bark Pretreated With Liquid Ammonia, Ph.D. thesis, McGill University, Montreal, 1953.

32. **Bryce, J. R. G.**, A Study of the Alkaline Extract of White Spruce Bark Pretreated With Liquid Ammonia, Ph.D. thesis, McGill University, Montreal, 1954.

33. **Timell, T. E.**, Constitution of a water soluble galactoglucomannan from the bark of amabilis fir *(Abies amabilis), Sven, Papperstidn.*, 64, 843, 1962.

34. **Timell, T. E.**, The structure of an arabino-4-0-methylglucuronoxylan from the bark of amabilis fir *(Abies amabilis), Sven. Papperstidn.*, 64, 748, 1961.

35. **Timell, T. E.**, Constitution of a glucomannan from the bark of amabilis fir *(Abies amabilis), Sven. Papperstidn.*, 64, 744, 1961.

36. **Ramalingam, K. V. and Timell, T. E.**, Polysaccharides present in the bark of Engelmann spruce *(Picea engelmanni), Sven. Papperstidn.*, 67, 512, 1964.

37. **Fu, Y. -L. and Timell, T. E.**, Polysaccharides in the secondary phloem of scots pine *(Pinus sylvestris)*. III. The constitution of a galactoglucomannan, *Cellul. Chem. Technol.*, 6, 744, 1972.

38. **Fu, Y. -L. and Timell, T. E.**, Polysaccharides in the secondary phloem of scots pine *(Pinus sylvestris)*, *Cellul. Chem. Technol.*, 6, 513, 1972.

39. **Fernandez, E. C.**, Douglas-Fir Bark: Structure and Alkaline Degradation of a Glucomannan, Ph.D. thesis, Oregon State University, Corvallis, 1977.

40. **Mian, A. J. and Timell, T. E.** Isolation and properties of a 4-0-methyl-glucuronoxylan from the inner bark of white birch *(Betula papyrifera), Tappi*, 43, 775, 1960.

41. **Jiang, K. S. and Timell, T. E.**, Polysaccharides in the bark of aspen *(Populus tremuloides).* III. The constitution of a galactoglucomannan, *Cellul. Chem. Technol.*, 6, 503, 1972.

42. **Jiang, K. S. and Timell, T. E.**, Polysaccharides in the bark of aspen *(Populus tremuloides).* II. Isolation and structure of an arabinan, *Cellul. Chem. Technol.*, 6, 499, 1972.

43. **Toman, R.**, Polysaccharides from the bark of white willow (*Salix alba* L.), structure of a xylan. III, *Cellul. Chem. Technol.*, 7, 351, 1973.

44. **Toman, R., Karacsonyi, S., and Kovacik, V.**, Polysaccharides from the bark of white willow (*Salix alba* L.), structure of a galactan, *Carbohydr. Res.*, 25, 371, 1972.

45. **Karacsonyi, S., Toman, R., Janecek, F., and Kubackova, M.**, Polysaccharides from the bark of white willow (*Salix alba* L). Structure of an arabinan, *Carbohydr. Res.*, 44, 285, 1975.

46. **Thornber, J. P. and Northcote, D. H.**, Changes in the chemical composition of a cambial cell during its differentiation into xylem and phloem tissues in trees. II, *Biochem. J.*, 81, 455, 1961.

47. **Bhattacharjee, S. S. and Timell, T. E.**, A study of the pectin present in the bark of amabilis fir *(Abies amabilis), Can. J. Chem.*, 43, 758, 1965.

48. **Timell, T. E. and Mian, A. J.**, A study of the pectin present in the inner bark of white birch (*Betula papyrifera* Marsh), *Tappi*, 44, 788, 1961.

49. Toman, R., Karacsonyi, S., and Kubackova, M., Studies on pectin present in the bark of white willow (*Salix alba* L.). Structure of the acidic and neutral oligosaccharides obtained on partial hydrolysis, *Cellul. Chem. Technol.,* 10, 561, 1976.

50. Scharfetter, H., Pizzi, A., and Du -T. Roosow, D., Some new ideas on tannin adhesives for wood. Paper presented to Internationl Union Forest Research Organization Conference, Merida, Venezuela, 1977.

51. Lewis, H. F., Brauns, F. E., Buchman, M. A., and Kurth, E. F., Chemical composition of redwood bark, *Ind. Eng. Chem.,* 36, 759, 1944.

52. Kurth, E. F., The chemical composition of barks, *Chem. Rev.,* 40, 33, 1946.

53. Kurth, E. F., Hubbard, J. K., and Humphrey, J. D., Chemical composition of ponderosa and sugar pine barks, *Proc. For. Prod. Res. Soc.,* 3, 276, 1949.

54. Hergert, H. L. and Kurth, E. F., The chemical nature of the cork from Douglas-fir bark, *Tappi,* 35, 59, 1952.

55. Kiefer, H. J. and Kurth, E. F., The chemical composition of the bast fibers of Douglas-fir bark, *Tappi,* 36, 14, 1953.

56. Kurth, E. F. and Smith, J. E., The chemical nature of the lignin of Douglas-fir bark, *Pulp Pap. Mag. Can.,* 55, 125, 1954.

57. Sarkanen, K. V. and Hergert, H. L., Classification and distribution, in *Lignins,* Sarkanen, K. V. and Ludwig, C. H., Eds., John Wiley & Sons, New York, 1971, chap. 3.

58. Hata, K. and Sogo, M., Chemical studies on the bark. III. On the lignin of Japanese red pine (*Pinus densiflora* S. et. Z.), *J. Jpn. Wood Res. Soc.,* 4, 5, 1958.

59. Sogo, M., Chemical studies on the bark. XIV. Comparison of phenolic acid, tannin, and Bjorkman lignin isolated from bark of some gymnosperm and dicotyledonus trees, *J. Jpn. Wood Res. Soc.,* 12, 293, 1966.

60. Sogo, M., On the permananate oxidation of bark phenolic acid, tannin, and Bjorkman lignin isolated from bark of some gymnosperm and angiosperm trees, *Tech. Bull. Fac. Agric. Kagawa Univ.,* 18, 158, 1967.

61. Sogo, M. and Hata, K., Chemical studies on the bark, XV. On the permanganate oxidation of bark phenolic acid, tannin, and Bjorkman lignin isolated from bark of some tree species, *J. Jpn. Wood Res. Soc.,* 13, 300, 1967.

62. Sogo, M., Hata, K., and Hirata, A., Chemical studies on the bark. X. Dissolution of lignin of red pine bark by cooking with sulfite solutions of various pH, *J. Jpn. Wood Res. Cos.,* 9, 194, 1963.

63. Sogo, M. and Hata, K., Chemical studies on the bark. XI. Properties of lignosulfonic acids from the outer bark of *Pinus densiflora, J. Jpn. Wood Res. Soc.,* 10, 36, 1964.

64. Sogo, M. and Hata, K., Chemical studies on the bark. XII. Molecular weight and viscosity behavior of lignosulfonates from *Pinus densiflora* outer bark, *J. Jpn. Wood Res. Soc.,* 10, 136, 1964.

65. Sogo, M., Ishihara, T., and Hata, K., Chemical studies on the bark. XIII. On the hydrogenolysis of the outer bark lignin of *Pinus densiflora, J. Jpn. Wood Res. Soc.,* 12, 96, 1966.

66. Sogo, M. and Hata, K., Chemical studies on the bark. XVI. On the aromatic nuclei of undissolvable bark lignin, *J. Jpn. Wood Res. Soc.,* 14, 334, 1968.

67. Higuchi, T., Ito, Y., Shimada, M., and Kawamura, I., Chemical properties of bark lignins, *Cellul. Chem. Technol.,* 1, 585, 1967.

68. Sogo, M. and Kawahara, M., The alcoholysis of the outer bark lignin of Japanese red pine (*Pinus densiflora,* S. et. Z.), *Tech. Bull. Fac. Agric. Kagawa Univ.,* 12, 269, 1961.

69. Hata, K., Schubert, W. J., and Nord, F. F., Fungal degradation of the lignin and phenolic acids of the bark of western pine, *Arch. Biochem. Biophys.,* 113, 250, 1966.

70. Nord, F. F. and Hata, K., Fungal degradation of pine bark lignin, *Current Aspects of Biochemical Energetics,* Kaplan, N.D. and Kennedy, E.P., Eds., Academic Press, New York, 1966, 315.

71. Swan, E. P., A study of western red cedar bark lignin, *Pulp Pap. Mag. Can.,* 67, T456, 1966.

72. Andersson, A., Erickson, M., Fridh, H., and Miksche, G. E., Zur Struktur des Lignins der Rinde von Laub und Nadelholzern, *Holzforschung,* 27, 189, 1973.

73. Smelstorius, J. A. and Stewart, C. M., Determination of total lignin, extraneous substances and polysaccharides in bark and wood of *Acacia penninervis, Holzforschung,* 28, 160, 1974.

74. Sogo, M., Nitrobenzene oxidation products from bark lignins of broad-leaved trees and coniferous trees, *Tech. Bull. Fac. Agric. Kagawa Univ.,* 10, 46, 1959.

75. Hata, K. and Sogo, M., Chemical studies on the bark. IV. Comparison of lignin between bark of gynmosperm trees and that of dicotyledonous trees, *J. Jpn. Wood Res. Soc.,* 4, 85, 1958.

76. Clermont, L. P., Study of lignin from stone cells of aspen poplar inner bark, *Tappi,* 53, 52, 1970.

77. Hata, K. and Sogo, M., Chemical studies on the bark. VII. On the lignin of the outer bark of "Mizumara", (*Quercus crispula* Blume), *J. Jpn. Wood Res. Soc.,* 6, 71, 1960.

78. Hergert, H. L., Secondary lignification in conifer trees, *Cellul. Chem. Technol. Symp.,* ACS Symp. Series, 48, 227, 1977.

79. **Hemingway, R. W. and Davies, G. W.**, Chemical anatomical and pulp properties of *Eucalyptus obliqua* bark, paper presented to 27th Appita Conference, Rotorura, New Zealand, 1973.
80. **Litvay, J. D. and Krahmer, R. L.**, Wall layering in Douglas-fir cork cells, *Wood Sci.*, 9, 167, 1977.
81. **Sette, P.**, Der Feinbau verkorkter Zellwande, *Microskopie*, 10, 178, 1955.
82. **Segall, G. H. and Purves, C. B.**, Chemical composition of wood barks, *Pulp Pap. Mag. Can.*, Convention issue, 149, 1946.
83. **Harwood, V. D.**, An Investigation of Ether and Ligroin Extracts of White Spruce Bark, Ph.D. thesis, McGill University, Montreal, 1949.
84. **Kurth, E. F. and Kiefer, H. J.**, Wax from Douglas-fir bark, *Tappi*, 33, 183, 1950.
85. **Kurth, E. F. and Hubbard, J. K.**, Extractives from ponderosa pine bark, *Ind. Eng. Chem.*, 43, 896, 1951.
86. **Kurth, E. F.**, The composition of conifer bark waxes and corks, *Tappi*, 50, 253, 1967.
87. **Hergert, H. L. and Kurth, E. F.**, The chemical nature of the cork from Douglas-fir bark, *Tappi*, 35, 59, 1952.
88. **Hergert, H. L. and Kurth, E. F.**, The chemical nature of the extractives from white fir bark, *Tappi*, 36, 137, 1953.
89. **Hergert, H. L.**, Chemical composition of cork from white fir bark, *For. Prod. J.*, 8, 335, 1958.
90. **Hall, J. A.**, Utilization of Douglas-fir bark, Pacific Northwest Forest and Range Experiment Station, U.S. Department of Agriculture Forest Service, Portland, Ore., 1971.
91. **Clark, I. T., Hicks, J. R., and Harris, E. E.**, Extractives of Douglas-fir and Douglas-fir lignin residues, *J. Am. Chem. Soc.*, 69, 3142, 1947.
92. **Clark, I. T., Hicks, J. R., and Harris, E. E.**, Constituents of extractives from Douglas-fir lignin residues, *J. Am. Chem. Soc.*, 70, 3729, 1948.
93. **Adamovics, J. A., Johnson, G., and Stermitz, F. R.**, Ferulates from cork layers of *Solanum tuberosum* and *Pseudotsuga menziesii, Phytochemistry*, 16, 1089, 1977.
94. **Laver, M. L., Loveland, P. M., Chen, M., Fang, C. -H., Fang, H. H. -L., Zerrudo, J. V., and Liu, Y. C. L.**, Chemical constituents of Douglas-fir bark: a review of more recent literature, *Wood Sci.*, 10, 85, 1977.
95. **Loveland, P. M. and Laver, M. L.**, Monocarboxylic and dicarboxylic acids from *Pseudotsuga menziesii* bark, *Phytochemistry*, 11, 430, 1972.
96. **Loveland, P. M. and Laver, M. L.**, ω-Hydroxyfatty acids and fatty alcohols from *Pseudotsuga menziesii* bark, *Phytochemistry*, 11, 3080, 1972.
97. **Rogers, I. H. and Grierson, D.**, Extractives from grand fir (*Abies grandis*, (Dougl.), Linal.), bark, *Wood Fiber*, 4, 33, 1972.
98. **Hata, K. and Sogo, M.**, Chemical studies on the bark. I. On the extractive, especially on wax from the outer bark of Japanese red pine (*Pinus densiflora*, S et. Z), *J. Jpn. Wood Research Soc.*, 39, 102, 1957.
99. **Rowe, J. W., Bower, C. L., and Wagner, E. R.**, Extractives of jack pine bark, occurrence of *cis-* and *trans*-pinosylvan dimethylether and ferulic acid esters, *Phytochemistry*, 8, 235, 1969.
100. **Hergert, H. L.**, personal communication, 1972.
101. **Hemingway, R. W.**, unpublished results.
102. **Pearl, I. A.**, The water-soluble and petroleum ether-soluble extractives of loblolly and slash pine barks, *Tappi*, 58, 143, 1975.
103. **Pearl, I. A.**, Ether-soluble and ethanol-soluble extractives of loblolly and slash pine barks, *Tappi*, 58, 135, 1975.
104. **Pearl, I. A.**, Variations of loblolly and slash pine bark extractive components and wood turpentine components on a monthly basis, *Tappi*, 58, 146, 1975.
105. **Hartman, L. and Weenick, R. O.**, A note on the fatty acid composition of the lipids from the bark of *Pinus radiata, N. Z. J. Sci.*, 10, 636, 1967.
106. **Weissmann, G.**, Rindenwachse von Kiefer und Fichte, *Seifen Oele Fette, Wachse*, 103, 399, 1977.
107. **Zinkel, D. F. and Rowe, J. W.**, A rapid method for the quantitative separation without alteration of ether-soluble acid and neutral materials, *Anal. Chem.*, 36, 1160, 1964.
108. **Hossfeld, R. L. and Hunter, W. T.**, The petroluem ether extractives of aspen bark, *Tappi*, 41, 359, 1958.
109. **Abramovitch, R. A., Coutts, R. T., and Knaus, E. E.**, Some extractives from the bark of *Populus balsamifera, Can. J. Pharm. Sci.*, 2, 71, 1967.
110. **Streibl, M., Konecny, K., and Mahdalik, M.**, Chemical composition of beech bark, Investigation of lipids, *Cellul. Chem. Technol.*, 3, 653, 1969.
111. **Streibl, M., Mahdalik, M., and Konecny, K.**, Composition of the lipoid portion of beech bark, *Holzforsch. Holzverwert.*, 22, 1, 1970.
112. **Streibl, M. and Mahdalik, M.**, Chemical composition of some hydrocarbons of the lipoid fraction from beech bark, *Holzforsch. Holzverwert*, 23, 32, 1971.

113. Streibl, M., Stransky, K., and Herout, V., *n*-Alkanes of some tree barks, *Collect. Czech. Chem. Commun.*, 43, 320, 1978.

114. Sears, K. D., Casebier, R. L., Hergert, H. L., Stout, G. H., and McCandlish, L. E., Catechinic acid, a base rearrangement product of catechin, *J. Org. Chem.*, 39, 3244, 1974.

115. Kolattukudy, P. E., Lipid polymers and associated phenols, their chemistry, biosynthesis, and role in pathogenesis, in *Recent Advances in Phytochemistry*, Loewus, F. A. and Reunckles, E., Eds., Plenum Press, New York, 1977, chap. 6.

116. Zhunchenko, A. G., Cherkasova, A. I., Chemical composition of birch bark, *Bibliography series 191*, (Suppl. 2), Weiner, J. and Pollock, V., Eds., Institute of Paper Chemistry, Appleton, Wis., 1973, 55.

117. Jensen, W., Study of outer bark of birch *(Betula vervucosa)* and cork *(Quercus suber)* by scanning electron microscopy, *An. Quim.*, 68, 871, 1972; *Abstr. Bull. I.P.C.*, 44, 3, 289, 1973.

118. Gonzales, A., Cork acids, *Acta Cient. Compostelana*, 5, 57, 1968; *Bibliography Series 191*, (Suppl. 2), Weiner, J. and Pollock, V., Eds., Institute of Paper Chemistry, Appleton, Wis., 1973, 23.

119. Seoane, E., Serra, M.C., and Aquillo, C., Two new epoxy acids from the cork of *Quercus suber*, *Chem. and Ind. (Lond)*, 662, 1977.

120. Guillemonat, A. and Triaca, M., Chemical composition of cork, Preliminary study of the *Kielmeyera coriacea* cork. *Bibliography Series 191*, (Suppl. 2), Weiner, J. and Pollock, V., Eds., Institute of Paper Chemistry, Appleton, Wis., 1973, 99.

121. Trocino, F. S., Method of extracting wax from bark, U.S. Patent 3,789,058, 1974.

122. Trocino, F. S., personal communication, 1977.

123. Trocino, F. S., Method of separating bark components, U.S. Patent 3,781,187, 1973.

124. Trocino, F. S., Extenders for thermosetting resins, U.S. Patent 3,616,201, 1971.

125. Kurth, E. F., Extraction of valuble products from bark, U.S. Patent 2,662,893, 1953.

126. Zenczak, P., Process for recovery of products from bark, U.S. Patent 2,781,336, 1953.

127. Zenczak, P., Method of producing, separating, and recovering reaction products from bark, U.S. Patent 2,947,764, 1960.

128. Swan, E. P., Alkaline ethanolysis of extractive free western red-cedar bark, *Tappi*, 51, 301, 1968.

129. Swan, E. P. and Naylor, A. F. S., Alkaline ethanolysis of western conifer barks, *Can. Dep. Fish For. Bi-Mon. Res. Notes*, 25, 32, 1969.

130. Heritage, C. C. and Dowd, L. E., Alkaline extraction of chemical products from bark, U.S. Patent 2,890,231, 1959.

131. Brink, D. L., Dowd, L. E., and Root, D. F., Separating wax products from aqueous akaline tree bark extracts and products, U.S. Patent 3,234,202, 1966.

132. Dowd, L. E., Brink, D. L., Gregory, A. S., and Esterer, A. K., Fractionation of alkaline extracts of tree barks, U.S. Patent 3,255,221, 1966.

133. Gygi, E. H. and Root, D. F., Fractionation of alkaline extracts of tree barks, U.S. Patent 3,560,536, 1971.

134. Miroshnichenko, E. V. and Fedorishchev, T. I., New method for the extraction of suberin from birch bark and surface-active substances from suberin, cited from Abstract Bulletin of the Institute of Paper Chemistry, 48, 306, 1977.

135. Smedman, L. A., Zavarin, E., and Teraniski, R., Chemotaxonomy of the genus *Abies*. III. Composition of oxygenated monoterpenes and sesquiterpene hydrocarbons from the cortical oleoresin of *Abies magnifica*, *Phytochemistry*, 8, 1457, 1969.

136. Smedman, L. A., Snajberk, K., and Zavarin, E., Chemotaxonomy of the genus *Abies*. IV. Oxygenated monoterpenoids and sesquiterpenoid hydrocarbons of the cortical turpentine from different *Abies* species, *Phytochemistry*, 8, 1471, 1969.

137. Zavarin, E., Snajberk, E., Reichart, T., and Tsien, E., Chemotaxonomy of the genus *Abies*. V. Geographic variability of the monoterpenes from the cortical blister oleoresin of *Abies lasiocarpa*, *Phytochemistry*, 9, 377, 1970.

138. Zavarin, E., Critchfield, W. B., and Snajberk, K., Composition of the cortical and phloem monoterpene of *Abies lasiocarpa*, *Phytochemistry*, 10, 3229, 1971.

139. Zavarin, E. and Snajberk, K., Geographic variability of monoterpenes from *Abies balsamea* and *Abies fraseri*, *Phytochemistry*, 11, 1407, 1972.

140. Lee, C. J., Snajberk, K., and Zavarin, E., Chemical composition of the cortical essential oil from *Abies balsamea*, *Phytochemistry*, 13, 179, 1974.

141. Zavarin, E., Snajberk, K., and Critchfield, W. B., Monoterpene variability of *Abies amabilis* cortical oleoresin, *Biochem. Syst.*, 1, 87, 1973.

142. Zavarin, E., Snajberk, K., and Fisher, J., Geographical variability of monoterpenes from cortex of *Abies, concolor, Biochem. System. and Ecol.*, 3, 191, 1975.

143. Balogh, B., Wilson, D. M., Burlingame, A. C., Lee, C. J., Snajberk, K., and Zavarin, E., 4,4-Dimethyl-2-cyclohepten-1-one in the cortical oleoresin of *Abies balsamea*, *Phytochemistry*, 11, 1481, 1972.

144. Zavarin, E., Snajberk, K., and Critchfield, W. B., Relation of cortical monoterpenoid composition to tree age and size, *Phytochemistry,* 16, 770, 1977.

145. Hanover, J. W., Inheritance of 3-carene concentration in *Pinus monticola, For. Sci.,* 12, 447, 1966.

146. Hanover, J. W., Comparative physiology of eastern and western white pine: Oleoresin composition and viscosity, *For. Sci.,* 21, 214, 1975.

147. Gerhold, H. D. and Plank, G. H., Monoterpene variations in vapors from white pines and hybrids, *Phytochemistry,* 9, 1393, 1970.

148. Marpeau, A., Baradat, P., and Bernard-Dagan, C., Terpenes of maritime pine. Biological and genetic aspects: heritability of the content of two sesquiterpenes lonigolene and caryophyllene, *Ann. Sci. For.,* 32, 185, 1975.

149. Simpson, R. F. and McQuilken, R. M., Terpenes of the bark oil of *Pinus radiata, Phytochemistry,* 15, 328, 1976.

150. Tobolski, J. J. and Hanover, J. W., Genetic variation in the monoterpenes of scotch pine, *For. Sci.,* 17, 293, 1971.

151. Norin, T. and Winell, B., Extractives from the bark of Scots pine *(Pinus sylvestris), Acta Chem. Scand.,* 26, 2297, 1972.

152. Squillace, A. E., Inheritance of monoterpene composition in cortical oleoresin of slash pine, *For. Sci.,* 17, 381, 1971.

153. Franklin, E. C. and Snyder, E. B., Variation and inheritance of monoterpene composition in longleaf pine, *For. Sci.,* 17, 178, 1971.

154. Wilkinson, R. C., Hanover, J. W., Wright, J. W., and Flake, R. H., Genetic variation in the monoterpene composition of white spruce, *For. Sci.,* 17, 83, 1971.

155. Rottink, B. A. and Hanover, J. W., Identification of blue spruce cultivars by analysis of cortical oleoresin monoterpenes, *Phytochemistry,* 11, 3255, 1972.

156. Wilkinson, R. C. and Hanover, J. W., Geographical variation in the monoterpene composition of red spruce, *Phytochemistry,* 11, 2007, 1972.

157. Norin, T. and Winell, B., Extractives from the bark of common spruce, *Picea abies, Acta Chem. Scand.,* 26, 2289, 1972.

158. Norin, T. and Winell, B., Neutral constituents of *Larix decidua* bark, *Phytochemistry,* 13, 1290, 1974.

159. Snajberk, K., Lee, C. J., and Zavarin, E., Chemical composition of volatiles from cortical oleoresin of *Pseudotsuga menziesii, Phytochemistry,* 13, 185, 1974.

160. Zavarin, E. and Snajberk, K., Geographical variability of monoterpenes from cortex of *Pseudotsuga menziesii, Pure Appl. Chem.,* 34, 411, 1973.

161. Snajberk, K. and Zavarin, E., Mono and sesqui-terpenoid differentiation of *Pseudotsuga* of the United States and Canada, *Biochem. System. and Ecol.,* 4, 159, 1976.

162. Zavarin, E. and Snajberk, K., Geographical differentiation of cortical monoterpenoids of *Pseudotsuga macrocarpa, Biochem. System. and Ecol.,* 4, 93, 1976.

163. Rowe, J. W. and Scroggins, J. G., Benzene extractives of lodgepole pine bark. Isolation of new diterpenes, *J. Org. Chem.,* 29, 1554, 1964.

164. Conner, A. H. and Rowe, J. W., Differentiating manool and 13-epimanool with NMR chiral shift reagents, *Phytochemistry,* 15, 1949, 1976.

165. Rowe, J. W. and Shaffer, G. W., Structures of contortadiol (agathadiol), contortolal (agatholal), and hydroxyepimanool (epitorulosol), *Tetrahedron Lett.,* 30, 2633, 1965.

166. Rowe, J. W., Ronald, R. C., and Nagasampagi, B. A., Terpenoids of lodgepole pine bark, *Phytochemistry,* 11, 365, 1972.

167. Bower, C. L. and Rowe, J. W., Extractives of jack pine bark. Occurrence of (+)-13-epimanoyl oxide and related labdane diterpenes, *Phytochemistry,* 6, 151, 1967.

168. Rowe, J. W., Nagasamagi, B. A., Burgstahler, A. W., and Fitzsimmons, J. W., Derivatives of nor-dehydroabietane from pine bark, *Phytochemistry,* 10, 1647, 1971.

169. Norin, T. and Winell, B., Nor-diterpenoids from the bark of *Pinus sylvestries, Acta Chem. Scand.,* 25, 611, 1971.

170. Zinkel, D. F. and Evans, B. B., Terpenoids of *Pinus strobus* cortex tissue, *Phytochemistry,* 11, 3387, 1972.

171. Rogers, I. H. and Rozen, L. R., Neutral terpenes from the bark of sitka spruce *(Picea sitchensis), Can. J. Chem.,* 48, 1021, 1970.

172. Quon, H. H. and Swan, E. P., Deoxypicropodophyllin and sugiol from *Thuja plicata* bark, *Can. Dept. of For., Biomon. Res. Notes,* 28, 23, 1972.

173. Quon, H. H. and Swan, E. P., Norditerpene alcohols in the barks of *Thuja plicata, Can. J. Chem.,* 47, 4389, 1969.

174. Fraser, H. S. and Swan, E. P., 19-Norisopimara-8(14),15-dien-3-one in *Thuja plicata* bark, *Can. Dept. For. Biomonth. Res. Notes,* 29, 12, 1973.

175. Hata, K., Sogo, M., and Yusuki, K., Chemical studies on bark. IX. On the extractives from the outer bark of *Chamaecyparis obtusa* (S. et. Z), *J. Jpn. Wood Res. Soc.*, 8, 167, 1962.

176. Zinkel, D. F., Pine resin acids as chemotaxonomic and genetic indicators, *Proc. 1977 Tappi For. Biol/Wood Chem. Cong.*, Atlanta, 1977.

177. Arya, V. P., Enzell, L., Erdtman, H., and Kuboto, T., Communic acid, a new diterpene acid from *Juniperous communis* L., *Acta Chem. Scand.*, 15, 225, 1961.

178. Arya, V. P., Isolation of communic acid from some juniper barks, *J. Sci. Ind. Res. (India)*, 21B, 201, 1962.

179. Zinkel, D. F. and Spalding, B. P., Strobic acid, a new resin acid from *Pinus strobus*, *Tetrahedron Lett.*, 27, 2459, 1971.

180. Zinkel, D. F., Toda, J. K., and Rowe, J. W., Occurrence of anticopalic acid in *Pinus monticola*, *Phytochemistry*, 10, 1161, 1971.

181. Spalding, B. P., Zinkel, D. F., and Roberts, D. F., New labdane resin acids from *Pinus elliotti*, *Phytochemistry*, 10, 3289, 1971.

182. Zinkel, D. F., Diterpene resin acids of *Pinus densiflora* needles and cortex, *Phytochemistry*, 15, 1073, 1976.

183. Rowe, J. W., Triterpenes of pine barks, Identity of pinusenediol and serratenediol, *Tetrahedron Lett.*, 34, 2347, 1964.

184. Rowe, J. W. and Bower, C. L., Triterpenes of pine barks, Naturally occurring derivatives of serratenediol, *Tetrahedron Lett.*, 32, 2745, 1965.

185. Rowe, J. W., Ronald, R. C., and Nagasampagi, B. A., Terpenoids of lodgepole pine bark, *Phytochemistry*, 11, 365, 1972.

186. Weston, R. J., Neutral extractives from *Pinus radiata* bark, *Aust. J. Chem.*, 20, 2729, 1973.

187. Kutney, J. P., Rogers, I. H., and Rowe, J. W., Neutral triterpenes of the bark of *Picea sitchensis*, (sitka spruce), *Tetrahedron*, 25, 3731, 1969.

187a. Uyeo, S., Ikada, J., Matsunaga, S., and Rowe, J. W., The structure and stereochemistry of *Abies* lactone, *Tetrahedron*, 24, 2859, 1968.

188. Rowe, J. W., The sterols of pine bark, *Phytochemistry*, 4, 1, 1965.

189. Laver, M. L. and Fang, H. H. L., *n*-Hexane-soluble components of *Pseudotsuga menziesii* bark, *Phytochemistry*, 10, 3292, 1971.

190. Seshadri, T. R. and Vedantham, T. N. C., Chemical examination of the barks and heartwoods of *Betula* species of American origin, *Phytochemistry*, 10, 897, 1971.

191. Rimpler, H., Kuhn, H., and Leuckert, C., Triterpenes of *Betula penda* (Roth) and *Betula pubescens* (Ehrh.), *Arch. Pharm.*, 299, 422, 1966.

192. Hejno, K., Jarolim, V., and Sorm, F., Some components of the white portion of birch bark, *Collect. Czech. Chem. Commun.*, 30, 1009, 1965.

193. Lindgren, B. O. and Svahn, C. M., Lupan-3-β-20-diol and lupan-3-β-20,28-triol in bark from birch *Betula verrucosa* (Erh.), *Acta Chem. Scand.*, 20, 1720, 1966.

194. Khan, M. A. and Rahman, A. U., Karachic acid, a new triterpenoid from *Betula vtilis*, *Phytochemistry*, 14, 789, 1975.

195. Zellner, J. and Weiss, L., Zur Chemi der Rinden. V. Schwarzerle (*Alnus glutinosa* L.), *Monatsh.*, 46, 312, 1925.

196. Kurth, E. F. and Becker, E. L., The chemical nature of the extractives from red alder, *Tappi*, 36, 461, 1953.

197. Matyukhina, L. G., Shmukler, V. S., and Ryabinin, A. A., Triterpenes of the bark of *Alnus subcordata*, *Bibliography Series 191*, (Suppl. 1), Roth, L. and Weiner, J., Eds., Institute of Paper Chemistry, Appleton, Wis., 1968, 42.

198. Pasich, B., Kowalesski, Z., and Grzybet, P., Triterpenoid compounds in plant materials, (Part 12), isolation of glutinone, taraxerol, and β-Sistosterol from the bark of the alder tree, *(Alnus glutinosa)*, *Bibliography Series 191*, (Suppl. 1,) Roth, L. and Weiner, J., Eds., Institute of Paper Chemistry, Appleton, Wis., 1968, 48.

199. Matyukhina, L. G., Triterpenes from the bark of Alnus barbata, *Bibliography Series 101*, (Suppl. 1), Roth, L. and Weiner, J., Eds., Institute of Paper Chemistry, Appleton, Wis, 1968, 42.

200. Pasich, B. and Kowalewski, Z., Triterpenoid compounds in plant material, (Part 10), isolation of lupenone from the bark of Alnus glutinosa, *Bibliography Series 191*, (Suppl. 1), Roth, L. and Weiner, J., Eds., Institute of Paper Chemistry, Appleton, Wis., 1968, 48.

201. Hart, N. K. and Lamberton, J. A., Morolic acid (3-hydroxyolean-18-en-28-oic acid) from the bark of *Eucalyptus papuana*, F. Muell., *Aust. J. Chem.*, 18, 115, 1965.

202. Aplin, R. T., Halsall, T. G., and Norin, T., The chemistry of triterpenes and related compounds. III. The constituents of the bark of *Platanus x hydriad* Brot. and the structure of platanic acid, *J. Chem. Soc.*, 3269, 1963.

203. Rowe, J. W. and Conner, A. H., Extractives in eastern hardwoods: a review, general technical report FPL 18, Forest Products Laboratory, For. Serv., U.S. Department of Agriculture, Madison, Wis., 1979.

204. Ryan, A. S., A review of the chemistry of quercetin, in Hall, J. A., *Utilization of Douglas-fir Bark,* Pac. Northwest For. and Range Exp. Sta., For. Serv. USDA, Portland, Oregon 1971. 109.

205. McClure, J. W., Physiology and functions of flavonoids, in *The Flavonoids,* Harborne, J. B., Mabry, T. J., and Mabry, H., Eds., Academic Press, New York, 1975, chap. 18.

206. Hergert, H. L., Economic importance of flavonoid compounds: wood and bark, in *The Chemistry of Flavonoid Compounds,* Geissmann, T.A., Ed., Macmillan, New York, 1962, chap. 17.

207. Hergert, H. L., Chemical composition of tannins and other polyphenols from conifer wood and bark, *For. Prod. J.,* 10, 610, 1960.

208. Lepetit, R. and Satta, C., Quercetin from the bark of *Pinus pinaster, Atti. Acad. Lincer.,* 25, 322, 1916.

209. Kurth, E. F., and Chan, F. L., Dihydroquercetin as an antioxidant, *J. Am. Oil Chem. Soc.,* 28, 433, 1951.

210. Kurth, E. F. and Chan, F. L., Extraction of tannin and dihydroquercetin from Douglas-fir bark, *J. Am. Leather Chem. Assoc.,* 48, 20, 1953.

211. Kurth, E. F., Quercetin from fir and pine bark, *Ind. Eng. Chem.,* 45, 2096, 1953.

212. Kurth, E. F., Hergert, H. L., and Ross, J. D., Behavior of certain 3-hydroxy-flavanones and basic salts of the alkali metals and ammonia, *J. Am. Chem., Soc.,* 77, 1621, 1955.

213. Kurth, E. F., Producing pure dihydroquercetin, U.S. Patent 2,744,919, 1956.

214. Kurth, E. F., Producing quercetin from barks of trees, Canadian Patent 532,806, 1956.

215. Kurth, E. F., Producing pure dihydroquercetin, U.S. Reissue Patent 25,135, 1962.

216. Gregory, A. S., Brink, D. L., Dowd, L. F., and Ryan, A. S., Douglas-fir bark as a source of quercetin, *For. Prod. J.,* 7, 135, 1957.

217. Roberts, J. R. and Gregory, A. S., U.S. Patent 2,832,765, 1958.

218. Gregory, A. S., Extraction Method, U.S. Patent 2,890,225, 1958.

219. Esterer, A. K. and Dowd, L. E., Processing of aqueous extracts of bark, U.S. Patent 3,131,198, 1964.

220. Brink, D. L., Method for fractionating aqueous extracts from barks of trees. U.S. Patent 3,189,596, 1965.

221. Esterer, A. K., Isolation of chroman derivatives from aqueous solutions, U.S. Patent 3,197,460, 1965.

222. Hergert, H. L., The flavonoids of lodgepole pine bark, *J. Org. Chem.,* 21, 534, 1956.

223. Hergert, H. L., Process for recovering flavonoids from bark, U.S. Patent 2,870,165, 1959.

224. Hergert, H. L. and Kurth, E. F., The isolation and properties of catechol from white fir bark, *J. Org. Chem.,* 18, 521, 1953.

225. Kurth, E. F., Ramanathan, V., and Venkataraman, K., The coloring matters of ponderosa pine bark, *J. Sci. Ind. Res. (India),* 15B, 139, 1956.

226. Tindale, M. D. and Roux, D. G., A phytochemical survey of the Australian species of *Acacia, Phytochemistry,* 8, 1713, 1969.

227. Clark-Lewis, J. W. and Porter, L. J., Phytochemical survey of the heartwood flavonoids of *Acacia* species from arid zones of Australia, *Aust. J. Chem.,* 25, 1943, 1972.

227a. Tindale, M. D. and Roux, D. G., An extended phytochemical survey of Australian species of *Acacia:* chemotaxonomic and phylogenetic aspects, *Phytochemistry,* 13, 829, 1974.

228. Drewes, S. E. and Roux, D. G., Condensed tannins. XV. Interrelationships of flavonoid components in wattle-bark extract, *Biochem. J.,* 87, 167, 1963.

229. Doskotch, R. W., Chatterjii, S. K., and Peacock, J. W., Elm bark derived feeding stimulants for the smaller European elm bark beetle, *Science,* 167, 380, 1970.

230. Doskotch, R. W., Mikhail, A. A., and Chatterjii, S. K., Structure of the water soluble feeding stimulant for *Scolytus multistriatus:* a revision, *Phytochemistry,* 12, 1153, 1973.

231. Hufford, C. D. and Lasswell, W. L., Uvaretin and isouvaretin, two novel cytotoxic C-benzylflavanones from *Uvaria chamae, J. Org. Chem.,* 41, 1297, 1976.

232. Hufford, C. D. and Lasswell, W. L., ^{13}C NMR studies of C-benzylated flavanones, *Lloydia,* 41, 152, 1978.

233. Kupchan, S. M., Siegel, C. W., Knox, J. R., and Udayamuthy, M.S., Tumor inhibitors. XXXVI. Eupatin and eupatoretin, two cytotoxic flavonols from *Eupatorium semiserratum, J. Org. Chem.,* 34, 1460, 1969.

234. Kupchan, S. M. and Bauerschmidt, E., Cytotoxic flavonols from *Baccharis sarothroides, Phytochemistry,* 10, 664, 1971.

235. Wattenberg, L. W. and Leong, J. C., Inhibition of the carcinogenic action of benzolalpyrene by flavones, *Cancer Res.,* 30, 1922, 1970.

236. Desphande, V. H., RamaRao, A. V., Venkataraman, K., and Wakhorkar, P. V., Wood phenolics of *Morus* species. III. Phenolic constituents of *Morus rubra* bark, *Indian J. Chem.*, 12, 431, 1974.

237. Harborne, J. B., Mabry, T. J., and Mabry, H., Eds., *The Flavonoids*, Academic Press, New York, 1975.

238. Pearl, I. A., Water extractives of American *Populus* pulpwood species bark: a review, *Tappi, 52*, 428, 1969.

239. Pearl, I. A. and Darling, S. F., The structures of salicortin and tremulacin, *Phytochemistry*, 10, 3161, 1971.

240. Pearl, I. A. and Darling, S. F., Further studies on the isolation of glucosides from the barks and leaves of *Populus tremuloides* and *Populus grandidentata, Tappi,* 47, 377, 1964.

241. Pearl, I. A. and Darling, S. F., Studies on the hot water extractives of the bark and leaves of *Populus deltoides, Can. J. Chem.,* 49, 49, 1971.

242. Ericksen, R. L., Pearl, I. A., and Darling, S. F., Further investigations on the hot water extractives of *Populus grandidentata* Michx. bark, *Tappi,* 53, 240, 1970.

243. Pearl, I. A. and Darling, S. F., Hot water phenolic extractives of the bark and leaves of diploid *Populus tremuloides, Phytochemistry,* 10, 483, 1971.

244. Pearl, I. A. and Darling, S. F., Hot water extractives of the leaves of *Populus heterophylla, J. Agric. Food Chem.,* 25, 730, 1977.

245. Thieme, H. and Richter, R., Isolation of a new phenol glycoside from *Populus tremula, Pharmazie,* 21, 251, 1966.

246. Pearl, I. A. Darling, S. F., DeHass, H., Loving, B. A., Scott, D. A., Turley, R. H., and Werth, R. W., Preliminary evaluation for glycosides of barks of several species of the genus *Populus, Tappi,* 44, 475, 1961.

247. Pearl, I. A. and Pottenger, C. R., Studies on the hot water extractives of the green bark of *Populus balsamifera, Tappi,* 49, 152, 1966.

248. Pearl, I. A., Justman, O., Beyer, D. L., and Whitney, D., Further studies on the hot water extractives of *Populus grandidentata* bark, *Tappi,* 45, 663, 1962.

249. Pearl, I. A., Darling, S. F., and Heller, S. F., Glucosides from the bark and leaves of triploid varieties of Populus species, *Tappi,* 49, 278, 1966.

250. Faber, H. B., The methanol extractable aromatic materials in the inner bark of *Populus tremuloides, Tappi,* 43, 406, 1960.

251. Pearl, I. A. and Darling, S. F., Studies on the barks of the family *Salicaceae.* XVIII. Studies on the hot water extractives of the brown bark of *Populus tricocarpa, Tappi,* 51, 537, 1968.

252. Estes, T. K. and Pearl, I. A., Studies on the bark of the family *Salicaceae.* XIII. Hot water extractives of the green bark of *Populus tricocarpa, Tappi,* 50, 318, 1967.

253. Pearl, I. A. and Darling, S. F., Tremuloidin, a new glucoside from the bark of *Populus tremuloides, J. Org. Chem.,* 24, 771, 1959.

254. Pearl, I. A. and Darling, S. F., Benzoates and salicyloylsalicin from the bark and leaves of *Populus grandidentata* and *Populus tremuloides, Tappi,* 48, 506, 1965.

255. Pearl, I. A. and Estes, T. K., Studies on the hot water extractives of the brown bark of *Populus tremuloides, Tappi,* 48, 532, 1965.

256. Pearl, I. A. and Darling, S. F., Studies on the bark of the family *Salicaceae.* XIV. Further studies on the barks of the triploid *Populus tremuloides, Tappi,* 50, 324, 1967.

257. Pearl, I. A. and Darling, S. F., Studies on the bark of the family *Salicaceae* XVI. The structure of salireposide, *Phytochemistry,* 7, 821, 1968.

258. Pearl, I. A. and Darling, S. F., Studies on the barks of the family *Salicaceae.* II. Salireposide from the bark of *Populus tremuloides, J. Org. Chem.,* 24, 1616, 1959.

259. Loeschcke, V. and Franksen, H., Tricocarpin, a new phenol glycoside from poplar bark significant as a resistance factor, *Naturwissenschaften,* 51, 40, 1964.

260. Pearl, I. A. and Darling, S. F., Variations in the hot water extractives of *Populus balsamifera* bark, *Phytochemistry,* 7, 1855, 1968.

261. Pearl, I. A. and Darling, S. F., Studies on the barks of the family *Salicaceae* (17) Tricoside, a new glucoside from the bark of *Populus tricocarpa, Phytochemistry,* 7, 825, 1968.

262. Pearl, I. A. and Darling, S. F., Investigations of the hot water extractives of *Populus balsamifera* bark, *Phytochemistry,* 8, 2393, 1969.

263. Ericksen, R. L., Pearl, I. A., and Darling, S. F., Populoside and grandidentatoside from the bark of *Populus grandidentata, Phytochemistry,* 9, 857, 1970.

264. Thieme, H. and Benecke, R., Isolierung eines neuen Phenolglycosides aus *Populus nigra* L., *Pharmazie,* 22, 59, 1967.

265. Jurd, L., The hydrolysable tannins, in *Wood Extractives,* Hillis, W.E., Ed., Academic Press, New York, 1962, chap. 6.

266. Hart, J. H. and Hillis, W. E., Inhibition of wood-rotting fungi by ellagitannins in the heartwood of *Quercus alba, Phytopathology,* 62, 620, 1972.

267. Mayer, W., Kunz, W., and Loebich, F., Die Struktur des Hamamelitannins, *Liebigs Ann. Chem.*, 638, 232, 1965.
268. Seikel, M. K., Hostettler, F. D., and Niemann, G. J., Phenolics of *Quercus rubra* wood, *Phytochemistry*, 10, 2249, 1971.
269. Mayer, W., Laver, K., Gabler, W., Panther, U., and Riester, A., Castalagin und Vesculagin, zwei neue, Kristallisierte Ellagengerbstoffe aus dem Holz der Edelkastanie und Eiche, *Naturwissenschaften*, 46, 669, 1959.
270. Mayer, W., Gabler, W., Riester, A., and Korger, H., Die Isolierung von Castalagen Vescalagin, Castalin und Vescalin, *Liebigs Ann. Chem.*, 707, 177, 1967.
271. Mayer, W., Einwiller, A., and Jochims, J. C., Die Struktur des Castalins, *Liebigs Ann. Chem.*, 707, 182, 1967.
272. Mayer, W., Sietz, H., and Jochims, J. C., Die Struktur des Castalagins, *Liebigs Ann. Chem.*, 721, 186, 1969.
273. Mayer, W., Kuhlmann, F., and Schilling, G., Die Struktur des Vescalins, *Liebigs Ann. Chem.*, 747, 51, 1971.
274. Mayer, W., Seitz, H., Jochims, J.C., Schauerte, K., and Schilling, G., Struktur des Vescalagins, *Liebigs Ann. Chem.*, 751, 60, 1971.
275. Schmidt, O. Th., Wurtele, L., and Harreus, A., Pedunculagin, ein 2,3,4,6-di-[(-)-Hexahydroxy-Diphenoyl]-Glucose aus Knoppern, *Liebigs Ann. Chem.*, 690, 150, 1965.
276. Mayer, W., Bilzer, W., and Schauerte, K., Isolierung von Castalagin und Vescalagin aus Valoneagabstoff, *Liebigs Ann. Chem.*, 754, 149, 1971.
277. Mayer, W., Gunther, A., Busath, H., Bilzer, W., and Schilling, G., Valolaginic Acid, *Liebigs Ann. Chem.*, 982, 1976.
278. Mayer, W., Schick, H., and Schilling, G., Isovalolaginic Acid, *Liebigs Ann. Chem.*, 2169, 1976.
279. Mayer, W., Bilzer, W., and Schilling, G., Catavaloninic Acid, Isolierung und Strukturermittlung, *Liebigs Ann. Chem.*, 876, 1976.
280. Yazaki, Y. and Hillis, W. E., Polyphenols of *Eucalyptus globulus, E. regnans,* and *E. Deglupta, Phytochemistry*, 15, 1180, 1976.
281. Hillis, W. E. and Yazaki, Y., Properties of some methylellagic acids and their glycosides, *Phytochemistry*, 12, 2963, 1973.
282. Hillis, W. E. and Carle, A., The formation of phenolic substances in *Eucalyptus gigantea* and *Eucalyptus sieberiana, Biochem. J.*, 74, 607, 1960.
283. King, H. G. C. and White, T., Tannins and polyphenols of *Schinopsis* (Quebraco) spp. Their genesis and interrelationships, *J. Soc. Leather Trades Chem.*, 41, 368, 1957.
284. Seikel, M. K. and Hillis, W. E., Hydrolysable tannins of *Eucalyptus delegetensis* wood, *Phytochemistry*, 9, 1115, 1970.
285. Hillis, W. E., The contribution of polyphenolic wood extractives to pulp color, *Appita*, 23, 89, 1969.
286. Hillis, W. E. and Yazaki, Y., Wood polyphenols of *Eucalyptus polyanthemos, Phytochemistry*, 12, 2969, 1973.
287. Jurd, L., Plant polyphenols. III. The isolation of a new ellagitannin from the pellicle of the walnut, *J. Am. Chem. Soc.*, 80, 2249, 1958.
288. Cunningham, J., Haslam, E., and Haworth, R. D., The constitution of piceatannol, *J. Chem. Soc.*, 2875, 1963.
289. Andrews, D. H., Hoffman, J. C., Purves, C. B., Quon, H. H., and Swan, E. P., Isolation, structure, and synthesis of a stilbene glucoside from the bark of *Picea glauca* (Moench.) Voss., *Can. J. Chem.*, 46, 2525, 1968.
290. Manners, G. D. and Swan, E. P., Stilbenes in the barks of five Canadian *Picea* species, *Phytochemistry*, 10, 607, 1971.
291. Gromova, A. S., Lutskii, V. I., and Tyukavkina, N. A., Hydroxystilbenes of the phloem of *Picea koraensis, Khim. Prirod. Soed.*, 6, 778, 1974.
292. Gromova, A. S., Lutskii, V. I., and Tyukavkina, N. A., Hydroxystilbenes of the inner and outer bark of *Picea ajanensis, Khim. Prir. Soedin.*, 11, 82, 1975.
293. Aritomi, M. and Donnelly, D. M. X., Stilbene glucosides in the bark of *Picea sitchensis, Phytochemistry*, 15, 2006, 1976.
294. Pearson, T. W., Kritz, G. S., and Taylor, R. J., Absolute identification of hydroxystilbenes. Chemical markers in *Engelmann spruce, Wood Sci.*, 10, 93, 1977.
295. Sano, Y. and Sakaibara, A., Studies on chemical components of *Akaezo matsu, (Picea glehnii)* bark. II. Isolation of stilbenes, *Res. Bull. Coll. Agr. Hokkaido Univ.*, 34, 287, 1977.
296. Markham, K. R. and Porter, L. J., Extractives of *Pinus radiata* bark. I. Phenolic components, *N.Z. J. Sci.*, 16, 751, 1973.
297. Hemingway, R. W. and McGraw, G. W., unpublished results
298. Hemingway, R. W. and McGraw, G. W., Polyphenols in *Ceratocystis minor* infected *Pinus taeda*: fungal metabolites, phloem, and xylem phenols, *J. Agric. Food Chem.*, 25, 717, 1977.

299. Yazaki, Y. and Hillis, W. E., Polyphenolic extractives of *Pinus radiata* bark, *Holzforschung*, 31, 20, 1977.

300. Hathway, D. E., The use of hydroxystilbenes as taxonomic tracers in the genus *Eucalyptus, Biochem. J.*, 83, 80, 1962.

301. Hart, J. H. and Hilliš, W. E., Inhibition of wood-rotting fungi by stilbenes and other polyphenols in *Eucalyptus sideroxylon, Phytopathology*, 64, 939, 1974.

302. Karchesy, J. J., Laver, M. L., Barofsky, D. F., and Barofsky, E., Structure of Oregonin, a natural diarylheptanoid-xyloside, *J. Chem Soc. Chem. Commun.*, 649, 1974.

303. Venkataraman, K., Flavones, in *The Flavonoids*, Harborne, J. B., Mabry, T. J., and Mabry, H., Eds., Academic Press, New York, 1975, chap. 6.

304. Farkas, L., Gabor, M., and Kallay, F., *Flavonoids and Bioflavonoids, Current Research Trends*, Elsevier Scientific. New York, 1977.

305. Anon., OSHA issues tentative carcinogen list, *Chem. Eng. News*, 56, 20, 1978.

306. Arcos, J. C., An overview. II. Cancer: chemical factors in the environment, *Am. Lab.*, 10, 29, 1978.

307. Morton, J. F., Further associations of plant tannins and human cancer, *Q. J. Crude Drugs Res.*, 12, 1829, 1972.

308. Bhargava, U. C., Pharmacology of Ellagic Acid From Black Walnut, Ph.D. thesis, University of Mississippi, University, 1967.

309. Bhargava, U. C. and Westfall, B. A., Antitumor activity of *Juglans nigra*, (black walnut) extractives, *J. Pharm. Sci.*, 57, 1674, 1968.

310. Turner, C. H., Aust. Pulp and Paper Manufact., Bernie, Tasmania, personal communication, 1970.

311. Hemingway, R. W. and Hillis, W. E., Behavior of ellagitannins, gallic acid, and ellagic acid under alkaline conditions, *Tappi*, 54, 933, 1971.

312. Hemingway, R. W. and Hillis, W. E., unpublished results, 1971.

313. Weinges, K., Bahr, W., Ebert, W., Goritz, K., and Marx, H. -D., Konstitution, Entstehung, und Bedeutung der Flavonoid Gerbstoffe, in *Fortschritte der Chemie Organisher Naturstoffe*, Vol. 27, Zeichmeister, L., Ed., Springer-Verlag, Basel, 1969, 159.

314. Roux, D. G., Recent advances in the chemistry and chemical utilization of the natural condensed tannins, *Phytochemistry*, 11, 1219, 1972.

315. Haslam, E., Natural procyanidins, in *The Flavanoids*, Harborne, J. B., Mabry, T. J., and Mabry, H., Eds., Academic Press, New York, 1975, chap. 10.

316. King, F. E. and Bottomly, W., The chemistry of extractives from hardwoods. XVIII. The occurrence of a flavan-3,4-diol (melacacidin) in *Acacia melanoxylon, J. Chem. Soc.*, 1399, 1954.

317. Clark-Lewis, J. W. and Mortimer, P. I., Flavan derivatives. II. Melacacidin and isomelacacidin from *Acacia* species, *J. Chem. Soc.*, 4106, 1960.

318. Clark-Lewis, J. W., Katekov, G. F., and Mortimer, P. I., Flavan derivatives. IV. Teracacidin, a new leucoanthocyanidin from *Acacia intertexta, J. Chem. Soc.*, 499, 1961.

319. Clark-Lewis, J. W. and Dainis, I., Flavan derivatives. XI. Teracacidin, melacacidin, and 7,8,4'-tri-hydroxy-flavanol from *Acacia sparsiflora* and extractives from *Acacia orites, Aust. J. Chem.*, 17, 1170, 1964.

320. Clark-Lewis, J. W. and Dainis, I., Flavan derivatives. XIX. Teracacidin and isoteracacidin from *Acacia obtusifolia* and *Acacia maidernii* heartwoods: phenolic hydroxylation patterns of heartwood flavonoids characteristic of sections and subsections of the genus *Acacia, Aust. J. Chem.*, 20, 2191, 1967.

321. Clark-Lewis, J. W. and Porter, L. J., Phytochemical survey of the heartwood flavonoids of *Acacia* species from arid zones of Australia, *Aust. J. Chem.*, 25, 1943, 1972.

322. Drewes, S. E. and Roux, D. G., Stereochemistry of flavan-3,4-diol tannin precursors: mollisacacidin, (−) leucotisetinidin, and (+) leucorobinetinidin, *Biochem. J.*, 90, 343, 1964.

323. Saaymon, H. M. and Roux, D. G., Configuration of quibourtacacidin and synthesis of isomeric racemates, *Biochem. J.*, 96, 36, 1965.

324. Drewes, S. E. and Roux, D. G., A new flavan-3,4-diol from *Acacia* auriculiformis by paper ionopho-resis, *Biochem. J.*, 98, 493, 1966.

325. DuPreez, I. C. and Roux, D. G., Novel flavan-3,4-diols from *Acacia cultriformis, J. Chem. Soc. C*, 1800, 1970.

326. Seshadri, T. R. and Venkaturamani, B., Leucocyanidin from mangroves, *J. Sci. Ind. Res. (India)*, 18B, 261, 1959.

327. Ganguly, A. K. and Seshadri, T. R., A study of leucoanthocyanidins of plants. II., *Tetrahedron*, 6, 21, 1959.

328. Nagarajan, G. R. and Seshadri, T. R., Leucocyanidins of arecunut and toon *(Cedrella toona)* wood, *J. Sci. Ind. Res. (India)*, 20B, 178, 1961.

329. Nagarajan, G. R. and Seshadri, T. R., Constitution of the leucocyanidin of the groundnut, *Curr. Sci. India*, 29, 178, 1960.

330. Chadha, J. S. and Seshadri, T. R., Leucocyanidins from the seeds of *Litchichinensis, Curr. Sci. India,* 31, 56, 1962.
331. Seshadri, T. R. and Vasishta, K., Polyphenols of *Psidium guaijava* plant, *Curr. Sci. India,* 32, 499, 1963.
332. Manson, D. W., The leucoanthocyanidin from black spruce inner bark, *Tappi,* 43, 59, 1960.
333. Baig, M. I., Clark-Lewis, J. W., and Thompson, M. S., Flavan derivatives. XXVII. Synthesis of a new racemate of leucocyanidin tetramethyl ether (2:3 *cis* - 3:4 *trans*) isomer: NMR spectra of the four racemates of leucocyanidin tetramethyl ether (5,7,3′,4′ tetramethoxy flavan-3,4-diol), *Aust. J. Chem.,* 22, 2645, 1969.
334. Bokadia, M. M., Brown, B. R., and Cummings, W., Polymerization of flavans. III. The action of lead tetra acetate on flavans, *J. Chem. Soc.,* 3308, 1960.
335. Betts, M. J., Brown, B. R., and Shaw, N. R., Reaction of flavonoids with mercaptoacetic acid, *J. Chem. Soc.,* 1178, 1969.
336. Freudenberg, K. and Weinges, K., Leuko- und Pseudoverbindungen der Anthocyanidine, Liebigs Ann. Chem., 613, 61, 1958.
337. Jacques, D., Opie, C. T., Porter, L. J., and Haslam, E., Plant proanthocyanidins. IV. Biosynthesis of procyanidins and observations on the metabolism of cyanidin in plants, *J. Chem. Soc. (Perkin Trans. 1,* 1637, 1977.
338. Roux, D. G. and Ferreira, D., α-Hydroxychalcones as intermediates in flavonoid biogenesis: the significance of recent chemical analogies, *Phytochemistry,* 13, 2039, 1974.
339. Haslam, E., Opie, C. T., and Porter, L. J., Procyanidin metabolism - a hypothesis, *Phytochemistry,* 16, 99, 1977.
340. Thompson, R. S., Jacques, D., Haslam, E., and Tanner, R. J. N., Plant proanthocyanidins. I. Introductions; the isolation, structure and distribution in nature of plant procyanidins, *J. Chem. Soc. Perkin Trans. 1,* 1387, 1972.
341. Hemingway, R. W. and McGraw, G. W., Progress in the chemistry of short leaf and loblolly pine bark flavonoids, *Appl. Polym. Symp.,* 28, 1349, 1976.
342. Hemingway, R. W. and McGraw, G. W., Southern pine bark polyflavonoids: Structure, reactivity, and use in wood adhesives, *Proc. Tappi For. Biol./Wood Chem. Symp.,* Atlanta, 1977.
343. Karchesy, J. J. and Hemingway, R. W., Loblolly pine bark polyflavonoids, Paper presented to Symp. Extractives Utilization Prob. Fine Chem. Res., Joint American Chemical Society/Chemical Society of Japan Chemical Congress, Washington, D.C., 1979.
344. Wong, E. and Birch, E. J., Biosynthesis of proanthocyanidins, *J. Chem. Soc. Chem. Commun.,* 979, 1975.
345. Drewes, S. E., Roux, D. G., Eggers, S. H., and Feeney, J., Three diastereo-isomeric-4,6-linked bileucofisetinidins from the heartwood of *Acacia mearnsii, J. Chem. Soc. C,* 1217, 1967.
346. Narayanan, V. and Seshadri, T. R., Chemical components of *Acer rubra* wood and bark: occurrence of procyanidin dimer and trimer, *Indian J. Chem.,* 7, 213, 1969.
347. Drewes, S. E., Roux, D. G., Saayman, H. M., Eggers, S. H., and Feeney, J., Some stereochemically identical biflavanols from the bark tannins of *Acacia mearnsii, J. Chem. Soc. C,* 1302, 1967.
348. Pelter, A., Amenechi, P. I., Warren, R., and Harper, S. H., The structure of two proanthocyanidins from *Julbernadia globiflora, J. Chem. Soc. C,* 2572, 1969.
349. DuPreez, I. C., Rowan, A. C., and Roux, D. G., A biflavonoid proanthocyanidin carboxylic acid and related biflavonoids from *Acacia luederitzii, J. Chem. Soc. Chem. Commun.,* 492, 1970.
350. Malan, E. and Roux, D. G., Flavonoids and tannins of the *Acacia* species, *Phytochemistry,* 14, 1835, 1975.
351. DuPreez, I. C., Rowan, A. C., and Roux, D. G., Hindered rotation about the Sp^2-Sp^3 hybridized C-C bond between flavonoid units in condensed tannins, *J. Chem. Soc. Chem. Commun.,* 315, 1971.
352. Ferreira, D., Hundt, H. K. L., and Roux, D. G., The stereochemistry of a tetraflavanoid condensed tannin from *Rhus lancea* L. f., *J. Chem. Soc. Chem. Commun.,* 1257, 1971.
353. Weinges, K. and Freudenberg, K., Condensed proanthocyanidins from cranberries and cola nuts, *J. Chem. Soc. Chem. Commun.,* 220, 1965.
354. Weinges, K., Kaltenhauser, W., Marx, H. -D., Nader, E., Nader, F., Perner, J., and Seiler, D., Procyanidine aus Fructen, *Liebigs Ann. Chem.,* 711, 184, 1968.
355. Weinges, K., Goritz, K., and Nader, F., Konfigurations Bestimmung von $C_{30}H_{26}O_{12}$ Procyanidinen und Strukturaufklarung eines neuen Procyanidin, *Liebigs Ann. Chem.,* 715, 164, 1968.
356. Weinges, K. and Perner, J., The structure of a $C_{30}H_{26}O_{12}$ procyanidin from cola nuts, *J. Chem. Soc. Chem. Commun.,* 351, 1967.
357. Weinges, K., Perner, J., and Marx, H. -D., Synthese des Octamethyl-Diacctyl-Procyanidins B-3, *Chem. Ber.* 103, 2344, 1970.
358. Fletcher, A. C., Porter, L. J., Haslam, E., and Gupta, R. K., Plant proanthocyanidins. III. Conformational and configurational studies of natural procyanidins, *J. Chem. Soc. Perkin Trans. 1,* 1628, 1977.

359. Mayer, W., Goll, L., VonArndt, E. M., and Mannschreck, A., Procyanidin A-2 (-)-Epicatechin ein zweizrmig Verknupftes, Kondensiertes Proanthocyanin aus *Aseculus hippocastanum, Tetrahedron Lett.*, 429, 1966.

360. Schilling, G., Weinges, K., Muller, O., and Mayer, W., ^{13}C-NMR-Spektroskopische Konstitutionser-mittlung der $C_{30}H_{24}O_{12}$ Procyanidine, *Liebigs Ann. Chem.*, 1471, 1973.

361. Jacques, D., Haslam, E., Bedford, G. R., and Greatbanks, D., Plant proanthocyanidins. II. Proan-thocyanidin-A2 and its derivatives, *J. Chem. Soc. Perkin Trans. 1, 2663*, 1974.

362. Weinges, K., ^{13}C-NMR spectroscopy as an expedient in the elucidation of the constitution of flavon-oids, in *Topics in Flavonoid Chemistry and Biochemistry*, Farkas, L., Gabor, N., and Kelley, F., Eds., Elsevier Scientific, New York, 1975, 64.

363. Geissman, T. A. and Dittman, H. J. K., A proanthocyanidin from avocado seed, *Phytochemistry*, 4, 359, 1965.

364. Porter, L. J., Extractives of *Pinus radiata* bark. II. Procyanidin constituents, *N. Z. J. Sci.*, 17, 213, 1974.

365. Hemingway, R. W., unpublished results, 1975.

366. Hergert, H. L., The chemistry and utilization of non-lignin phenolic polymers from conifers, Paper presented to 161st National American Chemical Society Meeting, Washington, D.C., 1971.

367. Betts, M. J., Brown, B. R., Brown, P. E., and Pike, W. T., Degradation of condensed tannins: structure of the tannin from common heather, *J. Chem. Soc. Chem. Commun.*, 1110, 1967.

368. Sears, K. D. and Casebier, R. L., Cleavage of proanthocyanidins with thioglycollic acid, *J. Chem. Soc. Chem. Commun.*, 1437, 1968.

369. Betts, M. J., Brown, B. R., and Shaw, M. R., Reaction of flavanoids with mercaptoacetic acid, *J. Chem. Soc. C*, 1179, 1969.

370. Brown, B. R. and Shaw, M. R., Reactions of flavanoids and condensed tannins with sulphur nucleo-philes, *J. Chem. Soc. Perkin Trans. 1, 2036*, 1974.

371. Sears, K. D. and Casebier, R. L., The reaction of thioglycolic acid with polyflavanoid bark fractions of *Tsuga heterophylla, Phytochemistry*, 9, 1589, 1970.

372. Karchesy, J. J., Loveland, P. M., Laver, M. L., Barofsky, D. F., and Barofsky, E., Condensed tannins from the barks of *Alnus rubra* and *Pseudotsuga menziesii, Phytochemistry*, 15, 2009, 1976.

373. Roux, D. G., Ferreira, D., Hundt, H. K. L., and Malan, E., Structure, stereochemistry, and reactiv-ity of natural condensed tannins as basis for their extended industrial application, *Appl. Polym. Symp.*, 28, 335, 1975.

374. Plomley, K. F., Tannin-formaldehyde adhesives for wood, Commonwealth Scientific and Industrial Research Organization Div. For. Prod. Technol. paper, 39, Melbourne, 1966.

375. Saayman, H. M. and Oatley, J. A., Wood adhesives from wattle bark extract, *For. Prod. J.*, 26, 27, 1976.

376. Pizzi, A., Hot-setting tannin-urea-formaldehyde exterior wood adhesive, *Adhes. Age*, 12, 27, 1977.

377. Van Der Westhuizen, P. K., Pizzi, A., and Scharfetter, H., A fast setting wattle adhesive for finger-jointing, Internation Union of Forest Research Organization Conf. Paper, Merida, Venezuela, 1977.

378. Kriebich, R. E., High speed adhesives for the wood-gluing industry, *Adhes. Age*, 17, 26, 1974.

379. McKenzie, A. W. and Yurritta, J. P., Starch tannin corrugating adhesives, *Appita*, 26, 30, 1972.

380. Saayman, H. M. and Brown, C. H., Wattle-base tannin-starch adhesives for corrugated containers, *For. Prod. J.*, 27, 21, 1977.

381. Saayman, H. M., Wood varnishes from wattle bark extract, *J. Oil Colour Chem. Assoc.*, 57, 114, 1974.

382. Manuel, W. A. and Lewis, H. A., Redwood tannin: use as a depressant for calcite and quartz in the flotation of fluorspar., *Ind. Eng. Chem.*, 36, 1169, 1944.

383. MacLean, H. and Gardner, J. A. F., Bark extracts in adhesives, *Pulp Pap. Mag. Can.*, 53, 111, 1952.

384. Hillis, W. E. and Urbach, G., The reaction of (+) catechin with formaldehyde, *J. Appl. Chem.*, 9, 474, 1959.

385. Hillis, W. E. and Urbach, G., Reaction of polyphenols with formaldehyde *J. Appl. Chem.*, 9, 665, 1959.

386. Hemingway, R. W. and McGraw, G. W., Formaldehyde condensation products of model phenols for conifer bark tannins, *J. Liq. Chrom.*, 1, 163, 1978.

387. Herrick, F. W. and Bock, L. H., Thermosetting exterior plywood adhesives from bark extracts, *For. Prod. J.*, 8, 269, 1958.

388. Brandts, T. G. and Lichtenberger, J. A., Resins from bark, Canadian Patent 708,936, 1965.

389. Hsa, T. -H., Dimethylolamide modified bark alkali product and method of making same, U.S. Patent 3,654,200, 1972.

390. Booth, H. E., Herzberg, W. J., and Humphreys, F. R., *Pinus radiata* (Don) bark tannin adhesives, *Aust. J. Sci.*, 21, 19, 1958.

391. Herzberg, W. J., *Pinus radiata* tannin formaldehyde resin as an adhesive for plywood, *Aust. J. Appl. Sci.*, 11, 462, 1960.
392. Steiner, P. R. and Chow, S., Factors influencing western hemlock bark extracts as adhesives, Report No. VP-X-153, Western Forest Products Laboratory, Vancouver, B.C., 1975.
393. Hartman, S., Alkali treated bark extended tannin aldehyde resinous adhesive, U.S. Patent 4,045,386, 1977.
394. Hall, R. B., Leonard, J.H., and Nicholls, G. A., Bonding particle-boards with bark extracts, *For. Prod. J.*, 10, 263, 1960.
395. Anderson, A. B., Wu, K. -T., and Wong, A., Utilization of ponderosa pine bark and its extract in particle-board, *For. Prod. J.*, 24, 48, 1974.
396. Anderson, A. B., Wong, A., and Wu, K. -T., Douglas-fir and western hemlock bark extracts as bonding agents for particle-boards, *For. Prod, J.*, 25, 45, 1975.
397. Anderson, A. B., Bark extracts as bonding agents of particle-boards, in, Wood Technology: Chemical Aspects, Goldstein, I.S., Ed., American Chemical Society, Washington, D.C., 1976, 235.
398. Herrick, F. W. and Conca, R. J., The use of bark extracts in cold setting water proof adhesives, *For. Prod. J.*, 10, 361, 1960.
399. Martin, W. N., Epoxy resin intermediates derived from bark, U.S. Patent 3,519,653, 1970.
400. Allan, G. G., Esters of bark phenolic acids, U.S. Patent 3,511,874, 1970.
401. Allan, G. G., Polyurethanes derived from esters of bark phenolic acids, Canadian Patent 839,847, 1970.
402. Miller, R. W. and VanBeckum, W. G., Bark and fiber products for oil well drilling, *For. Prod. J.*, 10, 193, 1960.
403. Hergert, H. L., Van Blaricom, L. E., Steinberg, J. C., and Gray, K. R., Isolation and properties of dispersants from western hemlock bark, *For. Prod. J.*, 15, 485, 1968.
404. Van Blaricom, L. E. and Gray, K. R., Drilling mud compositions, U.S. Patent 2,964,469, 1960.
405. Gray, K. R. and Van Blaricom, L. E., Bark treatment process and products, U.S. Patent 2,999,108, 1961.
406. Van Blaricom, L. E., and Tokos, G. M., Process for forming a chemical product from bark, U.S. Patent 2,975,126, 1961.
407. Herrick, F. W. and Hergert, H. L., Utilization of chemicals from wood, retrospect and prospect, in *Recent Advances in Phytochemistry*, Loewus, F. A. and Runeckles, V. C., Eds., Plenum Press, New York, 1977, chap. 11.
408. Van Blaricom, L. E. and Johnston, F. A., Iron complexed sulfonated polyflavonoids and their preparation, U.S. Patent 3,270,003, 1966.
409. Van Blaricom, L. E., Dewegert, H. R., and Smith, N. H., Grouting composition, U.S. Patent 3,490,933, 1970.
410. Pulkkinen, E. and Peltonen, S., Cationic flocculant from a phenolic acid fraction of conifer tree bark, *Tappi*, 61, 97, 1978.
411. Allan, G. G., Reaction products of lignin and bark extracts and process for same, U.S. Patent 3,470,148, 1969.
412. Mater, J., Marketing bark, agricultural, and horticultural products, Forest Products Research Society, Madison, Wis., 1971.
413. Allison, R. C. and Jordon, H. C., Use of hardwood bark as a poultry litter, in *Techniques and Processing of Bark and Utilization of Bark Products*, Mater, J., Ed., Forest Products Research Society Madison, Wis., 1973, 21.
414. Hartman, S., Polyurethane foams from the reaction of bark and diisocyanate in *Wood Technology: Chemical Aspects*, Goldstein, I.S., Ed., American Chemical Society, Washington, D.C., 1977, chap. 11.
415. Weldon, D., Processing yellow pine bark for use as an oil pollution scavenger, in *Techniques and Processing Bark and Utilization of Bark Products*, Mater, J., Ed., Forest Products Research Society, Madison, Wis., 1973, 31.
416. Randall, J. M., Variations in effectiveness of barks as scavengers for heavy metal ions, *For. Prod. J.*, 27, 51, 1977.
417. Martin, R. E. and Crawford, J. L., Sorption of sulfate mill odors by bark, in *Techniques and Processing Bark and Utilization of Bark Products*, Mater, J., Ed., Forest Products Research Society, Madison, Wis., 1973, 42.
418. Allan, G. G., Beer, J. W., Cousin, M. J., and Powell, J. C., Wood chemistry and forest biology, partners in more effective reforestation, *Tappi*, 61, 33, 1978.

Chapter 11

FOLIAGE

George M. Barton

TABLE OF CONTENTS

I. INTRODUCTION

Interest in foliage utilization has been accelerating in North America and elsewhere throughout the world for four main reasons. First and most important, there are fears of a global food shortage. In 1976, for example, the *Financial Post* pointed out that the worlds reserve food supply was only 35 days, in contrast to 89 days, 6 years earlier.[113] Although there may be short-term improvements during bumper harvest years, the fact remains that several poor growing years could cause a world-wide famine. Obviously then, if any part of the food chain normally fed to cattle, pigs, or poultry can be replaced by foliage, an equivalent amount of food or an equivalent amount of land to grow food would be available for human beings. There is reason to hope that foliage can be used for fodder, and this subject will be more fully developed later in this chapter. Second, there is considerable interest in complete tree harvesting techniques to improve the yield of lumber and pulp. This trend will mean that foliage will arrive with the main bole and branches to central sites, from which it may be processed, thus eliminating much of the labor costs of collecting it. Third, there has been heavy pressure in many areas to reduce air pollution from slash burning, and in some cases this has led to legislation and subsequent legal restrictions against this method of disposal. Thus, industry is faced and will be faced with what to do with vast amounts of foliage from their harvesting operations. Finally, foliage is being recognized as a source of unique chemicals that has present and future application in medicine, pharmacology, perfumery, cosmetics, and adhesives.

A. Nomenclature

It is important to define early in this chapter what is meant by foliage, since the word means different things to different people. Also, in deference to the fact that most of the work in the practical development of foliage use in poultry and animal fodder was initiated, prompted, and applied in the U.S.S.R., some Russian terms will be used throughout this text.

Foliage — Foliage will be defined here as a general term to include not only pure needles or leaves, but also varying ratios of twigs and small branches. Foliage from coniferous, deciduous, or mixtures from either may be included.

Needles — This term refers to pure needles only, and is confined to coniferous trees.

Leaves — These refer to pure leaves, deciduous trees only.

Technical or commercial foliage — Technical or commercial foliage (a term preferred in the U.S.S.R.) are considered synonymous, and are defined as all needles, twigs, leaves, shoots, and branches up to 0.6 cm (0.24 in.) in diameter. This diameter was arbitrarily chosen as representing optimum recovery in the conversion of foliage to fodder or chemicals. As increasing diameter of branches occur, wood and bark extractives become significant contaminants.

Muka — The Russian term muka (meaning flour) will be understood to mean foliage that has been dried and ground for use as a fodder vitamin supplement for poultry and animals. For the sake of brevity, muka used alone will refer to coniferous muka. Deciduous muka derived from broad leaves, as aspen* muka, for example, will be so designated. For reasons of optimum recovery, muka is usually made from technical foliage.

B. Historical Background

Historically, and presumably prehistorically, mankind has used foliage or substances extracted from it for medicinal, nutritive, or religious purposes. Pima Indians of Ari-

---

* Botanical names are given at the end of this chapter in order of appearance in the text.

zona, for example, used a boiled extract of willow leaves to relieve fever.[1] This was long before chemists had isolated the active febrifuge, salicin from its leaves and bark. Many other medicinal uses of foliage among native people are on record,[1] but one of the most interesting concerns the treatment of scurvy by Quebec Indians during a colonizing voyage of the French explorer, Jacques Cartier.[2] "On his second voyage to Newfoundland in 1535 he quartered his men for the winter near an Indian village, Stavacoma, in Quebec. During the winter, both the Indians and Cartier's crew came down with scurvy; 25 of Cartier's men died. Toward the middle of March, Cartier happened to notice an Indian who had been deathly sick with scurvy about ten days before. Now he was well and sound. By what native magic had he accomplished this, Cartier wondered. The Indian replied that he had taken the juice and sap from the leaves of a tree which he called Ameda. (This is now thought to be the American spruce). As this was a life or death matter, Cartier wanted to see how the brew was made. He asked the Indian to bring the branches and show him how the bark and leaves should be boiled together. The resulting concoction was such a powerful healing force that Cartier's men speedily recovered. He wrote in his diary: 'It wrought so well, that if all the physicians of Montpelier and Louaine had been there, with all the drugs of Alexandria, they would not have done as much in one year as that tree did in six days, for it did so prevail, that as many as used it, by the Grace of God recovered their health'."

Although the cure for scurvy was "reinvented" many times before biochemists identified vitamin C as the common, active component, the fact remains that these early North Americans had cured and prevented scurvy because of their instinctive ingestion of crude extracts of vitamin C from foliage.

Although a few examples of foliage utilization by native people were accepted by the new immigrants, many were lost while waiting to be reinvented. However, one characteristic property of fresh coniferous foliage, namely its pleasant fragrance, continued to intrigue mankind. In view of the relative ease by which this fragrance can be removed from foliage by steaming, it is understandable that even today, most of the worlds supply of foliage oils (essential oils) are prepared in cottage type industries using simple, inexpensive equipment.

C. Present Status

For the reasons already stated in the opening paragraph, interest in foliage utilization is being expressed throughout the world wherever trees grow in abundance. While the more important centers of interest are identified in this chapter, new ones are being reported as this book goes to press.

U.S.S.R. — To date the most extensive effort toward the development of marketable products from foliage has been made in the U.S.S.R. Although many of these products have been used by the cosmetic and pharmaceutical industries, the most important is the use of muka in animal and poultry feeds. The first Russian muka plant, with a capacity of approximately 100 t/year, started production in Latvia in 1955. Since that time, a vast amount of work has been done in developing more effective harvesting and transport systems, improved equipment for muka processing, and optimum plant size. The present total production in the U.S.S.R. is approximately 100,000 t/year. It is expected that this will rise to 200,000 t/year by 1985, and in order to meet this production level, some 300 additional muka plants are planned. Much of this work has been done at the Latvian Scientific Institute of Forest Problems in Riga, and also in the Laboratory on Living Tree Elements at the Kirov Academy of Forest Technology in Leningrad.

Canada — In Canada, foliage studies are well advanced, both in the East at Domtar

Research and Development headquarters, Senneville, Quebec, and in the West at the Western Forest Products Laboratory, Vancouver, B.C. Work at Domtar has concentrated on the separation of coniferous and deciduous foliage from chipped branches. It would probably be correct to say that Domtar has the first semicommercial muka plant in Canada. The species of immediate interest to Domtar are eastern maples, black spruce, and balsam fir. Currently, Domtar is completing a contract proposal with the government of Canada in an attempt to answer the question, "Can animal feed supplements from Canadian tree foliage be marketed profitably?" Results of this contract should be available early in 1979, and will include the cost of building and operating a muka plant, evaluation of muka as a poultry and animal feed stuff, market potential analyses, and selling price. In the West, the Western Forest Products Laboratory has been actively pursuing a foliage utilization program on several fronts, including harvesting and separation techniques, biomass studies, chemical analyses, and the use of foliage as adhesive fillers and extenders. A special joint program with the Canada Department of Agriculture, Experimental Station, Agassiz, B.C. is evaluating western coniferous muka as a fodder for poultry and ruminants. The trees of immediate importance to the Western Forest Products Laboratory are lodgepole pine and white spruce. These trees (or closely related trees) were chosen because they are well represented commercially throughout Canada, and because whole-tree logging is more likely to be practiced first on these trees than on mountainous or coastal tree species.

U.S. — There are at least four areas of activity in the U.S. One of these was a natural consequence of long-time interest and considerable expertise in full-forest utilization at the Complete-Tree Institute, University of Maine at Orono. Muka has already been prepared from Maine tree species, and both poultry and sheep feeding trials have been completed. In the South, at the Southern Forest Experiment Station, Pineville, La. the utilization of southern pine foliage is being studied. Poultry feeding trials have been completed in collaboration with the University of Louisiana. Also, interest in foliage has come from U.S. cattlemen. The president of Arrowhead Products Company, St. Louis, Mo., a company engaged in raising cattle for export, is actively engaged in a program to prepare muka from local puckerbrush. More recently, Weyerhaeuser Company has been evaluating Loblolly pine muka for cattle fodder.

Sweden — A new research group on foliage studies has been set up at the Swedish Forest Products Laboratory, Stockholm.

Finland — Both the Finnish Forest Research Institute and the Enso-Gutzeit Pulp and Paper Company have expressed interest in proceeding with the separation of foliage from full-forest chips and the subsequent utilization of the foliage.

South Africa — The technical director of South African Pulp and Paper Industry Company (SAPPI), Transvaal, has recommended to the management of SAPPI, a program of foliage utilization.

The purpose, then, of this chapter will be to identify major foliage chemicals and discuss their properties and uses. Because of the importance of food a special section of foliage for fodder will be included. Answers to questions relating to foliage biomass inventories or harvesting and processing techniques have either been explored in this book or have been covered in other foliage reviews.[3-5,7-9]

II. FOLIAGE CHEMICALS: THEIR STRUCTURES, PROPERTIES, AND USES

Foliage, unlike other tree components such as wood and bark, contains many unique chemicals directly or indirectly associated with the photosynthetic life support system. Since nitrogen-containing chemicals, e.g., chlorophyll, are essential to photosynthesis,

Table 1
CHEMICAL ANALYSES OF SLASH
PINE TREE PARTS

	Lignin[a] (%)	α-Cellulose (%)	Hemicellulose (%)
Needles	37.7	42.6	22.3
Bark	50.0	23.7	24.9
Branches	35.0	36.9	33.7
Top	32.5	41.4	31.2
Roots	31.3	44.6	25.6
Stem	27.8	51.1	26.8

[a] Figures are averages of three trees and are expressed as percentages of extractive-free ovendry weight.

Adapted from Howard, E. T., Physical and chemical properties of slash pine tree parts, *Wood Sci.,* 5(4), 312, 1973.

it is not surprising to discover that nearly all the nitrogen content of trees (approximately 1%) is to be found in needles and leaves. By comparison, bark contains approximately 0.2%, whereas wood contains less than 0.1% nitrogen. Although structural strength of foliage comes from its cellulose and lignin content, some of the lignin values reported in the literature are undoubtedly due to artifacts produced by reactions of phenolic monomers during the analytical isolation procedure. In addition, since most analytical data is given for technical foliage, these lignin and cellulose figures will reflect the varying amounts of wood and bark present in the mix and will be different than those for pure needles or leaves.[10] (Table 1.)

The variety of isolated and identified chemicals from foliage is enormous, with new ones being reported daily. Many are present in trace amounts and have been identified largely due to the excellence of modern chromatographic techniques. Clearly, this chapter must confine itself to those that occur in reasonable yield and have a proved or potential utility. For clarity, the format of a previous review[8] will be used, and chemicals will be classified under six main headings: essential oils, nonpolar compounds, chlorophylls, carotenoids, protein, and other extractives.

A. Essential Oils

Historically, foliage oils or essential oils obtained by steam distilling technical conifer foliage were among the first foliage chemicals utilized. The term essential is derived from a traditional belief that the fragrance of a substance contained the very essence of it, i.e., the vitally important ingredient. These essential oils usually have a pleasant scent and some are highly prized as perfume ingredients.

The oil is simply prepared by passing steam through finely divided needles and twigs, collecting the distillate, and separating the oil from the water. The equipment is not complex and is comparatively inexpensive[11] However, because of the labor intensive nature of collection, production of essential oils in North America has been sporadic and confined to part-time operations, mainly during the labor surplus periods of winter and spring.

Essential oils consist mainly of terpenes. These are monoterpenes ($C_{10}H_{16}$), sesquiterpenes ($C_{15}H_{24}$), and diterpenes ($C_{20}H_{32}$). All are formed in nature from the precursor Δ^3-isopentenyl pyrophosphate.[12] The monoterpenes are the most abundant and are further subdivided into the groups acyclic, monocyclic, and bicyclic (Figure 1).

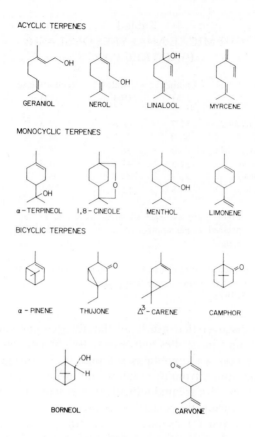

FIGURE 1.   Examples of acyclic, monocyclic, and bicyclic terpenes.

The yield of terpenes varies with species, season of collection, and interval between collection and processing. Yields of up to 2% (green leaf basis) are commercially obtainable from conifers. The three major constituents of the terpenes from essential oils are α-pinene, β-pinene, and Δ^3-carene, in descending concentrations. Exceptions occur as in the case of western red cedar oil that consists of 80 to 85% ℓ-thujone.[18]

One of the most abundant and ubiquitous of terpenes is α-pinene, which is not only the major component of conifer foliage, but also of wood turpentine. Both α-pinene and its isomer, β-pinene, are characterized by the ease that they can be converted into other terpene structures by simple chemical techniques. This can be very attractive commercially, since the market value of the converted derivative is often many times that of the original precursor. Examples of these transformations and the comparative values of the products are presented in Chapter 9.

1. Chemotaxonomy

It is hardly possible to leave the subject of essential oils without reference to their diagnostic value in taxonomy. Terpenes meet Harborne's four basic requirements necessary for chemosystematic studies, namely, sufficient chemical complexity and structural variety, physiological stability, widespread distribution, and quick and easy identification. The most recent, definitive work on the use of terpenes for chemotaxonomy has been done by von Rudloff. In a timely review,[14] he has compiled an analysis of volatile leaf oils of the families Pinaceae and Cupressaceae. An example of how this technique can be used is illustrated in his study of North American hemlocks.

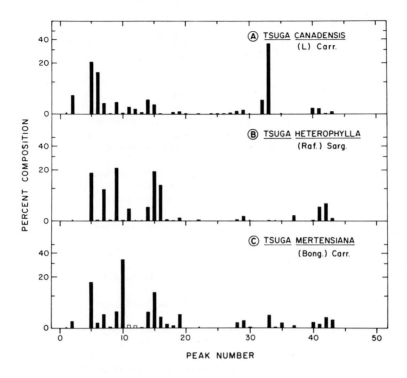

FIGURE 2. Bar histograms showing mean needle-oil terpene compositions of three species of hemlocks.

The genus *Tsuga* is represented in North America by the eastern, the western, and the mountain hemlock. The eastern hemlock has a wide range, being found from the Atlantic coast of Nova Scotia westward to southwestern Ontario and southward in the Appalachian Mountains to Georgia. No overlap with other *Tsuga* species exists and no varieties are recognized. Western hemlock, on the other hand, is found along the Pacific coast from Alaska to Oregon and northern California, as well as in the mountainous regions of the interior of British Columbia, Washington, Idaho, and Montana. At higher elevations above 1,000 m (3,300 ft), its range overlaps with the range of mountain hemlock. Natural hybrids of these species have been reported.

In Figure 2, von Rudloff clearly shows the similarities and differences in foliage composition among the above three hemlocks. Ignoring the trace components, it can be seen that all three hemlocks have in common α- and β-pinene (peaks #5 and #7), myrcene (#9), limonene (#14), β-phellandrene (#15), and α-terpineol (#29). Distinguishing features of eastern hemlock are tricyclene (#2), camphene (#6), and bornyl acetate (#33). Distinguishing features of western hemlock are myrcene (#9), *cis*-ocimene (#16), the virtual absence of camphene (#6), and no bornyl acetate (#33). Mountain hemlock has a leaf oil composition intermediate between eastern and western, except for the large amount of Δ³-carene (#10). In studies such as this, where the typical parent population is carefully chosen from protected geographical locations, the terpenoid data are taken as reference points, and various intermediate types can be defined according to their degree of quantitative intermediacy. In this way, the important study of hybrids can be greatly facilitated.

B. Nonpolar Compounds

Because of their historical and industrial importance, essential oils that are also

Table 2

NONPOLAR CONSTITUENTS FROM SCOTCH PINE
TECHNICAL FOLIAGE[15-17]

Constituent	Comments	Yield % dry wt basis
Triglycerides	One of the largest lipid groups in the petroleum ether extract Esterfied mainly with unsaturated fatty acids.	2.5
Acidic components Figure 6	40% Common resin acids of abietane, pimarane, and isopimatane 58% Diterpene acids of the labdane type	2.4
Steryl esters, sterols Figure 6	84% β-Sitosterol ester with some campesterol and stigmasterol Esterified with fatty acids varying in chain length from C_{12} to C_{24}. A small amount of free β-sitosterol was present also	1.1
Pinoprenyl acetates	Found only in the needles and as acetates	0.9
Diterpene alcohols Figure 6	The dominating constituent among these compounds was identified as abienol	0.6
Essential oils Figure 1	Mono and sesquiterpenes, already discussed	0.5
Wax esters	Mainly long chain aliphatic alcohols (arachinol, behenol, and lignocerol) esterified with fatty acids (oleic, lauric, and octacosanoic)	0.2
Diterpene aldehydes Figure 6	The aldehydes were of the abietane and pimarane types and related to the corresponding resin acids	0.1
Secondary fatty alcohols	*n*-nonacosan-10-ol was identified in the unsaponifiable fraction of the petroleum ether solubles	0.1

classified as nonpolar compounds have been discussed under a separate heading. However, as can be seen from Table 2 and Figure 3, this fraction contains a large number of other constituents, several of which are present in greater yields than the essential oils.[15-17] By combinations of selective solvent extraction, crystallization, and distillation procedures, purified components from this nonpolar fraction can be isolated. They have a wide range of uses. Fatty acid, for example, are widely used in protective coatings, intermediate chemicals, detergents, and as flotation agents. Resin acids, on the other hand, are used mainly in sizing paper to control water absorptivity, as emulsifying and tackifying agents in synthetic rubber manufacturing, as well as components in adhesives, surface coatings, printing inks, and chewing gum. Collectively, this nonpolar group of chemicals, representing as it does 12 to 13% of the total weight of technical foliage, could be a future source of chemical feedstocks. This group may also be responsible for some of the desirable properties of foliage as an adhesive extender. This subject will be explored later in this chapter.

C. Chlorophylls

The second group of foliage chemicals to be discussed are the chlorophylls, the pigment or pigment precursors of green foliage that are essential to the photosynthetic

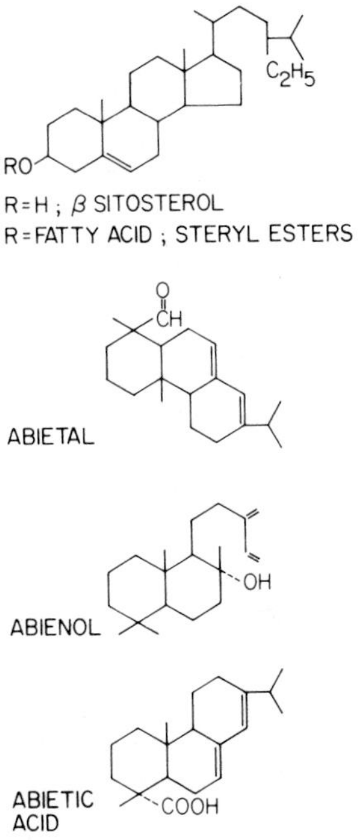

R=H ; β SITOSTEROL
R=FATTY ACID ; STERYL ESTERS

ABIETAL

ABIENOL

ABIETIC
ACID

FIGURE 3. Structures of selected nonpoplar constituents from scotch pine foliage[15-17]

process. Five closely related chlorophylls, designated a to e, occur in higher plants. The best known of these are chlorophyll a (blue-green pigment) and chlorophyll b (yellow-green). The chlorophylls are contained in lipoprotein bodies in the plastids. In green plants, chlorophylls a and b constitute 15 to 20% of the dry weight of these bodies known as chloroplasts. Vascular green plants contain chlorophylls a and b in the ratio of 3 to 1, and other chlorophylls are present in small or trace amounts. Except where specified in this review, the term chlorophyll will be understood to refer to the natural mixture of chlorophyll a and b.

The chlorophylls belong to the family of tetrapyrroles that include the open-chain bile pigments and the large-ring compounds of porphyrins. Their characteristic features show a chelated dihydroporphyrin nucleus, a cyclopentanone ring, and a phytyl ester grouping.

The structures of chlorophyll a and b, and related products are given in Figure 4. Pheophytin, also known as phaeophytin, is a bluish-black waxy pigment obtained from chlorophyll by a mild acid treatment. It is similar to chlorophyll, except that the magnesium atom has been replaced by hydrogen atoms. Pheophorbid is obtained by removal of both the magnesium and the phytyl group. Chlorophyllin, also known as sodium chlorophyllin, is a product resulting from the controlled action of alcoholic potassium or sodium hydroxide on alcoholic foliage extracts. The methyl and phytyl groups have been replaced by alkali, but the magnesium has not been replaced. Phytol is an alcohol obtained by the decomposition of chlorophyll and is used in the synthesis of vitamins E and K. The structure of the important blood pigment hemin is given in Figure 5 for comparison, and to emphasize the close chemical similarity between it and chlorophyll.

The use of chlorophyll and its derivatives in food coloring and as additives in lotions, creams, soaps, deodorants, and toothpastes is well known both in North America and

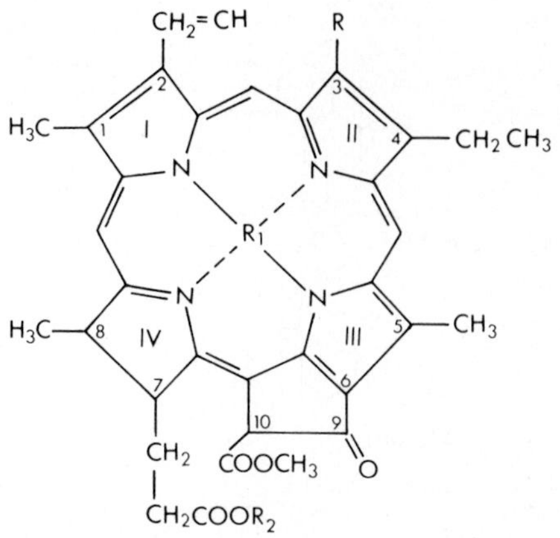

FIGURE 4.   Structure of chlorophyll and related products — phytyl:

$$R_2 = CH_3-CH-CH_2-CH_2-CH-CH_2-CH_2-CH_2-CH-CH_2-CH_2-CH_2C = CH-$$
$$\quad\quad\quad\quad | \quad\quad\quad\quad\quad\quad\quad | \quad\quad\quad\quad\quad\quad\quad\quad\quad\quad | \quad\quad\quad\quad\quad\quad\quad |$$
$$\quad\quad\quad CH_3 \quad\quad\quad\quad\quad CH_3 \quad\quad\quad\quad\quad\quad\quad\quad CH_3 \quad\quad\quad\quad\quad CH_3$$

chlorophyll a: $R = CH_3$, $R_1 = Mg$, $R_2 = $ phytyl, chlorophyll b: $R = CHO$, $R_1 = Mg$, $R_2 = $ phytyl, pheophytin: $R = CH_3$, $R_1 = 2H$, $R_2 = $ phytyl, and pheophorbid: $R = CH_3$, $R_1 = 2H$, $R_2 = H$.

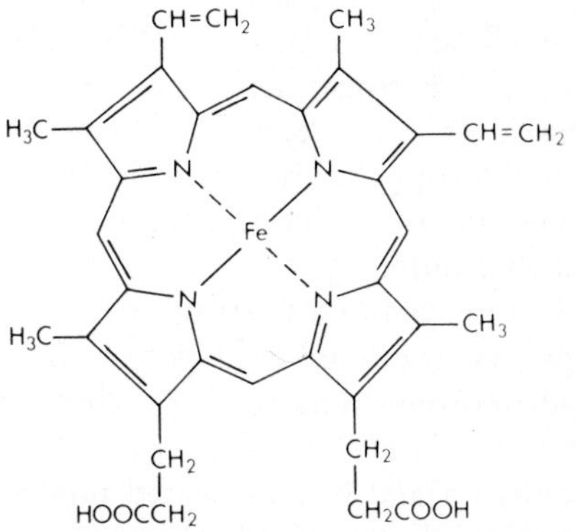

FIGURE 5.   Structure of hemin,

the U.S.S.R. However, their use in medicine and pharmacology may be less fa Research in the U.S.S.R.[6] has shown that chlorophyll preparations stimulate muscle, and skin tissues. More importantly, chlorophyll was found to stimulat action among experimental animals. Intraperitoneal doses of sodium chlorophy to cats and dogs increased the number of blood leucocytes. The interconver tagged chlorophyll to hemoglobin has been confirmed, thus opening the way treatment of human blood disorders by the partial in vivo synthesis of hem from adminstered chlorophyll preparations. In addition, sodium chlorophyl found to protect white mice and rabbits exposed to harmful irradiation. Anot

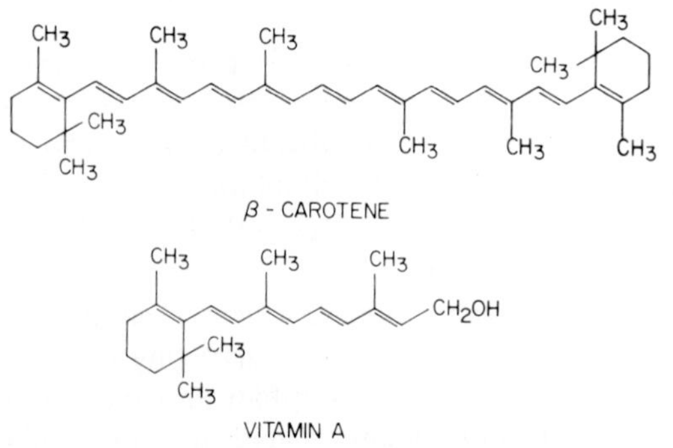

FIGURE 6.   Structures of β-carotene and vitamin A.

rivative of chlorophyll, pheophytin, was found to reduce the effect of benzene anemia in rabbits by accelerating leucocyte regeneration. In a study of the effect of pheophytin on hexenol narcosis in rats, it was found that the duration of narcosis was considerably reduced in comparison with control animals that received no pheophytin. It was postulated that pheophytin accelerated recovery of the liver function.

Studies in the U.S.S.R. are underway on the use of chlorophyll and its derivatives in the treatment of a diversity of diseases, including arterial sclerosis, ulcers, intestinal disorders, tuberculosis, neuritis, gallstones, and tooth decay. As a toothpaste additive, chlorophyll was particularly effective since it not only acted as a deodorant in the case of tooth decay, but also decreased swelling and hemorrhaging of the gums and improved saliva flow.

D. Carotenoids

Invariably associated in foliage with the chlorophylls is a class of labile, easily oxidizable yellow-red pigments called carotenoids. The term carotenoids is derived from carotene, the most plentiful pigment in carrots. Because of their solubility in fat solvents (e.g., petroleum ether and/or chloroform), they are also known as lipochromes, and like chlorophyll, are found mainly in the chromoplasts of leaves and needles.

From a chemical point of view, carotenoids are highly unsaturated (polyene) compounds containing a chromophoric system of alternate single and double bonds formed by condensation of isoprene units. Their color varies with the number and placement of the double bonds. In foliage, carotene is the most important example and occurs in three isomeric forms: α-, β-, and γ-carotene. The difference between them is the position of the double bonds. α-carotene is optically active, whereas β- and γ-carotene are optically inactive. The structure of β-carotene is given in Figure 6. Another carotenoid, xanthophyll, is also found in forest foliage. It is a yellow chloroplast pigment and is the dihydroxy-derivative of the optically active α-carotene. Xanthophyll can be separated from the carotenes by distribution between petroleum ether and 90% methanol. Carotene remains entirely in the petroleum ether phase, while xanthophyll remains in the methanol layer.

Undoubtedly the value of the carotenoids for nutrition depends on the ease with which a molecule of β-carotene can be converted to the important vitamin A (Figure 6). Thus, vitamin content of tree foliage is another example of pharmacological activity associated with these versatile constituents.

Historically and presumably prehistorically, both people and animals have included tree foliage in their diets. The value of this instinctive practice has been verified as a result of the finding that deciduous and coniferous foliage is an important source, not only of vitamin A, but of vitamins C, E, and K, provitamin D, and riboflavin.

E. Protein

Although coniferous foliage contains 7 to 8% protein, some deciduous species, such as poplar hybrids, average 19.5% on a moisture-free basis.[18] This compares very favorably with traditional fodder such as alfalfa ($\simeq$ 18% protein), and suggests that poplar leaves could provide protein for animal and human consumption. A compensation of the lower protein coniferous foliage is that it can be harvested even in winter when alfalfa and hay are unavailable.

More than half of crude foliage protein consists of simple proteins, with albumins and glutelins predominating. A Russian study involving pines[19] identifies 19 essential amino acids, the limiting ones being tryptophan and methionine. Lysine as well as methionine were identified as limiting amino acids in some preliminary studies involving spruce and pine needles at the Canada Agriculture Experiment Station, Agassiz, B.C.[20]

Although protein-rich juice can be liberated from leaf pulp by simple mechanical presses, there are difficulties in separating the protein from accompanying tannins and phenolic substances in tree foliage.[21] The present Soviet practice of using foliage products directly as an animal feed supplement after drying and grinding, would appear to be the most efficient utilization of foliage protein at the present time.

F. Other Foliage Extractives

It was previously noted that most of the research associated with foliage has been directed toward the conifers. An important exception to this is the monumental work of Pearl and his co-workers on chemical constituents from leaves of the family Salicaceae.[22] Initially, Pearl worked on poplar barks from which he isolated important and potentially important pharmaceuticals. These included: glucosides (such as benzoic and ϱ-hydroxybenzoic), and phenols (such as salicyl alcohol and pyrocatechol). He was among the first to recognize that these and many others, such as salicortin, 1-O-ϱ-coumaroyl-β-D-glucoside, populoside, ω-salicyloylsalicin, chrysin-7-glucoside, and 2-O-salicyloylsalicin, occurred also in the leaves.

Salicin is best known as an analgesic and antipyretic. As noted in the introduction, its use as a fever-reducing agent was well known to early American Indians and it has been used as a pharmaceutical since 1830. Tremuloidin, because of its unique benzoyl substituent in the 2-O-glucose position, is peculiarly suited as a starting material for the production of 3,4,6-tri-O-methyl-D-glucose, a very valuable compound for carbohydrate and sugar research. A demand for this sugar derivative would result in immediate commercial production of either the starting material, temuloidin, or the methylated sugar itself. Chemical structures of some of these important foliage chemicals of deciduous trees are shown in Figure 7 as follows:

- Populin: $R_1 = R_2 = R_3 = R_5 = R_6 = H$ and $R_4 = C_6H_5CO$
- Tremuloidin: $R_1 = C_6H_5CO$ and $R_2 = R_3 = R_4 = R_5 = R_6 = H$
- Salicyloylsalicin-2-benzoate: $R_1 = C_6H_5CO$, $R_2 = R_3 = R_4 = R_6 = H$, and $R_5 = o\text{-HOC}_6H_4CO$
- Salicyloylsalicin: $R_1 = R_2 = R_3 = R_4 = R_6 = H$ and $R_5 = o\text{-HOC}_6H_4CO$
- Salicin: $R_1 = R_2 = R_3 = R_4 = R_5 = R_6 = H$
- Triploside: $R_1 = C_6H_5CO$, R_2, R_3, R_4, or $R_5 = CH_3CO$, and $R_6 = H$
- Salireposide: $R_1 = R_2 = R_3 = R_5 = H$, $R_4 = C_6H_5CO$, and $R_6 = OH$.

It would be presumptuous to assume that any review of foliage chemicals is complete. Each tree species produces unique chemical compounds with a wide range of properties. Accordingly, a selection of some recent references to essential oils, chloro-

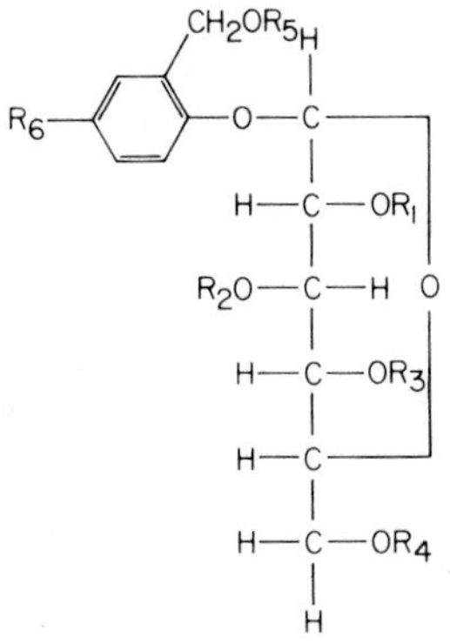

FIGURE 7. Structure of some important foliage chemicals of deciduous trees.

phyll, carotene, pigment constituents, and miscellaneous constituents have been summarized in Tables 3, 4 and 5.

III. FOLIAGE AS ADHESIVE FILLERS AND EXTENDERS

Until now, the properties and uses of isolated foliage chemicals have been discussed, however, for some applications, such as adhesive fillers and extenders, whole foliage has interesting possibilities. Studies[103] at the Western Forest Products Laboratory, Vancouver, B.C. have demonstrated that finely ground, dry foliage can replace traditional extenders and fillers such as wheat flour, corn cobs, walnut shells, and bark. Also, it can act as a partial replacement for phenol in both liquid and powdered phenol-formaldehyde resins.

Wheat flour has been the adhesive extender of choice in the wood-bonding industry. The purpose of adding wheat flour to adhesives is to maintain viscosity and obtain an adequate open-assembly time by moisture preservation (humectant). The quantity of wheat flour used in plywood industry in British Columbia alone is about 3600 to 4500 t/year (8 to 10 million lb/year). In North America, the total amount used is more than 45,000 t (100 million) and in Far East countries, is more than 227,000 t/year (500 million lb/year). Also, wheat flour prices have tripled in the recent years to 0.22 to $0.26/kg (to about 0.10 to $0.12/lb) and supply is getting difficult.

In contrast to extenders that are expected to provide useful chemical properties to the glue mix, fillers are usually added for physical bulk. Examples of commercially used fillers are powdered corn cobs, tree barks, and walnut shells. The quantity consumed in the plywood industry in British Columbia is about 4 times the wheat flour consumption (approximately 18,000 t/year, 40 million lb/year). Their supply is inconsistent and their price has also increased greatly.

An explanation of foliages humectant properties is probably to be found in its large nonpolar group of chemicals already discussed, and especially the triglycerides, fatty acids, and sterylesters. On the other hand the presence of low molecular weight phenols, flavonoids, and tannins would account for its reactivity with formaldehyde to give adhesive properties to the resin. In addition, technical foliage contains varying amounts of cellulose, lignin, and bark, which when finely ground, act as fillers. In concert, these foliage chemicals combine to make foliage an attractive alternate as an

Table 3
RECENT, SELECTED REFERENCES ON ESSENTIAL OIL TYPE FOLIAGE CONSTITUENTS

Species	Constituent(s)	Comment	Ref.
Juniperus sabina	Terpenes	Antimicrobial activity	23
Eucalyptus spp.,	Essential oils	General study	24
Betula costata	Triterpenoids	Identification	25
Juniper spp.	Monoterpenes	Comparison among three taxa	26
Abies alba	Monoterpenes	Composition of needles	27
Coniferous spp.	Terpene metabolites	Analysis of different organs	28
Pinaceae spp.	*n*-Heptane	Chromatographic study	29
Siberian fir	Needle oil	Review, emphasizing fragrance	30
Eucalyptus tereticornis	Essential oils	Composition	31
Quercus spp.	Triterpenes	Structure of cyclobalanone	32
Quercus spp.	Triterpenes	Identification	33
Pine spp.	Essential oils	Composition	34
Larch spp.	Monoterpenes	Seasonal variation	35
Larch spp.	Essential oils	Gas chromatographic analysis	36
Maritime pine	Terpene hydrocarbons	Comparative study	37
Pine spp.	Essential oils	Chemical composition	38
Abies spp.	Monoterpenes	Effect of age and size	39
Spruce, fir	Volatile oil	Insect resistance correlations	40
Larch spp.	Sesquiterpene hydrocarbons	Composition and seasonal variation	41
Picea excelsa	Volatile oils	Comparison of oil from healthy and insect infested trees	42
Juniper spp.	Essential oils	Composition, properties, and uses	43
Birch spp.	Triterpenoids	Quantitative determination using thin layer chromatography	44
Abies sibirica	Essential oils	Phenol content	45
Fir spp.	Monoterpenes and resin acids	Chemical composition	46
Betula costata	Triterpenoids	Identification of new structures	47
Eucalyptus camaldulensis	Essential oils	Composition	48
Eucalyptus globulus	Sesquiterpenes	Identification	49
Eucalyptus spp.	Essential oils	Study of introduced species	50
Eucalyptus spp.	Essential oils	Yield and citral content	51
Siberian conifers	Monoterpenes	Identification	52
Siberian conifers	Essential oils, phenols	Composition	53
Larch spp.	Essential oils	Chemical composition	54
Sitka spruce	Terpenes	Variations in leaf oil composition	55
Amabilis fir	Terpenes	Variations in leaf oil composition	56

adhesive filler and extender. Although much work remains to be done to examine influence of particle size, aging, and other factors on the durability of the resulting wood-glue bond, the following experiments indicate encouraging results. Because of the uniqueness of this approach in adhesives, these experiments are described in considerable detail.

A. Use in Plywood

In order to examine the feasibility of using foliage as a plywood adhesive additive, a commercial plywood phenol-formaldehyde resin that makes use of wheat flour and alder bark (Modal®) as extender and filler was selected for comparative purposes. The percentage of the additive in proportion to the phenol-formaldehyde resin solid weight was 55.4% which was equivalent to 35.7% to total solid weight of the glue.

Douglas-fir and 50/50 mix white spruce/lodgepole pine foliage were used for the

Table 4

RECENT, SELECTED REFERENCES ON CHLOROPHYLL, CAROTENE
AND ASSOCIATED PIGMENT FOLIAGE CONSTITUENTS

Species	Constituent(s)	Comment	Ref.
Larch	Chlorophyll	Seasonal dynamics	57
Pinus sylvestris	Chlorophyll	Effect of lead and zinc content	58
Deciduous spp.	Chlorophyll	Binding effects with protein complex	59
Pinus sylvestris	Chlorophyll and pigments	Seasonal changes	60
Deciduous spp.	Chlorophylls and carotenoids	Changes in summer leaves	61
Coniferous spp.	Phenophytin	Precipitation from alcohol extracts	62
Picea abies	Chlorophyll	Changes due to treatment with 2,4,5-T and dalapon	63
Deciduous spp.	Chlorophylls and carotenoids	A review of autumn leaf colors	64
Sitka spruce	Chlorophylls and carotenoids	Changes in response to shade and season	65
Deciduous spp.	Chlorophylls and carotenoids	Changes in summer leaves	66
General spp.	Chlorophyll and pheophytin	Effect of industrial pollution	67
Spruce	Chlorophyll	Effect of inorganic fertilizers	68
Conferous spp.	Chlorophylls and carotenoids	Improvement in extraction	69
Pine spp.	Chlorophylls and carotenoids	Effect of ecological patterns	70
Pine spp.	Protochlorophyllide	Photochemical biosynthesis	71
Pine and spruce spp.	Chlorophyll derivitives	Production and use of water-soluble chlorophyll derivitives	72
Pine spp.	Green pigment	Daily changes in seedlings	73
Coniferous spp.	Pigments	Changes due to disease	74

experiments. For the Douglas-fir foliage-phenol-formaldehyde mixture, the glue was prepared by completely replacing all wheat flour and 0, 57, and 100% replacement of the bark filler (Modal®). The white spruce/lodgepole pine foliage-phenol-formaldehyde glue mix was prepared by complete replacement of all wheat flour in all cases and 33, 67, and 100% replacement of bark filler. The viscosities of the glues, other than the one with 100% bark replacement (that was found to be lower than the others) was similar to that of the control glue.

Plywood, 1.59 mm (5/8 in.) thick, was made with white spruce veneers, 3.2 mm (1/8 in.) thick. Glue spread was 26.8 k/100 m² of double glueline (55 lb/1000 ft.² The plywood assemblies were pressed at 141°C (285 K) and 1.21 MPa (175 psi) pressure, using different press and open assembly times, as shown in Table 6. For each manufacturing condition, three 1.22 × 1.22 m (4 × 4 ft) plywood sheets were made. From each panel, more than 100 standard shear specimens were cut from the center portion of the panel and 40 samples were randomly selected for testing.

From the Table 6 results, it appears that total replacement of commercial glue filler and extender with foliage in phenol-formaldehyde resin is feasible. Further experiments, using cyclic wet-dry conditions to more fully examine the durability properties of this glue, are being conducted.

In another experiment, a series of adhesives were made by mixing different quantities of foliage powder (Douglas-fir) with liquid phenol-formaldehyde resin. The phenolic resin and Douglas-fir foliage powder were mixed in the proportions of 0 to 80% foliage content, based on solid weight of the adhesive-foliage mixture, at 10% intervals. In the high foliage-content range (greater than 50%), the viscosity of the adhesive became relatively high because water was absorbed into the foliage powder. Additional water was added to the adhesive for viscosity adjustment in this case.

Each of the adhesive-foliage mixtures spread on white spruce veneers at a spread of

Table 5

RECENT, SELECTED REFERENCES ON MISCELLANEOUS FOLIAGE CONSTITUTENTS

Species	Constituent(s)	Comment	Ref.
Cedrus atlantica	Diterpene phenols	General studies	75
Picea abies	Peroxidase	Identification in needles	76
Pinaceae spp.	Lignians	Isolation and identification	77
Liriodendron tulipifera	Sesquiterpene lactone	Characterization of peroxyferolide	78
Pinus sylvestrix	Diterpene alcohol	Identification of isoabienol	79
Larix spp.	Phenolics	Analysis of flavonoids by chromatography (HPLC)	80
Eucalyptus spp.	Antifungal substances	General studies	81
Pinus contorta	Glycosides	Isolation and identification	82
Pinus sylvestris	Phenolic glycosides	Isolation and identification of some flavonoid glucosides	83
Larix leptolepis	Phenolics	Seasonal variation of flavonoids	84
Picea sitchensis	Gibberellin	Isolation and identification	85
Pinaceae spp.	Phenolic acids and glycosides	Isolation and identification	86
Pinus sylvestris	Flavonoids	Isolation and identification	87
Pinus spp.	Protein	Protein composition vs. needle age	88
Norway spruce	Amino acids	Extraction and analysis	89
Pine spp.	Protein, amino acids	Composition	90
Pinus radiata	Protein	Composition of foliage leachates	91
Pinaceae spp.	Acetophenones	Identification and distribution	92
Picea abies	Glutathione and glutathione reducatse	Seasonal variation	93
Coniferous spp.	Phytol and vitamin E	Preparation	94
Siberian larch	Flavonoids	Seasonal variation	95
Coniferous spp.	Vitamins	Preparation of water-soluble vitamins	96
Spruce spp.	Proteins and chlorophyll paste	A new extraction method	97
Salicaceae family	Hot water extractives	Studies involving isolation and identification	98
Picea abies	Nonsaponifiables	Composition in needle fractions	99
Pine and spruce spp.	Needle constituents general	Differences in composition from green needles to forest litter	100
Pinus sylvestris	Phenolic glycosides	Isolation and identification	101
Pinus sylvestris	Polyisoprenols	Isolation and identification	102

24.4 and 14.7 kg/100 m² (50 and 30 lb/1000 ft,² respectively) of double gluelines was made into 5-ply plywood. The open-assembly time was 5 min, followed by pressing at 149°C (300 K) for 10 min at 1.38 MPa (200 psi).

The test boards were cut into 25.4 × 25.4 mm (1 × 1 in) specimens for testing according to the torsion-shear method[104] and checked by the standard internal-bond (IB) testing procedure. The results of IB tests are shown in Figure 8. Because the torsion-shear results were very similar to that of the standard IB test, and showed the same trend, the torsion-shear results are not included in Figure 8.

Figure 8 results indicate that, with the inclusion of foliage up to 50% of 60% solid content of the adhesive mixture, strength remains relatively constant at about 1.03 MPa (150 psi). With further addition, the strength dropped sharply, but even at 80% foliage content, the internal bond strength was still maintained at 0.41 to 0.48 MPa (60 and 70 psi, respectively), well above the Canadian Standards Association (CSA) - 0188 minimum requirement of 0.28 MPa (40 psi). These internal bond strength results indicate that foliage powder has potential for use as a plywood, hardboard, or particleboard liquid-resin extender.

Table 6
COMPARISON OF BOND QUALITY OF PLYWOODS MADE OF FOLIAGE ADHESIVES AND CONTROL PHENOL-FORMALDEHYDE GLUE

Douglas-Fir Foliage Adhesives

		Control		Bond quality					
				0% Additive[a]		57% Additive[a]		100% Additive[b]	
Pressing time (min)	Open assembly time (min)	psi (MPa)	WF (%)	psi (MPa)	WF (%)	psi (MPa)	WF (%)	psi (MPa)	WF (%)
4	15	139 (0.96)	65	150 (1.03)	66	174 (1.20)	54	145 (1.00)	34
5	15	175 (1.21)	83	160 (1.10)	85	190 (1.31)	86	191 (1.32)	52
6	15	167 (1.15)	88	166 (1.14)	93	193 (1.33)	83	191 (1.32)	91
7	15	194 (1.34)	97	161 (1.11)	96	188 (1.30)	89	212 (1.46)	90

White Spruce/Lodgepole Pine Foliage Adhesives

		Control		Bond quality					
				0% Additive[a]		57% Additive[a]		100% Additive[b]	
Pressing time (min)	Open assembly time (min)	psi (MPa)	WF (%)	psi (MPa)	WF (%)	psi (MPa)	WF (%)	psi (MPa)	WF (%)
4	15	199 (1.37)	71	201 (1.39)	88	197 (1.36)	76	174 (1.20)	80
4	25	166 (1.14)	81	236 (1.63)	91	186 (1.28)	92	185 (1.28)	84
6	15	188 (1.30)	96	178 (1.23)	89	199 (1.37)	94	197 (1.36)	92
6	25	183 (1.26)	94	206 (1.42)	86	188 (1.30)	92	191 (1.32)	85

Note: The specimens were tested after CAS-0122 vacuum-pressure water-soak treatment. psi = shear strength and WF = wood failure.

[a] All wheat flour was replaced.
[b] All wheat flour and bark filler was replaced.

From Chows, S., Foliage as adhesive extender: a progress report, presented at the 11th Washington State University Symposium on Particleboard, Pullman, Wash., 1977. With permission.

B. Use in Particle and Waferboard

A 50/50 mixture of Douglas-fir foliage and a powdered novolac resin, having a phenol-to-formaldehyde ratio of 1 to 0.08, was prepared. Douglas-fir planer shavings (moisture content 9%) and the resin-foliage mixtures at 2.5, 5, and 10% on the weight of wood, were pressed into particleboard at 193°C (380°F) for 10 min. The resulting

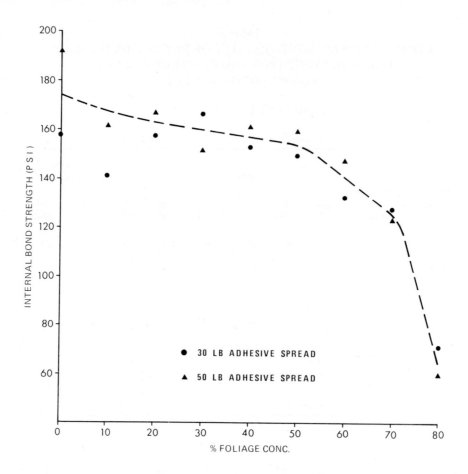

FIGURE 8.   Relationship between internal bond strength of plywood and percentage of foliage concentration in phenol-formaldehyde resin. (From Chow, S., Foliage as adhesive extender: a progress report, presented at the 11th Washington State University Symp. on Particleboard, Pullman, Wash., 1977. With permission.)

board thickness was 9.5 mm (3/8 in), and the specific gravity was 0.8. Similarly, control boards made with the same phenolic-powder resin alone were also prepared at 1.25, 2.5, and 5% resin-content levels, based on wood weight.

Two boards for each resin content were made. Three specimens each for standard IB and bending-strength tests were cut from each panel. Only dry tests were performed.

The results of the tests are shown in Table 7, and show that the bond quality for the 2.5 and 5% resin-foliage mixtures is similar to those for the 2.5 and 5% resin boards, indicating that the foliage is acting as an adhesive, and hence, providing a replacement for the resin.

In another experiment, the same powdered resin as above was used. A 50/50 mixture of white spruce foliage and the resin was prepared. Aspen flakes (industrially cut and dried) and resin-foliage mixtures of 2.2, 2.8, 3.4, and 6% (based on the weight of wood) were pressed into 12.7 mm-thick (½ in) waferboards at 202°C (395°F) for 7.5 min. The resulting board specific gravity was about 0.65 ± 0.01. Control boards made with the same resin were also prepared, having resin contents of 1.1, 1.4, 1.7, and 3%. Three boards were prepared for each resin content and four bending-strength specimens were cut from each board. Two of the boards were tested dry and the other two were tested after 2 hr of boiling in water to conform to the CSA standard. Results are

Table 7
COMPARISON OF INTERNAL BOND AND BENDING-STRENGTH TESTS OF PARTICLEBOARD MADE OF FOLIAGE-ADHESIVE MIXTURE AND CONTROL PHENOLIC RESIN

Phenolic-resin content (%)	Foliage extended		Resin alone	
	IB (psi) (MPa)	MOR (pai) (MPa)	IB (psi) (MPa)	MOR (psi) (M
1.25	18 (0.12)	1480 (10.20)	3 (0.02)	470 (3.24)
2.5	65 (0.45)	1810 (12.48)	21 (0.14)	1130 (7.79)
5.0	140 (0.97)	2560 (17.65)	83 (0.57)	2070 (14.27)

From Chows, S., Foliage as adhesive extender: a progress report, presented at the 11th Washington State University Symposium on Particleboard, Pullman, Wash., 1977. With permission.

Table 8
COMPARISON OF BENDING STRENGTHS OF WAFERBOARD MADE OF FOLIAGE-ADHESIVE MIXTURE AND CONTROL PHENOLIC RESIN

Resin content (%)	Control		Foliage adhesive mix	
	Dry (psi) (MPa)	Boiled (psi) (MPa)	Dry (psi) (MPa)	Boiled (psi) (MPa)
1.1	750 (5.17)	110 (0.76)	1500 (10.34)	280 (1.93)
1.4	760 (5.24)	170 (1.17)	2000 (13.79)	370 (2.55)
1.7	1830 (12.62)	610 (4.21)	3230 (22.27)	1150 (7.93)
3.0	3460 (23.86)	1360 (9.38)	3560 (24.55)	1460 (10.07)

Note: Each value is the average of six specimens from three boards.

From Chow, S., Foliage as adhesive extender: a progress report, presented at the 11th Washingon State University Symposium on Particleboard, Pullman, Wash., 1977. With permission.

shown in Table 8 and demonstrate the effectiveness of foliage in enhancing the bending strength of the board tested under these conditions.

Although much work remains to be done, particularly in the area of identifying the

Table 9
COMPARISON OF CONIFEROUS AND
DECIDUOUS MUKA WITH ALFALFA

Component	Spruce muka	Birch muka	Alfalfa meal
Protein (%)	8.79	8.0	18.3
Fats (%)	6.54	8.2	3.2
Cellulose (%)	35.6	18.0	26.2
Nitrogen-free extractives (%)	34.0	56.8	41.8
Ash (%)	4.4	4.2	9.6
Carotene (mg/kg)	139.0	380.0	172.0
Riboflavin (mg/kg)	6.0	4.0	13.2
Calcium (%)	0.72	0.78	1.13
Phosphorous (%)	0.17	0.26	0.31
Potassium (%)	0.44	0.73	1.34
Magnesium (%)	0.59	0.3	0.2
Iron (mg/kg)	158.5	101.0	212.0
Manganese (mg/kg)	292.0	30.0	29.0
Copper (mg/kg)	5.6	8.0	9.9
Zinc (mg/kg)	31.5	121.0	16.0
Cobalt (mg/kg)	158.0	90.0	360.0

From Keays, J. L. and Barton, G. M., Recent advances in foliage utilization. Can. For. Serv., West. For Prod. Lab., Inf. Rep. VP-X-137, 1975.

active foliage chemicals responsible for its desirable adhesive properties, it would appear that dried, pulverized coniferous foliage can partially replace phenol in both liquid and powdered phenol-formaldehyde resins. In addition, it has been successful as an extender and filler for plywood adhesives. Results of planned full scale mill studies as well as current economic studies on processing dried, pulverized foliage will determine the commercial potential of this unique use of foliage.

IV. FOLIAGE FOR FODDER

It is becoming increasingly evident that the world is moving at an accelerated pace from relative abundance to absolute scarcity. The nonrenewable resources are dwindling, and the cost of recovery and processing is increasing. The renewable resources, the seas, the farm lands, and the forests are rapidly losing their reserve capacity. World demand for these resources continues to increase, and there is little indication that world population or resource use will soon stabilize or decrease. In order to meet the world demand for clothing, shelter, and food, it is becoming increasingly important that all resources, especially those like foliage that have been burned or left to decay, be used effectively and efficiently in the future.

There are at least two persuasive arguments to consider foliage for fodder namely, foliage like alfalfa contains vitamins, protein, and minerals (Table 9) and foliage is being used as fodder supplements (muka) for poultry and animals in the U.S.S.R. In fact, at the recommended supplement levels outlined in Table 10, several important benefits are reported such as increased livestock weight and productivity, faster growth, healthier animals, and increased resistance to disease and infection.

Although these arguments are impressive, each country must investigate its own species for nutrient levels, and more importantly, for possible toxic effects. Ponderosa pine, for example, has been identified with anti-oestrogenic activity, and ingestion of its foliage has been linked with abortion in cattle.[105,106]

Table 10
RECOMMENDED DOSAGE AND REPORTED BENEFITS
FROM THE USE OF CONIFEROUS MUKA IN THE
U.S.S.R.[6,7]

Livestock	Recommended supplement level (%)	Acceptance levels (kg/head/day)	Benefits
Poultry	5	—	Reduced susceptibility to disease, increased (18%) weight, increased egg production
Cattle	5	1—2	Reduced susceptibility to disease, increased vitality, increased (15—20%) weight, Easier gestation
Milk cows	5	1—2	Healthier cows, increased (12%) milk production
Pigs	5	—	Increased (15%) weight,
Sheep	—	0.25—0.50	improved growth, increased
Goats	—	0.25—0.50	productivity, healthier
Horses	—	1—2	animals, improved reproductive capacity

Adapted from Keays, J. L. and Barton, G. M., Can. For. Serv., West, For. Prod. Lab., Inf. Rep. VP-X-137, Vancouver, B.C., 1975; Keays, J. L., *Appl. Polym. Symp.*, 28, 445, 1976.

Several analytical studies[107-110] on North American species have now been completed and nutrient levels similar to those reported in the Russian review[6] have been found. (Table 11.) However, whereas U.S.S.R. studies had emphasized rapid (40%) losses in carotene content in foliage after felling, a study[107] on lodgepole pine and white spruce showed that except for the essential oils, there was no appreciable loss of any organic constituent within the time span of the experiment (49 days in winter, 20 days in summer) (Table 12).

Although carotene and its associated vitamin A content are not so highly prized in North America because of the low cost of synthetic vitamins, loss of carotene after felling was considered a potential indicator of foliage freshness and ultimate value as an animal or poultry fodder. Thus, the findings shown in Table 12 are important to the commercial development of foliage in North America since they strongly suggest that no costly changes need to be made in the normal cycles of felling and harvesting practices already in use. Unlike the harvesting of hay or alfalfa, coniferous foliage can be collected and processed even under winter conditions.

This study[107] of lodgepole pine and white spruce needles also included analytical data on inorganic constituents such as calcium, phosphorus, magnesium, copper, iron, manganese, and zinc. Of these, calcium showed the greatest difference since it was consistently three times higher in spruce than in pine needles. No explanation was given other than to suggest that it represented a genetic difference between these species.

Several feeding trials involving muka from North American conifers have also been completed, and the results summarized in Table 13. Not unexpectedly, these experiments have not shown the improved growth responses demonstrated by the U.S.S.R. studies. This is believed mainly due to energy-rich North American control (basal) diets that have been optimized for maximum growth. Also, coniferous foliage products were lower in Total Digestible Nutrients (T.D.N.), on the basis of acid detergent fiber than alfalfa because of higher lignin contents. Varying amounts of lignin and cellulose from

Table 11
ANALYTICAL DATA ON NORTH AMERICAN CONIFEROUS FOLIAGE

Species	Form of foliage	Protein (%)	Carotene (mg/kg)	Chlorophyll (mg/kg)	Ether extractives (%)	Essential oils %	Lignin (%)	Acid detergent fiber (%)	Total digestible nutrients[a] (%)	Ref.
Lodgepole pine	Needles, winter	6.8	78	1283	—	0.76	—	37.9	54.7	107
	Needles, summer	6.3	69	1109	—	0.54	—	36.3	57.3	—
White spruce	Needles, winter	5.9	56	921	—	0.69	—	34.8	57.2	107
	Needles, summer	5.9	49	859	—	0.54	—	34.1	57.3	—
White spruce	Needles, green	7.2	67	1089	—	0.80	—	40.5	52.8	109
	Needles, steam-distilled	6.9	37	707	—	0.16	—	42.6	51.2	—
	Needles, oven-dried	6.6	45	758	—	0.45	—	40.8	52.6	—
Red spruce	Needles	6.41	—	—	8.06	—	16.78	35.97	56.1	108
Balsam fir	Needles	6.62	—	—	10.38	—	15.65	33.19	58.1	108
Loblolly pine	Needles	7.3	—	—	8.0	—	—	28.9	61.1	110
	Technical foliage	6.6	—	—	7.6	—	—	33.5	57.8	—
Timothy hay[b]	—	15.0	—	—	3.1	—	5.1	36.4	55.8	108

a   Calculated from acid detergent fiber.
b   Included for comparison.

Table 12
EFFECT OF TIME ON NUTRIENT CONSTITUENTS
IN LODGPOLE PINE AND WHITE SPRUCE
NEEDLES

Sample	Days	Carotene[a] (mg/kg)	Protein (%)	Chlorophyll (mg/kg)	Essential oils (%)
			Winter		
Pine A[b]	0	88	6.7	1330	0.88
	7	88	7.1	1348	—
	21	93	7.5	1512	—
	35	90	6.4	1324	—
	49	94	6.6	1456	0.68
Pine B	0	72	6.1	1057	0.79
	7	97	6.8	1394	—
	21	93	6.2	1378	—
	35	85	6.8	1211	—
	49	99	7.0	1321	0.63
Spruce C	0	48	5.8	746	0.75
	7	54	4.4	967	—
	21	53	4.3	925	—
	35	65	5.4	1299	—
	49	53	4.7	1168	0.71
Spruce D	0	69	5.9	952	0.45
	7	72	5.7	1257	—
	21	71	5.0	1122	—
	35	76	6.2	1138	—
	49	79	5.6	1288	0.40
			Summer		
Pine E	0	73	5.9	1186	0.57
	4	68	6.5	1146	—
	8	57	7.4	1042	—
	12	60	7.3	921	—
	20	50	7.2	1015	0.38
Pine F	0	55	5.2	816	0.55
	4	44	5.6	621	—
	8	46	5.9	608	—
	12	51	5.8	655	—
	20	50	5.5	724	0.29
Spruce G	0	61	5.7	994	0.56
	4	78	4.9	1150	—
	8	75	6.8	1056	—
	12	79	5.7	1168	—
	20	68	5.9	1174	0.45
Spruce H	0	34	5.0	567	0.75
	4	38	5.8	643	—
	8	39	5.2	685	—
	12	37	5.8	619	—
	20	34	5.5	622	0.68

[a] All results reported on moisture-free weight basis.
[b] All trees, both pine and spruce, were felled in the same area of British
Columbia within 3.2 km of each other.

From Hannus, K., Silvichemicals in technical foliage. III. The composi-
tion of the acid fraction in nonpolar extracts of technical foliage of pine
(*Pinus silvestris* L.), Pub. of Ser. A278, Inst. of Wood Chem. and Cell.
Tech., Abo Akedemi, Abo, Finland, 1974. With permission.

Table 13
SUMMARY OF MUKA FEEDING EXPERIMENTS IN NORTH AMERICA

Test animal	Foliage species	Form of foliage	Treatment of foliage prior to feeding	Details of feeding experiment	Results	Ref.
Chickens, broilers	Loblolly pine	Needles and technical foliage	Dried and ground	Rations were formulated at 2.5 and 5.0% levels and compared against alfalfa controls	Did not deleteriously affect growth, feed conversions or mortality, taste panel judged that the cooking quality of pine fed birds was fully equivalent to controls	110
Hens, laying	Loblolly pine	Needles and technical foliage	Dried and ground	Ration was formulated at 20% level and compared with standard laying ration at Louisiana State University	Egg quality was equivalent to that of the reference ration in all respects including taste, but production was lower because of the low metabolizable energy content	110
Chickens, broilers	Red spruce, balsam fir	Needles	Dried and ground	Ration was formulated at 5% and compared with a standard corn-based diet	The growth rate to 7½ weeks of age was depressed, feed consumption was not different from controls, but the feed to gain ratio was significantly poorer for the foliage fed birds, taste was not affected	108
Sheep, wether	Red spruce, balsam fir	Technical foliage	Ground, heated to drive off volatiles, equal quantities were blended	Ration was formulated at 25% level and compared with timothy hay	The addition of muka to timothy hay did not affect palatability as indicated by voluntary intake, however, the resulting mixture of 25% muka and 75% hay had significantly lower digestibility than the timothy hay alone	111

| Chickens, broilers | White spruce | Needles | Dried and ground compared with steamed, dried and ground | Rations were formulated at 2.5, 5.0, and 10% levels and compared with a standard maize diet, pelleted and unpelletted rations were compared. | Addition of muka, either steamed or untreated, reduced growth except at the 2.5% steamed level that was comparable to basal diet, pelleting the supplemental diets significantly stimulated growth, mortalities were normal, and no stress in vital organs was noted | 109 |

branchwood in technical foliage will thus affect T.D.N. levels, but in general, coniferous technical foliage will approximate T.D.N. values of 53% as found in a low- to medium-grade hay. A slight improvement in this value might be expected in deciduous technical foliage since lignin is approximately 20% lower in hardwoods.

While other tree species must be examined for toxicity and more feeding trials, especially those involving ruminants, need to be undertaken, a viewpoint of cautious optimism on the use of foliage for fodder can be expressed. Where foliage can be obtained from an integrated forest products operation close to livestock, it can compete economically, and to a large extent nutritionally, with hay. It may even have an advantage over hay in northern communities since coniferous foliage can be processed for fodder under winter conditions. Although muka has not shown the growth responses reported in the U.S.S.R., it is encouraging to find that no toxic factors, other than those reported for ponderosa pine, have been observed, nor were adverse taste problems experienced for either eggs or meat.

V. OPTIONS AND CONCLUSIONS

It is not the purpose of this chapter to provide detailed information on foliage processing options already discussed in previous reviews.[6,8] However, some repetition is necessary in order to provide future direction for foliage utilization.

There is growing evidence[112] that under the right set of conditions (for example, whole tree harvesting, private logging roads, integrated pulp and/or sawmill complexes, wood burning boilers, etc.), the cost of delivering foliage to a central location adjacent to a mill site is economically attractive. What is done with this foliage will depend on species, geographic location, and accessibility to livestock, as well as competition from traditional crops of alfalfa and hay.

In addition, future uses of foliage will depend on results of research studies such as those already described. Figure 9 presents some options for coniferous foliage already in use or under consideration. With the exception of the production of essential oils, similar options for deciduous foliage could be proposed. From a commercial point of view, it is obvious that those options that produce more than one marketable product are likely to be the most successful. It makes little sense, for example, to produce essential oils and waste spent foliage that still has value as fodder or adhesive. Neither should it be forgotten that as petrochemicals become scarcer and more expensive, demand for nonpolar foliage components, such as essential oils, triglycerides, resin acids, sterols, etc., for chemical feed stocks will increase. Ideally, before the spent foliage is finally used for fuel, remaining components such as the chlorophylls, carotenoids, and protein should be removed and utilized.

It may be concluded that foliage contains a wide variety of chemicals with a great range of properties. Although some progress has been made in utilizing these chemicals (such as essential oils), more effort must be made if this potential is to be realized. While no one can accurately predict the future, it is reasonable to assume that traditional food supplies will not be able to meet the demands of increasing population trends. It is, therefore, important to use foliage to feed poultry and animals in order to release equivalent land to grow food for human consumption. No longer can we be excused of wasting foliage because of ignorance. Future generations will judge us on how well we manage and utilize this valuable, renewable resource.

ACKNOWLEDGMENT

This chapter is dedicated to the memory of my colleague, the late John L. Keays

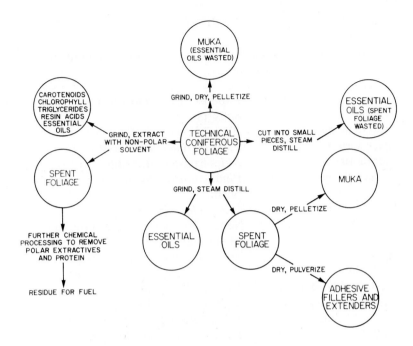

FIGURE 9.   Options for processing coniferous foliage.

who was the catalyst for foliage research in North America. As his friend, Harold Young expressed so well in his obituary "He was a leader ..... and succeeded in making the world listen". Neither will those who were privileged to call him friend forget his good natured humour, enthusiasm, or compassion.

APPENDIX

Common Names Used in the Text and Their Botanical Equivalents

Aspen	*Populus tremuloides* Michx.
Willow	*Salix* spp.
Eastern maples	*Acer rubrum* L.,
Black spruce	*Picea mariana* (Mill.) BSP
Balsam fir	*Abies balsamea* (L.) Mill.
Lodgepole pine	*Pinus contorta* Dougl. var. *latifolia* Engelm.
White spruce	*Picea glauca* (Moench) Voss
Puckerbrush	Noncommercial trees and shrub species that usually occur as successional species on every continent of the world.
Slash pine	*Pinus elliotii* var. *elliottii* Engelm.
Western red cedar	*Thuja plicata* Donn
Eastern hemlock	*Tsuga canadensis* (L.) Carr.
Western hemlock	*Tsuga heterophylla* (Ref.) Sarg.
Mountain hemlock	*Tsuga mertensiana* (Bong.) Carr.
Scotch pine	*Pinus silvestris* L.
Poplar hybrids	*Populus* spp.
Russian coniferous spp.	*Pinus silvestris, Pinus Koraninsis, Picea abies, Picea jezoensis, Abies sibirica*
Poplar	*Populus* spp.
Alder	*Alnus rubra* Bong.
Douglas-fir	*Pseudotsuga menziesii* (Mirb.)
Ponderosa pine	*Pinus ponderosa* Laws.
Red spruce	*Picea rubens* Sarg.
Loblolly pine	*Pinus taeda* L.,

REFERENCES

1. Vogel, V. J., *American Indian Medicine*, University of Oklahoma Press, Norman, Okla., 1970, 393.
2. Bailey, H., *The Vitamin Pioneers*, Rodale Books, Emmaus, Pa., 1968, 3.
3. Young, H. E., Forest biomass inventory: The basis for complete tree utilization, in For. Biol./Wood Chem. Conf., Madison, Wis., 119, 1977.
4. Keays, J. L., Complete-tree utilization. An analysis of the literature. Part II. Foliage, Can. For. Serv., West. For. Prod. Lab., Inf. Rep. VP-X-70, Vancouver, B.C., 1971.
5. Keays, J. L., Projection of world demand and supply for wood fiber to the year 2000, *Appl. Polym. Symp.*, 28, 29, 1975.
6. Keays, J. L. and Barton, G. M., Recent advances in foliage utilization, Can. For. Serv., West. For. Prod. Lab., Rep. VP-X-137, Vancouver, B.C., 1975.
7. Keays, J. L., Foliage. I. Practical utilization of foliage, *Appl. Polym. Symp.*, 28, 445, 1976.
8. Barton, G. M., Foliage. II. Foliage chemicals, their properties and uses, *Appl. Polym. Symp.*, 28, 465, 1976.
9. Barton, G. M., McIntosh, J. A., and Chow, S., The present status of foliage utilization, *Am Inst. Chem. Eng.*, in press.
10. Howard, E. T., Physical and chemical properties of slash pine tree parts, *Wood Sci.*, 5(4), 312, 1973.
11. Bender, F., Cedar leaf oils, Can. For. Serv., East. For. Prod. Lab., Publ., Ottawa, Rept. No. 1008, 1963.
12. Richards, J. H. and Hendrickson, J. B., *The Biosynthesis of Steroids, Terpenes and Acetogenins*, W.A. Benjamin, New York, 1964, 217.
13. von Rudloff, E., Gas-liquid chromtography of terpenes. VI. The volatile oil of *Thuja plicata* Donn, *Phytochemistry*, 1, 195, 1962.
14. von Rudloff, E., Volatile leaf oil analysis in chemosystematic studies of North American conifers, *Biochem. Syst.*, 2, 131, 1975.
15. Hannus, K. and Pensar, G., Silvichemicals in technical foliage. I. Water steam distilled oil from pine material, *Pap. Puu*, 55, 509, 1973.
16. Hannus, K. and Pensar, G., Silvichemicals in technical foliage. II. Nonpolar lipids in the technical foliage of Scots pine, *Finn. Chem. Lett.*, 1, 255, 1974.
17. Hannus, K., Silvichemicals in technical foliage. III. The composition of the acid fraction in nonpolar extracts of technical foliage of pine (*Pinus silvestris* L.), *Pub. of Ser. A 278, Inst. of Wood Chem. and Cell. Tech.*, Åbo Akademi, Åbo, Finland, 1974.
18. Dickson, R. E. and Larson, P. R., Muka from populus leaves: a high-energy feed suppliment for livestock, Tappi For. Biol. Wood Chem. Conf., Atlanta, 1977, 95.
19. Repyakh, S. M., Rakhmilevich, V. A., and Levin, E. D., Protein composition of pine needles of different ages, *Khim. Drev.*, 4, 111, 1976.
20. Hunt, J. R., personal communication, 1978.
21. Pirie, N. W., Leaf protein: a beneficiary of tribulation, *Nature (London)*, 253, 239, 1975.
22. Pearl, I. A., Extractives of hardwood wastes as sources of chemicals, *For. Prod. J.*, 18(2), 60, 1968.
23. Akimov, Yu. A., Kharchenko, G. I., and Krylova, A. P., Antimicrobial activity of terpenes from *Juniperus Sabina* L., *Prikl. Biokhim. Mikrobiol.*, 13(2), 185, 1977.
24. Calderon, G., DeNigrinis, S., and Calle, A., Study of Columbian essential oils. III. Essential oils of the *Eucalyptus, Rev. Colomb. Cienc. Quim. Farm.*, 3(1), 95, 1976.
25. Uvarova, N. I., Malinosvskaya, G. V., and Elkin, Yu. N., Triterpenoids from *Betula costata* leaves, *Khim. Prir. Soedin.*, 6, 757, 1976.
26. Hoerster, H., Csedo, K., and Racz, G., Comparison of essential oils of three Juniper taxa of the oxycedrus section, *Pharmazie*, 31(12), 888, 1976.
27. Scheffer, J. J. C., Koedam, A., and Cijbels, M. J. M., Trace components of essential oils isolated by combined liquid-solid and gas-liquid chromatography. I. Monoterpene hydrocarbons in the essential oil of *Abies alba* Miller needles, *Pharm. Weekbl.*, 111(52), 1309, 1976.
28. Stepanov, E. V., Volatile terpenic metabolites of different organs of forest forming coniferous species, *Ekologiya*, 5, 35, 1976.
29. Kolesnikova, R. D., Lafysh, V. G., and Chernodubov, A. I., Chromatographic study of heptane in essential oil of representatives of the family Pinaceae, *Khim. Prir. Soedin.*, 5, 613, 1976.
30. Opkyke, D. L. J., Monographs on fragrance raw materials. Fir needle oil, Siberian, *Food Cosmet. Toxicol.*, 13(4), 450, 1975.
31. DeRiscala, E. C., Juliani, H. R., and Fumarola, M. J., Essential oils of *Eucalyptus terticornis (E. umbellata), Essenze Deriv. Agrum.*, 46(2), 176, 1976.
32. Kammano, Y., Tachi, Y., and Sawada, J., Studies on the constituents of *Quercus* Spp. VI. Triterpenes of *Quercus glauca* Thunb., Structure of cyclobalanone, *Yakugaku Zasshi*, 96(10), 1207, 1976.

33. Tachi, Y., Kamano, Y., and Sawada, J., Studies on the constituents of *Quercus* spp. VII. Triterpenes of *Quercus gilva* Blume, *Yakugaku Zasshi,* 96(10), 1213, 1976.

34. Chernodubov, A. I., Kolesnikova, R. D., and Deryuzhkin, R. I., Some features of the essential oil of common pine tree subspecies, *Izv. Vyssh. Uchebn. Zaved., Lesn. Zh.,* 19(5), 93, 1976.

35. Deryuzhkin, R. I., Krasnoboyarova, L. V., and Kolesnikova, R. D., Annual dynamics of the essential oil of some Larch species and the content of monoterpenoid hydrocarbons in it, *Rastit. Resur.,* 12(3), 418, 1976.

36. Krasnoboyarova, L. V., Latysh, V. G., and Kolesnikova, R. D., Gas chromatographic analysis of the essential oil of Larch, *Gidroliz. Lesokhim. Prom-St.* 6, 9, 1976.

37. Zimmermann-Fillon, C. and Bernard-Dagan, C., Qualitative and quantitative variations of terpene hydrocarbons during needle and shoot growth in the Maritime pine: comparative study of eight phenotypes, *Can. J. Bot.,* 55(8), 1009, 1977.

38. Kolesnikova, R. D., Chernodubov, A. I., and Latysch, V. G., Composition of the essential oils of some pine species from the Caucasus and Krymsk regions, *Rastit. Resur.,* 13(2), 351, 1977.

39. Zavarin, E., Snajberk, K., and Critchfield, W. B., Relation of cortical monoterpenoid composition of *Abies* to tree age and size. *Phytochemistry,* 16(6), 770, 1977.

40. Karasev, V. S., The role of volatile oil composition for trunk pest resistance in coniferous plants. Experiments on lumber., *Symp. Biol. Hung.,* 16, 115, 1976.

41. Kolesnikova, R. D., Krasnoboyarova, L. V., and Latysh, V. G., Composition and yearly dynamics of oxygen-containing oil of Larch species. *Rastit. Resur.,* 13(4), 669, 1977.

42. Madziara-Brovsiewicz, K. and Strzelecka, H., Conditions of spruce (*Picea excelsa*) infestation by the engraver beetle (*Ips typographus*) in mountains of Poland. I. Chemical composition of volatile oils from healthy trees, *Z. Angew. Entomol.,* 83(4), 409, 1977.

43. Akimov, Yu. A., Nilov, G. I., and Litvinenko, R. M., Essential oils of Junipers of the ancient Mediterranean area, composition, properties, and possible uses, *Tr. Nikitsk. Bot. Sada,* 69, 79, 1976.

44. Polonik, S. G., Pokhilo, N. D., Baranov, V. I., and Uvarova, N. I., Quantitative determination of triterpenoids of the dammarane series of densitometry of thin-layer chromatograms, *Khim. Prir, Soedin.,* 3, 349, 1977.

45. Gornostaeva, L. I., Repyakh, S. M., and Levin, E. D., Phenols from *Abies sibirica* essential oil, *Khim. Prir. Soedin.,* 3, 417, 1977.

46. Shmidt, E. N., Dubovenko, Z. V., and Tagil-tsev, Y. G., Chemical composition of monoterpenes and resin acids of oleoresins of far eastern species of fir trees, *Khim. Prir. Soedin.,* 1, 118, 1977.

47. Uvarova, N. I., Malinovskaya, G. V., and Elyakov, G. B., Some new triterpenoids from leaves of *Betula costata, Tetrahedron Lett.,* 50, 4617, 1976.

48. Tanker, M., Sener, B., and Soner, O., Essential oil of *Eucalyptus camaldulensis* planted at Datca, Ankara Univ., Eczacilik Fak. Mecm., 6(2), 181, 1976.

49. De Pascual Teresa, J., Urones, J. G., and Gonzalez Mateos, F., Sesquiterpenes from the essential oil fraction of *Eucalyptus globulus, An. Quim.,* 73(5), 751, 1977.

50. Abbasov, R. M., Mamedov, F. M., and Shikhiev, A. S., Study of essential oil from some introduced species of *Eucalyptus* under Apsheron peninsula conditions, *Izv. Akad. Nauk. Az SSR Ser. Biol. Nauk,* 2, 10, 1977.

51. Pinto, A. J. D., De Souza, C. J., and Donalisio, M. G. R., Selecting *Eucalyptus* with regard to essential oil yield and citral content, *Bragantia,* 35, 115, 1976.

52. Gornostaeva, L. I., Repyakh, S. M., and Levin, E. D., Monoterpenes of essential oils from Siberian coniferous species, *Khim. Prir. Soedin.,* 6, 784, 1977.

53. Gornostaeva, L. I., Repyakh, S. M., and Levin, E. D., Composition of essential oil phenols from Siberian conifers, *Khim. Drev.,* 1, 109, 1978.

54. Kolesnikova, R. D., Latysh, V. G., and Krasnoboyarova, L. V., Chemical composition of the essential oil of larch, *Khim. Prir. Soedin.,* 4, 456, 1976.

55. Von Rudloff, E., Chemosystematic studies in the genus picea. V. Variation in leaf oil terpene composition of sitka spruce, *Phytochemistry,* 17, 127, 1978.

56. Von Rudloff, E., Chemosystematic studies in the genus abies. III. Leaf and twig oil analysis of amabilis fir, *Can. J. Bot.,* 55, 3087, 1977.

57. Mezentseva, V. T., Deryuzhkin, R. I., and Skorobogatova, T. I., Seasonal dynamics of chlorophyll in the needles of different larch species and ecotypes, *Izv. Vyssh. Uchebn. Zaved. Lesn. Zh.,* 19(6), 132, 1976.

58. Swieboda, M., Chlorophyll content in pine *(Pinus sylvestris)* needles exposed to flue dust from lead and zinc works, *Acta Soc. Bot. Pol.,* 45(4), 411, 1976.

59. Davtyan, V. A., Kazaryan, V. V., and Movsesyan, G. M., Change in the content of chlorophyll and its binding strength with lipoprotein complex in some deciduous species, *Biol. Zh. Arm.,* 29(11), 57, 1976.

60. Tsoneva, P., Changes in the chlorophyllase activity and pigment content in *Pinus sylvestris* needles during one year, *God. Sofii. Univ., Biol. Fak.*, 67(2), 139, 1976.

61. Billore, S. K., Mehta, S. C., and Mall, L. P., Changes in chlorophylls A, B and carotenoids in summer leaves of tree species in a dry deciduous forest, *J. Indian Bot. Soc.*, 55(1), 56, 1976.

62. Solodkaya, G. F., Precipitation of pheophytin from alcohol extracts of tree foliage, *Lesokhim. Podsochka*, 8, 14, 1976.

63. Tonecki, J., Changes of respiration intensity and chlorophyll content in needles of Norway spruce (*Picea abies* L Karst) seedlings treated with 2,4,5-T and dalapon, *Acta Agrobot.*, 28(2), 177, 1977.

64. Hass, H. B., Leaves of autumn, *Chem. Technol.*, 7(9), 525, 1977.

65. Lewandowska, M. and Jarvis, P. G., Changes in chlorophyll and carotenoid content, specific leaf area and dry weight fraction in Sitka spruce, in response to shading and season, *New Phytol.*, 79(2), 247, 1977.

66. Billore, S. K., Metha, S. C., and Mall, I. P., Changes in chlorophyll and carotenoid in summer leaves of a tropical deciduous tree Buchanania lanzan Spreng, *Sci. Cult.*, 43(7), 324, 1977.

67. Gowin, T. and Goral, I., Chlorophyll and pheophytin content in needles of different age of trees growing under conditions of chronic industrial pollution, *Acta Soc. Bot. Pol.*, 46(1), 151, 1977.

68. Bocharova, L. V., Effect of inorganic fertilizers on the content and seasonal dynamics of chlorophyll in spruce needles in 10 year crops, Deposited Doc. 1402-75, 15, 1974.

69. Malyutina, L. A., Ushkova, E. V., and Vyrodov, V. A., Improvement in the process of extracting biologically active substances from coniferous needles, *Lesokhim Podsochka*, 3, 9, 1977.

70. Bopp, L. A., Content of basic pigments in the pine needles of different ecological origin, Deposited Doc. 1402-75, 37, 1974.

71. Vozilova, L. O. and Okuntsov, M. M., Photochemical biosynthesis of protochlorophyllide in the green needles of pine seedlings, *Tr. NII. Biol. i. Biofiz. pri. Tomsk. Un-te.*, 7, 159, 1976.

72. Baranova, R. A., Production and use of water-soluble derivatives of chlorophyll from pine, and spruce needles, *Vitamin. Rastiteln Resursy i Ikh Ispol'z*, 339, 1977.

73. Vozilova, L. O. and Stvolova, A. P., Daily changes of green pigments in pine seedling needles, *Vopr. Biol.*, 119, 1977.

74. Chernysheva, N. K. and Knyazeva, V. V., Changes in the composition of conifer needles injected with different diseases *Lesokhim Podsochka*, 12, 8, 1977.

75. Lercker, G., Casalicchio, C., and Conte, L. S., Study of some constituents of lipid fractions from soil. VIII. Some minor components of the cedar tree (*Cedrus atlantica* Man.) system, *Agrochimica*, 21(3), 207, 1977.

75a. Lercker, G., Casalicchio, C., and Conte, L. S., Study of some constituents of lipid fractions from soil. VIII. Some minor components of the cedar tree (*Cedrus atlantica* Man.) System, *Agrochimica*, 21(4), 207, 1977.

76. Esterbauer, H., Grill, D., and Zotter, M., Peroxidase in needles of *Picea abies* (L.) Karst, *Biochem. Physiol. Pflanz.*, 172(1), 155, 1978.

76a. Esterbaner, H., Grill, D., and Zotter, M., Peroxidase in needles of *Picea abies* (L.) Karst, *Biochem. Physiol. Pflanz.*, 172(2), 155, 1978.

77. Tyukavkina, N. A., Medvedeva, S. A., and Ivanova, S. Z., Lignan compounds in the needles of some species of the *Pinaceae* family, *Khim. Drev.*, 6, 94, 1977.

78. Doskotch, R. W. and El-Feraly, F. S., Isolation and characterization of peroxyferolide, a hydroperoxy sesquiterpene lactone from *Liriodendron tulipifera*, *J. Org. Chem.*, 42(22), 3614, 1977.

79. Ekman, R., Sjoholm, R., and Hannus, K., Isoabienol, the principal diterpene alcohol in *Pinus sylvestris* needles, *Acta Chem. Scand Ser. B*, 31(10), 921, 1977.

80. Niemann, G. J. and Koerselman-Kooy, J. W., Phenolic from Larix needles. XIII. Analysis of main *Larix* flavonoids by high pressure liquid chromatography, *Planta Med.*, 31(3), 297, 1977.

81. Edgawa, H., Tsutsui, O., Tatsuyama, K., and Hatta, T., Antifungal substances found in leaves of *Eucalyptus* Spp., *Experimentia*, 33(7), 889, 1977.

82. Higuchi, R. and Donnelly, D., *Pinus*. II. Glycosides from Pinus contorta needles, *Phytochemistry*, 16(10), 1587, 1977.

83. Popoff, T. and Theander, O., Phenolic glycosides from *Pinus sylvestris* L., *Appl. Polym. Symp.*, 28, 1341, 1976.

84. Niemann, G. J., Seasonal variation of main flavonoids in leaves of *L. leptolepis*, *Acta Bot. Neerl.*, 25(5), 349, 1976.

85. Lorenzi, R., Saunders, P. F., Heald, J.K., and Horgan, R., A novel gibberellin from needles of *Picea sitchensis*, *Plant Sci. Lett.*, 8(3), 179, 1977.

86. Medvedeva, S. A., Ivanova, S. Z., and Tyukavkina, N. A., Phenolic acids and their glycosides in the needles of some *Pinaceae* species, *Khim. Drev.*, 3, 93, 1977.

87. Kowalska, M., Isolation of additional flavonoid compounds from Scotch pine (*Pinus sylvestris*) needles, *Rocz. Akad. Roln. Poznaniu*, 92, 47, 1977.

88. Repyakh, S. M., Rakhmilevich, V. A., and Levin, E. D., Protein composition of pine needles of different ages, *Khim. Drev.,* 4, 111, 1976.

89. Lunderstaedt, J., Extraction and analysis of free and protein-bound amino acids from Norway spruce foliage, *Mod. Methods For. Genet., Pap. Int. Union For. Res. Organ. Biochem. Genet. Worshop,* 78, 1976.

90. Repyakh, S. M., Novikova, N. G., and Levin, E. D., Composition of the amino acids of protein isolated from the needles of the common pine, *Khim. Drev.,* 3, 104, 1977.

91. Franich, R. A. and Wells, L. G., Protein in foliage leachates of *Pinus radiata, Phytochemistry,* 15(11), 1595, 1976.

92. Ivanova, S. Z., Medvedeva, S. A., and Tyukavkina, N. A., Acetophenones from needles of some species of the family *Pinaceae, Khim. Drev.,* 1, 103, 1978.

93. Esterbauer, H. and Grill, D., Seasonal variation of glutathione and glutathiane reductase in needles of *Picea abies, Plant Physiol.,* 61(1), 119, 1978.

94. Chernysheva, N. K., Provitamin concentrate made from coniferous needles and preparation of phytol and vitamin E based on it, *Vitamin. Rastitel'n Resursy i Ikh Ispol'z,* 343, 1977.

95. Varaksina, T. N., Yearly dynamics of flavonoids in needles and phloem of the Siberian larch, *Obmen. Veshchestv. I Produktivn. Khvoin. Novosibirsk, "Nauka",* 170, 1977.

96. Fragina, A. I., Conifer needles as a source of water-soluble vitamins, *Vitamin. Rastitel'n Resursy i Ikh Ispol'z,* 337, 1977.

97. Cossette, C. and Law, K. N., Use of tree foliages. A method for the extraction of proteins and a chlorophyll paste contained in spruce needles, *Ann. Actas,* 44(2), 117, 1977.

98. Pearl, I. A. and Darling, S. F., Studies on the leaves of the family salicaceae. XVII. Hot-water extractives of the leaves of *Populus heterophylla, J. Agric. Food Chem.,* 25(4), 730, 1977.

99. Pensar, G., Hannus, K., and Lax B., Nonsaponifiable components in needle fractions from industrial wood product wastes of spruce *(Picea abies), Tutkimus Tek.,* 1, 28, 1975.

100. Theander O., Leaf litter of some forest trees. Chemical composition and microbiological activity, *Tappi,* 61(4), 69, 1978.

101. Popoff, T. and Theander, O., The constituents of conifer needles. VI. Phenolic glycosides from *Pinus sylvestris, Acta Chem. Scand. Ser. B,* 31(4), 329, 1977.

102. Hannus, K. and Pensar, G., Polyisoprenols in *Pinus silvestris* needles, *Phytochemistry,* 13, 2563, 1974.

103. Chow, S., Foliage as adhesive extender: a progress report, presented at 11th Washington State University Symposium on Particleboard, Pullman, Wash., 1977.

104. Shen, K. C. and Carroll, M., A new method for evaluation of internal strength of particleboard, *For. Prod. J.,* 19(8), 17, 1969.

105. Kamstra, L. D., Ponderosa pine needles bad for pregnant cows, *Chem. Eng. News,* 53(32), 32, 1975.

106. Cook, H. and Kitts, W. D., Anti-oestrogenic activity in yellow pine needles *(Pinus ponderosa), Acta Endocrinol.,* 45, 33, 1964.

107. Barton, G. M. and MacDonald, B. F., A new look at foliage chemicals, *Tappi,* 61, 45, 1978.

108. Gerry, R. W. and Young, H. E., A preliminary study of the value of conifer muka in a broiler ration, *Res. Life Sci.,* 24(1), 1, 1977.

108a. Gerry, R. W. and Young, H. E., A preliminary study of the value of conifer muka in a broiler ration, Res. Life Sci., 24(2), 1, 1977.

109. Hunt, J. R. and Barton, G. M., Nutritive value of spruce muka (foliage) for the growing chick, *Anim. Feed Sci. Technol.,* 3, 63, 1978.

110. Watts, A. B. and Manwiller, F. G., Evaluation of Southern pine needles and twigs as a potential poultry feed supplement, *For. Prod. J.,* submitted.

111. Apgar, W. P., Dickey, H. C., and Young, H. E., Estimated digestibility of conifer muka fed to sheep, *Res. Life Sci.,* 24(1), 1, 1977.

111a. Apgar, W. P., Dickey, H. C., and Young, H. E., Estimated digestibility of conifer muka fed to sheep, *Res. Life Sci.,* 24(2), 1, 1977.

112. Lightfoot, H. D., Animal feed supplement from Canadian tree foliage, Canadian Government Contract 06SZ.KL210-7-5022, 1979.

113. Brown, L. R., Why North America must decide who gets how much food and on what terms, *Financial Post,* Oct. 11, 1976.

Chapter 12

INTEGRATED PLANTS FOR CHEMICALS FROM BIOMASS

Irving S. Goldstein

TABLE OF CONTENTS

I. INTRODUCTION

In the preceding chapters, the technical feasibility of converting biomass into a variety of useful chemicals has been established. Although such technical feasibility is obviously a necessary condition for the utilization of biomass for chemicals, it is not a sufficient one, and some further reason for the use of this raw material whether it be economic, political, or social is required. These factors, especially the economics, are explored in the next and final chapter. However, one economic principle is so fundamental, yet has been ignored so often in the past, that it warrants special emphasis in a brief chapter of its own.

This principle is the need to maximize the yield of products from raw materials for greatest economic efficiency. It is especially important in the case of a multicomponent raw material like biomass. Early schemes for conversion of wood to chemicals considered only single products produced from a portion of the wood, and consequently carried a higher raw material cost as an economic burden. The raw material cost of a cellulose-derived chemical such as ethanol, for example, can double when it is a single product, compared to its cost when coproducts are derived from the remaining biomass components, the hemicelluloses and lignin. Yet, such single product plants have been designed and operated, and their failure to operate profitably (except in times of national emergency) has prejudiced many against the whole concept of chemicals from biomass. Even established industries could not thrive on a single product. Imagine the cost of gasoline as the only product from petroleum or steak as the only product from meat processing!

Biomass conversion plants should convert all the components of the starting material into useful products just as petroleum refineries and meat packing plants utilize all their raw materials. Examples of such integrated schemes for biomass processing are presented in Sections II to V. Some products fall into more than one category, for example, glucose can be used for either chemicals or food and ethanol for chemicals or fuel. The integrated schemes cited should not be considered an exhaustive listing, but rather as an illustration of what can be done.

II. PRODUCTION OF CHEMICALS ONLY

One possible integrated scheme for selectively converting each of the components of biomass into chemicals is shown in Figure 1. In this approach, the readily hydrolyzed hemicelluloses are first removed from the biomass and converted into simple sugars under mild conditions that leave the cellulose and lignin essentially unaffected. Subsequent hydrolysis of the cellulose provides almost pure glucose and a lignin residue that can be further processed to phenols and other aromatic compounds. The separate sugar streams from hydrolysis and prehydrolysis can be converted into various products.

The presence of lignin during the cellulose hydrolysis is a barrier to enzymatic hydrolysis, so the processing sequence shown in Figure 1 is better suited for acid hydrolysis of extensively lignified biomass. However, the lignin and hemicelluloses can be separated from the biomass before hydrolysis of the cellulose as shown in Figure 2. The residual cellulose is then amenable to any hydrolysis technique, or when isolated under carefully controlled conditions, it can be used for fiber as discussed in the following section.

Obviously, conventional pulping processes can effect this separation sequence, but for purposes of chemicals production, greater flexibility of process conditions can be tolerated since residual fiber strength is not important. The sequences in Figure 2 have been generalized from specific reaction descriptions.[1-3]

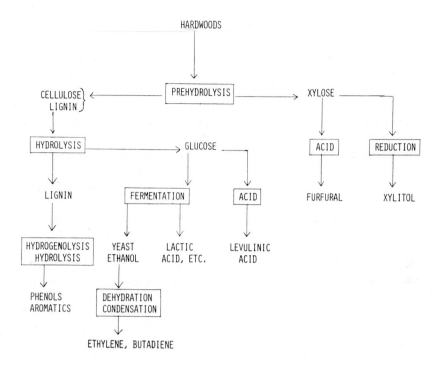

FIGURE 1.   An integrated scheme for conversion of wood into chemicals.

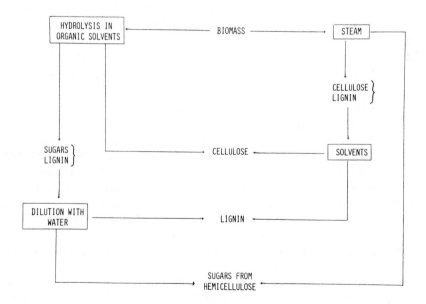

FIGURE 2.   Alternative sequences of biomass component separation.

Complete conversion of all the biomass to chemicals could possibly result in marketing problems when one product has much greater apparent utility than the others. For example, large volume ethylene production from cellulose would result in coproduction of larger volumes of phenol and furfural than present markets for these chemicals could absorb. New utilization patterns would be needed, such as conversion of phenols to other aromatics and conversion of furfural to other useful intermediates.

III. COPRODUCTION OF CHEMICALS AND FIBERS

Those chemicals that are already being produced in conjunction with existing conventional pulping operations are generally considered by-products rather than coproducts, because their volume is small in relation to the wood processed. Although the tall oil produced in the kraft pulping of southern pines is significant, much smaller quantities of dimethyl sulfoxide are derived from kraft pulping liquors, and of ethanol and vanillin from sulfite pulping liquors.

During kraft pulping most of the hemicelluloses are destroyed by conversion into saccharinic acids with consequent consumption of alkali. Especially when dissolving pulp is the desired product and all hemicelluloses must be removed anyhow, some mills are removing the hemicelluloses before pulping by prehydrolysis with steam or mild acid. The sugars so produced, mostly xylose in the case of hardwoods and mannose in the case of softwoods, could be processed to chemicals.

The lignin and saccharinic acids in kraft black liquor have been the subject of considerable research effort directed at their isolation and utilization for chemicals. Barriers to major use of these cell wall fragments for chemicals are their dilute state and admixture with pulping chemicals, as well as their present utility in providing the reducing environment for pulping chemicals recovery and the energy for pulp mill operation. They would have to be replaced with another carbon source in the overall pulp mill economy.

It seems more likely that any integrated plant for the coproduction of chemicals and fibers would be organized according to Figure 2. The processing conditions for the separation and isolation of hemicelluloses and lignin for subsequent conversion into chemicals would differ from present commercial processes, but would be carefully controlled to prevent extensive degradation of the cellulose in the residue that might make it unsuitable for applications as a fiber.

IV. COPRODUCTION OF CHEMICALS AND FOOD

A number of integrated systems for joint production of chemicals and food can be designed. Possible systems based on sugarcane and corn have been analyzed by Lipinsky,[4] while others can be based on sugar hydrolyzates from cell wall carbohydrates.

The corn system contains three main components: grain, stover, and cobs. All are conventionally used in either human or animal food or feed, and all are convertible to chemicals. The added flexibility and biomass utilization that can be gained by controlling product flows between food and chemicals to meet changing needs and conditions can provide more efficient use of the ultimate resource, the land. In the case of sugarcane, the options are not as great. Generally the juice would be processed to sucrose and the bagasse to chemicals. However, fermentation of at least part of the sucrose, especially crude material in molasses, could be advantageous at times.

It is obvious that in any process scheme involving hydrolysis of cellulose to glucose, the glucose can be used for food as well as chemicals. Depending on prices and markets, one use or the other could be favored. Still another important food coproduct is the yeast resulting from fermentation of sugars to ethanol. This material is a valuable protein supplement in animal rations.

V. COPRODUCTION OF CHEMICALS AND FUELS

In the broadest sense, any organic material can serve as a fuel whether it be in the form of solids such as coal or wood; liquids such as oils, low boiling hydrocarbons, or alcohols; or gases such as methane or carbon monoxide. Thus, chemicals derived from biomass can serve as fuels either in external markets or in the internal energy balance of the processing plant. In considering coproduction of chemicals and fuels, it is the marketing of biomass conversion products for fuel applications that defines this category.

Probably the most publicized potential fuel from biomass is ethanol, which can alternatively be processed to chemicals. A plant producing only ethanol for both fuel and chemicals apparently meets the definition of this section, but unless other chemicals were derived from the other biomass components, it would not be an integrated plant in the true sense. Wider marketing options for ethanol as chemical or fuel strengthen its high rank among potential biomass derivatives.

Gasification of biomass to carbon monoxide and hydrogen meets the criterion that all the raw material be converted to useful products. All or part of this synthesis gas can be used directly as a low Btu pipeline fuel, and the remainder converted to such products as ammonia, methanol, methane, or higher hydrocarbons. All of these but ammonia can also be considered as fuels.

Pyrolysis of biomass yields charcoal, pyrolysis oils, and synthesis gas. These liquid, solid, and gaseous fuel products could provide the economic base that would permit further exploitation of the chemical values of these materials to enhance overall process profitability. Chemical production would be geared to market demand with any residuals being relegated to fuel status.[5]

REFERENCES

1. Lora, J. H. and Wayman, M., Delignification of hardwoods by autohydrolysis and extraction, *Tappi*, 61(6), 47, 1978.
2. Chang, P. C. and Paszner, L., Comparative dissolution rates of carbohydrates and lignin during aqueous acidified organosolve saccharification of alcohol-benzene pre-extracted aspen and Douglas-fir heartwood, TAPPI Forest Biology/Wood Chemistry Conference, Atlanta, 1977.
3. Poster session at World Conference on Future Sources of Organic Raw Materials, Iotech Corporation, Kanata, Ontario, 1978.
4. Lipinsky, E. S., Fuels from biomass: integration with food and materials systems, *Science*, 199, 644, 1978.
5. Soltes, E. J. and Elder, T. J., Thermal degradation routes to chemicals from wood, *Proc. 8th World Forestry Congress, U.N.Food and Agriculture Org.*, Rome, in press.

Chapter 13

ECONOMIC AND OTHER CONSIDERATIONS

Irving S. Goldstein

TABLE OF CONTENTS

I. INTRODUCTION

In this final chapter, it seems appropriate to discuss those nontechnical factors that will ultimately determine what kind and what volume of chemicals will be derived from biomass in the future. The conceptual feasibility of chemicals from biomass as developed in the preceding chapters will be tempered by real world economic, political, and social factors. Although many of these are imponderable, others can be analyzed, allowing at least some basis for speculation and prognosis. Such economic factors as raw material costs, capital requirements, cost of chemicals from biomass, and cost and availability of fossil fuels are considered, as well as political and environmental factors and the inertial aspects of changing the resource base.

II. RAW MATERIAL COSTS

While raw material costs are important economic factors in most manufacturing operations, they are especially critical in the conversion of biomass into chemicals for two reasons. In the first place, the heterogeneous composition of biomass makes low yields of products inevitable when expressed as a percentage of total biomass even for an integrated process. Reduction in product weight is also to be expected in going from biomass with its high oxygen content to chemicals such as ethylene with no oxygen, as well as from loss of CO_2 during fermentation. The maximum possible yield of ethylene from glucose would be only 31%, and even of ethanol only 51%. With such yield limitations, the raw material costs for biomass must be kept low.

Because of its great abundance, crude biomass should be relatively inexpensive, and in fact, raw wood can be obtained for 0.01 to \$0.02/lb. However, the price of cellulose depends on its purity, and the extent to which it has been separated from the lignin and hemicelluloses with which it is naturally associated. Paper pulps increase in price from about \$0.10/lb for groundwood, to \$0.20/lb for bleached chemical pulps, and high purity chemical cellulose or dissolving pulp can reach \$0.25/lb. Obviously, substrates such as the latter two, which are susceptible to enzymatic hydrolysis, are much too expensive as a raw material for chemicals, except for special cases such as waste cellulose sludge from papermaking. Crystalline sucrose at \$0.15/lb is also too expensive for large volume applications.

Urban wastes have also been identified as potential raw materials over a wide range of costs from a negative value for unusable residues that must be disposed of to a value of several cents a pound for recyclable fibrous material. However, the most abundant sources of cheap, renewable raw biomass material for chemicals are whole-tree chips from low-grade hardwoods and agricultural residues. While grain is often abundant, its price fluctuates excessively.

Low-grade hardwoods present a utilization problem in all parts of the country, since they are not suitable for structural applications or for pulping because of size, species, defects, or bark content. Depending on locations they may consist of material remaining in high-graded stands of mixed hardwoods, the hardwood component of pine stands, or shrubby species. Whole-tree chips produced in the woods from such material can be delivered within a 50-mi radius for just under \$20/ton of dry wood or less than \$0.01/lb.[1]

For agricultural residues, a wide range of delivered costs have been cited depending on type and location.[2] Some representative values include field corn residue 30 to \$47/ton, wheat straw residue 11 to \$21/ton, and bagasse 3 to \$27/ton, dry basis within a 15-mile radius. These values are also, for the most part, below \$0.01/lb and generally below \$0.02/lb.

Table 1

VALUE OF ALTERNATIVE FUELS BASED ON EQUIVALENT
HEAT CONTENT

Fuel	Cost ($)						
Natural gas (per 1000 ft³)	1.50	1.75	2.00	2.25	2.50	2.75	3.00
#6 Fuel oil (per gal)	0.24	0.28	0.32	0.36	0.40	0.44	0.48
Bituminous coal (per ton)	44.25	51.62	58.99	66.37	73.74	81.12	88.50
Wood chips at 100% moisture content (per ton dry wood)	22.12	25.82	29.50	33.18	36.88	40.56	44.26

From Curtis, A. B., Jr., Fuel value calculator, Southeastern Area State and Private Forestry, U.S. Department of Agriculture, Atlanta, 1978.

In all probability the minimum value that will come to be placed on crude biomass raw material suitable for conversion into chemicals will be determined by its value as an alternative source of fuel by direct combustion in place of oil or coal. Table 1 compares the value of alternative fuels based on equivalent heat content at various natural gas costs. Factors other than equivalent heat content alone enter into the actual market value of alternative fuels. For example, the regulated cost of natural gas is less than the equivalent heat value from oil. Also, solid fuels such as wood and coal are not as convenient to handle and require more expensive combustion equipment than liquid and gaseous fuels, so their value will be reduced accordingly. Environmental protection costs also reduce the value of some fuels.

While for most applications the value of wood as a fuel will be determined by the cost of coal, in locations where wood is abundant and coal must be hauled from a great distance the direct substitution of wood for oil can be attractive with a resultant higher competitive price for wood as fuel. For example, replacement of oil at $1/gal with wood at $75/ton can still provide savings. However, with current coal prices of approximately $50/ton the fuel value of wood generally, and consequently its probable minimum value as a chemical raw material, is about $25/ton.

III. CAPITAL INVESTMENT

For a number of reasons, capital equipment costs in the conversion of biomass to chemicals are greater than in conventional petrochemical processing. Some of these causes include the greater difficulty in storing and handling solids compared to liquids, the lower bulk density of biomass materials, and stringent reaction conditions for lignin conversion analogous to coal liquefaction.

It is apparent from Table 2, which presents comparative 1975 plant investment costs, that biomass conversion facilities are approximately three times as expensive as petrochemical facilities, and that biomass and coal conversion facilities are comparable in cost.

The ethanol plant investment costs cited in Table 2 for wood as a raw material were based on a dilute acid hydrolysis process. Estimates for ethanol plant investment for enzymatic hydrolysis vary widely, but are equal to or greater than acid hydrolysis costs. For a 25 million gal/year ethanol plant, capital investment estimates range from $65 million[5] to $121 million.[6] In the latter estimate, 38% of fixed capital investment rep-

290 *Organic Chemicals from Biomass*

Table 2
PLANT INVESTMENT FOR METHANOL AND ETHANOL PRODUCTION

Product	Volume (10^6 gal/yr)	Raw material	Investment ($\$10^6$)
Methanol	50	Natural gas	23.1
	50	Coal	74.4
	50	Wood	64
	200	Natural gas	61
	200	Coal	178
	200	Wood	169
Ethanol (95%)	25	Ethylene	20
	25	Wood	70
	100	Ethylene	53
	100	Wood	185

From Katzen, R., Assoc., The feasibility of utilizing forest residues for chemicals,, PB-258 630, National Technical Information Service, U.S. Department of Commerce, Springfield, Va., 1976.

resents substrate pretreatment, enzyme production, and enzyme recovery facilities that are not required for the acid hydrolysis process. If these costs are subtracted, the remaining investment for hydrolysis, concentration, fermentation, and distillation facilities approximates the acid hydrolysis investment cost from Table 2. All the plants require large volume tanks and reactors because of the low dilutions involved.

Although the relative magnitude of capital investment requirements for biomass conversion to chemicals compared to petrochemical production is discouraging, one bright spot is that the individual plant size would be much smaller for biomass conversion. In contrast to the multi-billion dollar capital investment required for individual world-scale petrochemical plants and the large coal conversion plants considered to be necessary for economy of scale, a large chemicals from biomass industry would not be concentrated nor would individual installations be as large. The need to procure a biomass raw material supply within a reasonable hauling radius would lead to plants of relatively modest size widely dispersed throughout the resource base. A compromise between economy of scale and biomass procurement limitations would be in the range of about 2000 tons/day, dry raw material basis. This has proven to be a convenient size for many pulp mills. Estimated capital investment requirements for an integrated plant would be somewhat above $100 million[4], an order of magnitude lower than the billions of dollars needed for individual petrochemical and coal conversion plants. This relatively small capital investment for individual biomass conversion plants could favorably influence a decision in favor of chemicals from biomass for incremental or replacement plant capacity.

IV. COST OF CHEMICALS FROM BIOMASS

Based on raw material and capital cost estimates discussed above, the selling price of chemicals from biomass can be calculated. Some values for single product plants are given in Table 3. It is apparent that these are higher than 1978 market prices for the chemicals under consideration, especially since the values were based on 1975 dollars. However, it is also obvious that the cost of these chemicals from biomass is approaching current market prices. With refinements in biomass conversion technology still to be realized and the continuing escalation of crude oil prices, it is certain that even single product plants for biomass conversion will eventually become economically feasible.

Table 3
SELLING PRICE OF SINGLE PRODUCT CHEMICALS FROM BIOMASS

Chemical	Raw material	Raw material cost ($)	Estimated selling price ($)	Market price (1978)($)	Ref.
Methanol	Wood	25/ton	0.55—0.75/gal	0.44/gal	4[a]
Ethanol (95%)	Wood (acid)	25/ton	1.20—1.60/gal	1.12/gal	4[a]
Ethanol (95%)	Corn	3/bushel	1.20—1.40/gal		4[a]
Ethanol (95%)	Wood (enzyme)	20/ton	1.75/gal		5[b]
Ethanol (95%)	Urban waste (enzyme)	5/ton	1.43/gal		5[b]
Furfural	Hardwood	25/ton	0.52/lb	0.52/lb	4[a]

[a] Prices include overhead and 20% return on investment before taxes.
[b] Prices represent manufacturing cost and do not include overhead and profit.

Even now, an integrated plant for converting hardwoods into ethanol, furfural, and phenol could be profitable at current market prices. Figure 1 represents the necessary selling price for ethanol at various raw material costs to return 20% before taxes on $100 million (as valued in 1975) invested in a plant to produce from 1500 dry tons/day of hardwoods; 25 million gal of 95% ethanol, 75 million lb of furfural, and 52 million lb of phenol, annually. Coproduct credits of $0.30/lb for furfural (1978 market price $0.52/lb) and $0.22/lb for phenol (1978 market price $0.19/lb) were used in constructing this curve.[4]

With a $25/ton raw material cost for hardwood chips, a 20% return on investment could be realized at an ethanol selling price of only $0.80/gal, well below the current market price.

The wood hydrolysis process represented by Figure 1 is the Madison dilute acid percolation process, with a maximum potential glucose yield of 60%. Improvement of sugar yield from hydrolysis to 90% as from a strong acid process, and keeping all other factors constant would result in a 37.5 million gal annual production of ethanol with only modest additional capital requirements for slightly larger fermentation and distillation capacity. The selling price for ethanol could then be decreased to about $0.60/gal or about $0.09/lb. At this price level, the material cost for producing ethylene from ethanol (96% yield[7]) would be only $0.15/lb of ethylene, and for producing butadiene from ethanol (70% yield[7]) would be only $0.22/lb of butadiene. These projected raw material costs are only slightly higher than 1978 prices for ethylene of $0.13/lb and butadiene of $0.21/lb. Capital costs for these simple petrochemical type conversions should not be high, permitting selling prices for ethylene in the $0.20/lb range and for butadiene in the $0.30/lb. range, about 50% higher than 1978 prices.

V. COST AND AVAILABILITY OF FOSSIL FUELS

These factors are the most important considerations affecting the ultimate large scale conversion of biomass into chemicals and at the same time the most imponderable. There is wide disagreement about how fast oil prices will continue to rise and how soon supplies will become exhausted.

At the Houston Conference on Chemical Feedstock Alternatives,[8] one prediction placed the beginning of the decline of the petrochemical industry at about 1990 with 50% or more of the new feedstocks being added to the system coming from nonpetroleum sources. At the other end of the prediction scale for petroleum cost and availability, the middle of the next century was cited as the approximate date for the depletion of world oil reserves. Somewhere in the intervening period crude oil prices in

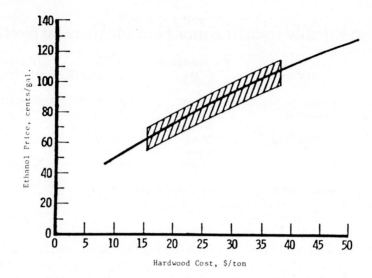

FIGURE 1.   Ethanol selling price for 20% return on investment from integrated plant producing ethanol, furfural and phenol from hardwoods. (From Katzen, R., Assoc., The feasibility of utilizing forest residues for chemicals, PB-258 630, National Technical Information Service, U.S. Department of Commerce, Springfield, Va., 1976, 189, with permission.)

current dollars will double, at which point the production of ethylene by dehydration of alcohol will probably become economically feasible.[9] This, of course, presumes that crude oil prices will continue to climb at a faster rate than biomass raw material costs. The rapid escalation of petroleum prices established by the OPEC countries in 1979 indicates that this will probably be so.

Coal can also serve as a raw material for the production of chemicals. Although it is also a finite resource and will ultimately be consumed leaving only biomass as a perpetually renewable carbon source, in the short range coal will probably play an important role in chemical production because development work is farther advanced than on chemicals from biomass, and because the massive use of coal to produce synthetic liquid fuels will also produce chemicals. The cost of these chemicals will depend on whether they would come from a free standing plant for chemicals from coal whose economics are not encouraging,[8] or as by-products of synthetic fuel plants. It has been predicted that chemicals from coal will also become competitive with petrochemicals when the price of low-sulfur crude oil doubles.[10]

Since much petrochemical production is based on natural gas, the deregulation of natural gas prices could have an important impact on the competitive position of chemicals from biomass. For example, at about $1.50/1000 ft^3 for natural gas, the selling price for methanol is about $0.40/gal, but at $3.00/1000 ft^3 for natural gas, the selling price for methanol from natural gas would have to be about $0.60/gal.[4] This latter price falls within the range projected for the production of methanol from biomass. Based on equivalent heat content, a natural gas price of $3.00/1000 ft^3 is not out of line with current fuel oil prices (Table 1).

The increasing costs for gas and oil directly result in increasing costs for the petrochemicals, for which they are the raw materials. These chemical costs will continue to converge with, and eventually exceed the costs of chemicals from substitute raw materials such as biomass by the year 2000.[11] At that time, feedstocks from all sources other than oil and gas will constitute 10 to 15% of the total.[11]

VI. POLITICAL AND SOCIAL CONSIDERATIONS

Classical laissez-faire economics based on supply, demand, and profitability are not necessarily the most important factors in determining a change in a resource base, as has been pointed out by Berg[12] in his analysis of the history of the switch from wood to coal, and then to oil and gas as primary energy resources. Although he emphasizes process technology advances as having been more important than the relative prices of the wood and coal, a similar role may be assigned to political and social decisions, as shown by the different rates of substitution of coal for wood in England and France.

In present times, it is also possible that government incentives and disincentives could influence decisions on the resource base to be used for the production of chemicals. Mechanisms such as subsidies, price supports, favorable tax treatment, price ceilings, import duties, etc. can completely reverse an unfavorable venture analysis based on marketplace economics alone. Motivation for such legislation could result from an unfavorable balance of international trade or severe unemployment. A country with abundant coal or biomass resources could choose to provide employment for its people in mining or harvesting the raw materials for production of chemicals, instead of buying petroleum from abroad with scarce foreign exchange. In fact, South Africa is converting coal into fuels and chemicals today, and similar decisions in favor of biomass are conceivable elsewhere.

VII. ENVIRONMENTAL CONSIDERATIONS

The increased use of biomass for conversion to chemicals will have environmental effects that will require attention, but do not appear to be cause for concern. The removal of agricultural residues from the land may require changes in agronomic practices to maintain productivity of the land. The increased harvesting of wood will have an impact on soils, watersheds, wildlife, etc. These same concerns exist in harvesting for conventional wood uses, and acceptable harvesting practices have been developed. However, less is known about the effect of whole-tree chipping and complete removal of biomass from the land. In one study on the recovery of a deforested ecosystem,[13] vegetational and biogeochemical recovery was relatively rapid, although loss of nutrients in drainage water was high. However, the increased availability of light coupled with high soil temperatures, increased soil moisture and dissolved nutrients provided high potential for the rapid regrowth of vegetation. The long-range effects of nutrient loss need to be evaluated.

Another environmental aspect is the possibility that a preservationist attitude by the public toward the forest environment could limit the availability of the additional wood that will be needed for conversion to chemicals. While this is more likely to apply to public lands, an affluent society might choose to allow the major portion of all the biomass that grows each year to be recycled by natural agents, rather than permitting the harvest of this annual growth as a chemical raw material.

VIII. INERTIAL ASPECTS OF CHANGING THE RESOURCE BASE

According to Davies[14] the chemical industry is now entering a resource and husbandry era, where the imperatives are to improve, cheapen, and assure supplies of wanted products by materials economy and substitution of more abundant feedstocks. According to this analysis, the chemical industry must turn toward the resource base to continue to thrive. Just as there was an earlier change from a coal tar base to a petroleum base, new feedstocks (including biomass) will be used in the future.

This will require a change in thinking among many of those who are now involved in the petrochemical industry. Chemical engineers find liquids and gases easier to handle than solids, and both the petroleum industry and the sector of the chemical industry involved in organic feedstocks are accustomed to handling a liquid raw material by tanker or pipeline. The prospect of collecting a solid raw material over an area of several thousand square miles is foreign and uncomfortable for them, so much that its feasibility has been questioned. On the other hand, the forest products industries, which regularly assemble even larger quantities of wood in their operations, are equally uncomfortable with the idea of producing chemicals rather than their familiar products.

These inertial barriers will have to be overcome, possibly by joint ventures between forest products and chemical or petroleum companies.

IX. CONCLUSION

Our present reliance on fossil hydrocarbons for organic chemicals cannot continue any longer than the availability of this finite, depletable resource. When this depletion will occur is still open to argument, but even before the total exhaustion of the world's oil and gas supplies, the cost of these resources will increase to a level at which chemicals from alternative resources including biomass will be able to compete with petrochemicals. This point will be reached sooner in some countries than others because of the local availability of alternative resources and governmental policy decisions affecting their economics.

While the exact shape of a chemicals from biomass industry cannot be foreseen, the various technical options described in the preceding chapters provide the framework from which it will develop. Although the ultimate cross-over point at which biomass-derived chemicals will be less expensive than petroleum-based products depends more on economic, political, and social factors than on technological aspects, improvements in technology can hasten its arrival. Important areas deserving emphasis include the development of an inexpensive pretreatment to improve enzyme accessibility of lignified biomass, process improvements in acidic cellulose hydrolysis to improve yields, utilization of lignin residues from hydrolysis, and process improvements to reduce capital costs in thermochemical reactions. Even though the future of chemicals from biomass cannot be predicted with certainty, they will make a significant contribution to the problems of organic chemicals supply that will be caused by the depletion of fossil hydrocarbons.

REFERENCES

1. Goldstein, I. S., Holley, D. L., and Deal, E. L., Economic aspects of low-grade hardwood utilization, *Forest Prod. J.,* 28(8), 53, 1978.
2. Benemann, J. R., Biofuels: a survey, EPRI ER-746-SR special report, Electric Power Research Institute, Palo Alto, California, 1978, 6-2.
3. Curtis, A. B., Jr., Fuel value calculator, Southeastern Area State and Private Forestry, U.S. Department of Agriculture, Atlanta, 1978.
4. Katzen, R., Assoc., The feasibility of utilizing forest residues for chemicals, PB-258 630, National Technical Information Service, U.S. Deptartment of Commerce, Springfield, Va., 1976.

5. **Anon.**, U.S. Army statement on alcohol fuels, submitted to U.S. House Committee on Advanced Energy Technologies, Washington, D.C., Sept. 20, 1978.

6. **SRI International,** Preliminary economic evaluation of a process for the production of fuel grade ethanol by enzymatic hydrolysis of an agricultural waste, Report to Fuels from Biomass Branch, Department of Energy, Menlo Park, Calif., 1978.

7. **Faith, W. L., Keyes, D. B., and Clark, R. L.,** *Industrial Chemicals,* 3rd ed., John Wiley & Sons, New York, 1965.

8. **Van Antwerpen, F. J., Ed.,** Proc. Conference Chem. Feedstock Alternatives, American Institute of Chemical Engineers, New York, 1977, 125, pp.

9. **Sarkanen, K. V.,** Renewable resources for the production of fuels and chemicals, *Science,* 191, 773, 1976.

10. **Richards, D.,** *Chem. Mark. Rep.,* 209(19), 1976.

11. **Wishart, R. S.,** Industrial energy in transition: a petrochemical perspective, *Science,* 199, 614, 1978.

12. **Berg, C. A.,** Process innovation and changes in industrial energy use, *Science,* 199, 608, 1978.

13. **Likens, G. E., Bormann, F. H., Pierce, R. S., and Reiners, W. A.,** Recovery of a deforested ecosystem, *Science,* 199, 492, 1978.

14. **Davies, D. S.,** The changing nature of industrial chemistry, *Chem. Eng. News,* 56(12), 30, 1978.

INDEX

O

P

Wood pulp, crystallinity, see also Pulping
    processes, 12

X

Xylans, 128, 129
  chemicals derived from, 129
  furfural from, 133
  in angiosperms, 126
  in angiosperms and gymnosperms, 127
  in cedar, 126
  preparation for hydrolysis, 130—131
  technologies for utilization, 29
  unbranched β-1,3″-linked, 10
Xylem ray cells, 191
Xylitol, 130—132
Xylose, 128—131, 147
Xylose-derived chemicals

ethanol from, 34
furfural, 132—135
hemicellulose hydrolysis and, 127
xylan, 128
xylitol and polyols, 131—132
xylose, 128—131

Y

Yeast(s)
  fermentation by, 30, 35—36
  from wood, 147
  immobilized cell technology, 36

Z

Zymomonas, 34